Technology, Design and Process Innovation in the Built Environment

T0251087

Spon Research

publishes a stream of advanced books for built environment researchers and professionals from one of the world's leading publishers.

Forthcoming:

Location-Based Management
System for Construction
Improving productivity using
flowline
978-0-415-37050-9
R. Kenley and O. Seppanen

Employee Resourcing in
Construction
978-0-415-37163-6
**A. Raiden, A. Dainty and
R. Neale**

Technology, Design and Process Innovation in the Built Environment

Edited by Peter Newton, Keith Hampson and Robin Drogemuller

Spon Press
an imprint of Taylor & Francis

LONDON AND NEW YORK

First published 2009
by Taylor & Francis
2 Park Square, Milton Park, Abingdon, Oxfordshire OX14 4RN

Simultaneously published in the USA and Canada
by Taylor & Francis
711 Third Avenue, New York, NY 10017

First issued in paperback 2016

Taylor & Francis is an imprint of the Taylor & Francis Group, an informa business

© 2009 Taylor and Francis

Typeset in Sabon by Wearset Ltd, Boldon, Tyne and Wear

This publication presents material of a broad scope and applicability.
Despite stringent efforts by all concerned in the publishing process,
some typographical or editorial errors may occur, and readers are
encouraged to bring these to our attention where they represent errors
of substance. The publisher and author disclaim any liability, in whole
or in part, arising from information contained in this publication. The
reader is urged to consult with an appropriate licensed professional
prior to taking any action or making any interpretation that is within
the realm of a licensed professional practice.

British Library Cataloguing in Publication Data
A catalogue record for this book is available from the British Library

Library of Congress Cataloging in Publication Data
Technology, design, and process innovation in the built
environment/edited by Peter Newton, Keith Hampson, and Robin
Drogemuller.
p. cm.
Includes bibliographical references and index.
1. Building–Technological innovations. 2. Buildings–Technological
innovations. 3. Building materials–Technological innovations. I.
Newton, P. W. (Peter Wesley), 1948– II. Hampson, Keith (Keith
Douglas) III. Drogemuller, Robin.
TH153.T436 2009
690–dc22 2008035396

ISBN 13: 978-1-138-98853-8 (pbk)
ISBN 13: 978-0-415-46288-4 (hbk)

Contents

Figures

Tables

Contributors

The editors

Professor Peter Newton is a Research Professor in the Institute for Social Research, Swinburne University of Technology, Melbourne. Prior to joining Swinburne in 2007 Dr Newton held the position of Chief Research Scientist in the Commonwealth Scientific and Industrial Research Organiation (CSIRO), where for over ten years he was Science Director of the Sustainable Built Environment Program, and from 2001 through 2006 was Director of the Sustainability Program in the Australian Cooperative Research Centre (CRC) for Construction Innovation.

Professor Keith Hampson is CEO of the CRC for Construction Innovation, with responsibility for crafting a blend of commercial and public-good outcomes in the built environment on behalf of industry, government and research partners nationally. Prior to this Professor Hampson was Director of Research in Queensland University of Technology's School of Construction Management and Property, and Coordinator of the Postgraduate Project Management Program. For 13 years prior to this he led a private engineering and property development consultancy and was active in design, construction and facility management roles.

Professor Robin Drogemuller is Professor of Digital Design in the School of Design, Queensland University of Technology, Brisbane. Prior to joining QUT in 2007 Professor Drogemuller was leader of the Urban Informatics group in the CSIRO, and from 2001 through 2006 was Director of the ICT Program in the CRC for Construction Innovation. His major research and development work over the last decade has been in the use of information technology to support decision-making in the design, construction and operation of buildings.

The authors

Paul Akhurst, Facilities Director, Sydney Opera House, Sydney (at the time of the research)

Michael Ambrose, Urban Systems Program, CSIRO Sustainable Ecosystems, Melbourne

Dr Nick Blismas, School of Property Construction and Project Management, RMIT University, Melbourne

Fanny Boulaire, Urban Systems Program, CSIRO Sustainable Ecosystems, Melbourne

Professor Terry Boyd, Faculty of Business and Informatics, Central Queensland University, Rockhampton (previously Queensland University of Technology, Brisbane)

Professor Peter Brandon, THINKLab, University of Salford, Manchester

Dr Stephen Brown, Urban Systems Program, CSIRO Sustainable Ecosystems, Melbourne

Stuart Bull, Arup, Sydney

Dr Ivan Cole, Office of the Chief, CSIRO Materials Science and Engineering, Melbourne

Dr Penny Corrigan, Surfaces, Thin Films and Interface Research Program, CSIRO Materials Science and Engineering, Melbourne

Dr Philip Crowther, School of Design, Queensland University of Technology, Brisbane

Dr Lan Ding, Urban Systems Program, CSIRO Sustainable Ecosystems, Sydney

Professor Robin Drogemuller, School of Design, Queensland University of Technology, Brisbane

Stephen Egan, Urban Systems Program, CSIRO Sustainable Ecosystems, Melbourne

Professor Martin Fischer, Center for Integrated Facility Engineering, Stanford University, Palo Alto

Professor John Gero, Volgenau School of Information Technology and Engineering, George Mason University, Washington (previously University of Sydney)

Dr Medha Gokhale, GHD, Brisbane, previously City Design, Brisbane City Council

Dr **Leman Figen Gul**, School of Architecture and the Built Environment, University of Newcastle, Newcastle

Professor **Keith Hampson**, Cooperative Research Centre for Construction Innovation, Brisbane

Mary Hardie, School of Engineering, University of Western Sydney, Sydney

Dr **Ralph Horne**, Centre for Design, RMIT University, Melbourne

Professor **Richard Hough**, Arup and University of NSW, Sydney

Delwyn Jones, Cooperative Research Centre for Construction Innovation and Ecquate Pty Ltd, Brisbane

Professor **Stephen Kajewski**, School of Urban Development, Queensland University of Technology, Brisbane

Steven Kenway, Urban and Industrial Water Program, CSIRO Land and Water, Brisbane

Professor **Arto Kiviniemi**, Granlund, Helsinki (previously ICT for Built Environment, VTT Technical Research Centre of Finland)

Professor **Arun Kumar**, School of Urban Development, Queensland University of Technology, Brisbane

Chris Linning, BIM Manager, Sydney Opera House, Sydney

Professor **Mary Lou Maher**, Key Centre of Design Computing and Cognition, University of Sydney and Human-Centered Computing Cluster, National Science Foundation, Washington

Dr **Karen Manley**, School of Urban Development, Queensland University of Technology, Brisbane

Kevin McDonald, Urban Systems Program, CSIRO Sustainable Ecosystems, Brisbane

Professor **Graham Miller**, School of Engineering, University of Western Sydney, Sydney

Steve Moller, Sustainable Built Environment Pty Ltd, Melbourne (previously CSIRO Manufacturing and Infrastructure Technology)

Professor **Lidia Morawska**, School of Physical and Chemical Sciences, Queensland University of Technology, Brisbane

Professor **Peter Newton**, Institute for Social Research, Swinburne University of Technology, Melbourne

Dr **Phillip Paevere**, Urban Systems Program, CSIRO Sustainable Ecosystems, Melbourne

Associate Professor Rabee Reffat, Department of Architecture, King Fahd University of Petroleum and Minerals, Dhahran

Dr Seongwon Seo, Urban Systems Program, CSIRO Sustainable Ecosystems, Melbourne

Associate Professor Sujeeva Setunge, School of Civil, Environmental and Chemical Engineering, RMIT University, Melbourne

Dr Ambalavanar Tharumaraja, Urban Systems Program, CSIRO Sustainable Ecosystems, Melbourne

P.C. Thomas, Team Catalyst, Sydney

Grace Tjandraatmadja, Urban and Industrial Water Program, CSIRO Land and Water, Melbourne

Dr Selwyn Tucker, Urban Systems Program, CSIRO Sustainable Ecosystems, Melbourne

Professor Terry Turney, Centre for Green Chemistry, Monash University, Melbourne (previously CSIRO Manufacturing and Infrastructure Technology)

Professor Ron Wakefield, School of Property Construction and Project Management, RMIT University, Melbourne

Professor Derek Walker, School of Property Construction and Project Management, RMIT University, Melbourne

Kendra Wasiluk, Sustainability Research Institute, The University of Leeds (previously Centre for Design, RMIT University, Melbourne)

Achi Weippert, School of Urban Development, Queensland University of Technology, Brisbane

Preface

The operating environment of the property, design, construction and facility management industry (also known as the AECO sector: architecture, engineering, construction and operations) in Australia and internationally is rapidly changing. In the space of less than a decade, a sector of industry previously lacking in research-based innovation, highly fragmented and litigious, and seemingly protected from international competition by the friction of distance and purported uniqueness of product, was changing. Globalization had intensified competition within Australia while opening up market opportunities internationally, a revolution in IT and communications was realizing opportunities for application in a previously overlooked sector, and the realities of a resource-constrained and carbon-constrained twenty-first century were beginning to register with the designers, constructors and operators of the built environment.

The receptivity of the AECO sector to research-based innovation originating in Australia is increasing, in part due to the efforts of R&D partnerships brokered through a uniquely Australian innovation: the Cooperative Research Centres Program (https://www.crc.gov.au).

The formation of the Cooperative Research Centre for Construction Innovation (CRC CI) in 2001 marked the beginning of a sustained commitment to construction research by a complementary grouping of industry, government and research bodies across Australia. The CRC itself evolved from collaborations initiated in 1994 through the Commonwealth Scientific and Industrial Research Organisation's former Division of Building, Construction and Engineering, and Queensland University of Technology's (QUT) former School of Construction Management and Property. To better serve the research needs of the Australian industry, this working relationship was formalized as the QUT/CSIRO Construction Research Alliance in 1998. The Alliance also brought together RMIT's Department of Building and Construction Economics, and the Construction Industry Institute Australia (CIIA). It drew on member institutions' national and international networks, and established a record of delivering valuable research outcomes to industry.

In this transition to improved industry innovation through the CRC CI, major roles were played by leading organizations (listed below) in driving new industry-relevant research, setting up programs to encourage communication between stakeholders, and commissioning and disseminating a series of industry development reports. Success from a combination of these initiatives was evident in the emergence of a cohesive 2020 industry vision and much greater cooperation between stakeholders in pursuing innovation and performance improvements that will benefit the entire industry.

CRC CI partners since 2001 have included the Australian Building Codes Board, Arup Australasia, Bovis Lend Lease, Brisbane City Council, Brookwater, Building Commission (Victoria), Commonwealth Scientific and Industrial Research Organization (CSIRO), Curtin University of Technology, DEM, John Holland, Kennards Hire, Leighton Contractors, Mirvac, Nexus Point Solutions, Parsons Brinkerhoff, Queensland Building Services Authority, Queensland Department of Main Roads, Queensland Department of Public Works, Queensland Department of State Development and Innovation, Queensland University of Technology, Rider Levett Bucknall, RMIT University, Sydney Opera House, Transfield Services Australia, University of Newcastle, University of Sydney, University of Western Sydney, Thiess, Western Australian Department of Housing and Works, and Woods Bagot.

The efforts of the partners to the CRC have been complemented by the activities of the following industry organizations:

Australian Construction Industry Forum (ACIF)
Australian Procurement and Construction Council (APCC)
Construction and Property Services Industries Skills Council (CPSISC)
Australian Sustainable Built Environment Council (ASBEC), which the CRC CI was instrumental in establishing in 2003.

This book consolidates key applied research outcomes from the CRC CI and research initiatives from its International Construction Research Alliance (ICALL) partners in the Center for Integrated Facility Engineering, Stanford University, USA; the Research Institute for the Built and Human Environment, University of Salford, UK; and the VTT Technical Research Centre, Finland. It also captures related and complementary work from Australian researchers who have made significant contributions in this key theme area of *Technology, Design and Process Innovation in the Built Environment*.

The underpinning objective of this theme was creation of new knowledge and innovative technologies from a deliberate convergence of computer science, design science, building science, materials science and environmental science, engineered by the directors of the CRC's Sustainable Built Assets Program and the ICT Program. The result has been the emergence of a Building Information Modelling platform that has provided the basis for an integrated, performance-based modelling of built environment systems, ranging from the scale of building element to the city precinct.

Structuring the parts of this book as Materials, Design, Construction, Facilities management and re-lifing, and Innovation: capture and implementation, reflects the logical progression through the applied research outcomes focused on the supply chain of this industry. The integration of these themes also reinforces the editors' belief that a key means of improving the performance of this diverse industry is to facilitate innovation across its traditional functional interfaces.

The Australian property, design, construction and facility management industry is on the cusp of a new era in realizing the benefits of closer industry, government and research relationships leading to improved innovation performance. Industry-led initiatives are breaking new ground in applied research and implementation, with the potential for a major boost to innovation in the built environment. Building information modelling for facility management and eco-efficiency assessment of a building in real time during the design process are but two such successes. Against an acknowledged historical backdrop of poor innovation in the sector, more effective solutions to long-felt industry challenges are now being achieved through national and international collaboration in the Australian CRC CI.

However, we have reached a stage in Australia which demands renewed and innovative efforts to improve this critical industry, and through it the built environment. We are at a time of global recognition of the need to firmly address climate change and a carbon-constrained future – a problem that is exacerbated by the challenges of ever-increasing population in a resource-constrained world. The sense of urgency is clear.

In late 2007, Australian industry confirmed its commitment to the establishment of the Sustainable Built Environment Research Centre as the successor to the CRC CI, which is due to cease operations in June 2009. Support for this new national centre was founded on the challenge to deliver applied research to transform Australia's infrastructure and building industry – environmentally, socially and economically.

Professor Peter W. Newton
Professor Keith D. Hampson
Professor Robin M. Drogemuller
July 2008

Part I
Introduction

1 Transforming the built environment through construction innovation

Peter Newton, Keith Hampson and Robin Drogemuller

Achieving sustainable urban development for a projected global population of 9.2 billion in 2050, 70 per cent of whom will be living in urban settings (United Nations 2008), represents one of the principal challenges of the twenty-first century. Australia, as one of the world's most urbanized societies, led this global transition 125 years ago. Its cities are classed among the world's most liveable. Liveability, however, does not equate to *sustainability*. Indeed the current trajectory of Australia's urban development has been classed as unsustainable (Newton 2006, 2007a).

Transforming buildings and infrastructure to become more sustainable elements of our built environment is a key challenge for the property, construction, planning, design and facilities management industry, as well as governments at all levels. The roadmap by which this built environment transformation can be driven is clear but complex (see Figure 1.1).

At the heart of the transition is the promise of *virtual building* – an ability to assess the performance of a proposed built asset (e.g., lifetime cost, environmental impact, social benefit, locality impact) prior to construction. Central to virtual building is the *building information model* (BIM), an integrative digital technology that permits information-sharing between disciplines. Together with the work of the OGC (Open Geospatial Consortium), BIM provides the basis for a more rigorous cross-disciplinary specification of information required for a *convergence* of building science, design science, engineering and construction, environmental science, management science and spatial science knowledge in a modelled representation of a complex system, which is the built environment. The *city of bits* is a powerful metaphor first introduced by Bill Mitchell (1995) that has stimulated our thinking about the manner in which a complex city can be conceived – as a collection of *material objects* with different attributes and behaviours that can be assembled and re-assembled in a myriad of ways to deliver our living and working built environment.

Developments in materials and manufacturing processes, *and in design, construction and facilities management processes*, are all providing the basis for a transformation in the built environment sector that will be required to meet the challenges of:

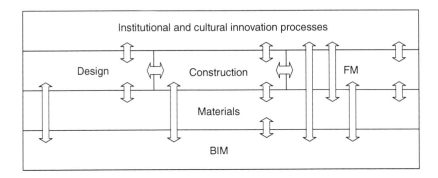

Figure 1.1 Framework for sustainable construction and the built environment.

- a rapidly growing population;
- increasing consumption;
- a resource-constrained world;
- a carbon-constrained world linked to greenhouse gas emissions and climate change;
- increasing urbanization in advanced industrial, newly industrializing and less developed countries – each with similar built environment goals, but different endowments in natural, human and financial capital;
- globalization and the competitiveness that is unleashed for industry efficiency.

An awareness that the built environment design, construction and facilities management industry was lacking in the levels of productivity, competitiveness and innovation apparent in other industrial sectors has led to a series of initiatives by the Australian government and industry seeking to identify how technology, product and process innovation in the IT, materials, design, construction and facilities management domains can be more successfully identified, diffused and implemented within an architecture, engineering, construction and operations (AECO) organization (again, see Figure 1.1).

To become more sustainable, the built environment will need to embody significantly higher levels of innovation – in its products and processes – than was characteristic of the previous century. The Cooperative Research Centre for Construction Innovation (CRC CI) was established in 2001 with a charter to assist the AECO industry deliver a more competitive and environmentally sustainable built environment. *Eco-efficiency innovation* was a key objective of applied R&D undertaken within the Sustainable Built Assets program of the CRC in close collaboration with its IT Platform – one of the key convergences it pioneered between design science and sustainability science.

The sections that follow focus on the significance, key challenges and principal transitions required in:

- the built environment;
- the AECO industry;
- innovation systems, including contributions made by the CRC CI to assist in the transition to a more sustainable built environment.

The built environment

Significance

The importance of the built environment is unquestionable. It is typically a nation's greatest asset (Newton 2006). It is where a nation's population lives and, in advanced industrial societies, where 95 per cent of the population works and where approximately 80 per cent of national GDP is generated: 'Its design, planning, construction and operation is fundamental to the productivity and competitiveness of the economy, the quality of life of all citizens, and the ecological sustainability of the continent' (Newton *et al.* 2001). The built environment also represents the myriad of enclosed spaces – homes, offices, shopping centres, entertainment venues, transport vehicles – where the population, on average, spends 97 per cent of its time (Newton *et al.* 1997).

Preference for urban (as opposed to rural) living is strong in Australia, where approximately 88 per cent of the population lives in centres of 1,000 or more residents (Newton 2008a). This is now a dominant global trend – and accelerating. The planning and management of sustainable urban settlement is possibly the greatest global challenge of the twenty-first century, especially when it is coupled with adaptation to the projected impacts of climate change and resource constraints.

Key challenges

The challenges faced by Australian built environments are well established (House of Representatives Standing Committee on Environment and Heritage 2005, 2007; Newton 2006, 2008b), and are summarized below.

Efficiency and competitiveness

In a globalized world, a nation's built environments are often assessed in terms of their contribution to international competitiveness (OECD 2008). Engineers Australia's Infrastructure Report Card (Hardwicke 2008) has assigned an overall rating of C+ (within an A–F range) for Australia's roads, rail, electricity, gas, ports, water and airports, in large part due to a significant backlog in infrastructure expenditure (Regan 2008). Costs of urban traffic congestion have been forecast to increase from approximately A\$9.4 billion in 2005 to an estimated A\$20.4 billion by 2020 (BTRE 2007). Infrastructure performance will be further tested by projected impacts of climate change (CSIRO *et al.* 2007).

Resilience to climate change

In relation to forecast impacts of climate change (Hennessy 2008), Australia stands to suffer more than any other developed country. Resilience will be tested in terms of built environment adaptability to:

- sea-level rises in combination with storm surges and their impact on coastal settlements and their infrastructures (Church *et al.* 2008);
- a rise in temperatures, especially numbers of days over 35°C, intensifying heat island effects in cities and peak demands for electricity for cooling buildings (Howden and Crimp 2008);
- urban flooding due to increases in localized extreme daily rainfall events (Abbs 2008);
- damage to life and property in peri-urban regions due to increasing risk of megafires (Leicester and Handmer 2008).

Mitigation of climate change will require a fundamental transition in energy supply (to renewables) and a de-carbonizing of the economy and population lifestyles in what is now a carbon-constrained world. As one of the largest per capita emitters of greenhouse gas, this will also represent a significant challenge to Australia's AECO sector in delivering a future carbon-neutral built environment

A resource-constrained built environment

The impact of peak oil and fuel security will be considerable as it reshapes mobility and accessibility for residents and businesses alike within cities (Newman 2008; Dodson and Sipe 2008), as will threats to the safe yield of traditional urban water supplies to many Australian cities and settlements. The dominant twentieth-century planning paradigm that assumed an abundant supply of resources – land, water, energy – would continue to be available into the future (Rees and Roseland 1991; Meadows *et al.* 1972) has lost all currency.

Liveability

Australia's major capital cities are regularly rated by international agencies such as the Economist Intelligence Unit as among the world's most liveable. Liveability, however, does not equate to sustainability: the ecological footprints of Australia's major settlements are all of the order of 7 hectares per capita – three times the global average (Newton 2007b). The challenge of winding back contemporary levels of resource consumption – at built environment, industry and household levels – is clear. Historically low levels of housing affordability and historically high levels of automobile dependence are but two indicators of where liveability is declining for many segments of the population. The manner in which our built environments have been planned, constructed and operated post-1950 has been a major contributing factor to this situation.

Key transitions

A number of key transitions are required in our built environment systems to achieve sustainable urban development *goals*: using resources more efficiently, using wastes as resources, restoring and maintaining urban environmental quality, enhancing human wellbeing, and implementing more efficient and effective urban and industrial planning and management systems (Newton 2007a). The key *transitions* are as follows (after Newton 2008b).

Energy – transitioning from a fossil fuel-based economy to one centred on distributed renewable energy (Jones 2008; Graham *et al.* 2008; Fell *et al.* 2008; Dicks and Rand 2008). Assisting with this longer-term transition will be significant improvements in the energy efficiency of building materials manufacturing, the built environment 'shell' (Centre for International Economics 2007), in energy-efficient sub-division design (see Ambrose, Chapter 13 in this volume) and in more integrated land use-transport planning.

Water – transitioning from a centralized divert–use–dispose system to a closed loop integrated urban water system which incorporates stormwater and recycled wastewater as additional sources of supply (Diaper *et al.* 2008; see also Kenway and Tjandraatmadja, Chapter 15).

Materials – transitioning from a system predicated on resource extraction–manufacture–use–dispose, to one based on industrial ecology and life cycle principles (Jones *et al.*, Chapter 3), cradle-to-cradle manufacture and construction (e.g., via off-site manufacturing; Blismas and Wakefield, Chapter 19) and dematerialization in construction via design for deconstruction (Crowther, Chapter 12) and re-lifing of building stock (Setunge and Kumar, Chapter 24). A revolution in material science (Turney, Chapter 2) provides opportunity for more functional facades, intelligent interiors and clever constructions.

Communications – transitioning from an era of low bandwidth systems supporting transmission of voice, data and low-resolution images across a digital divide (Newton 1995), to a wireless, high-bandwidth internet environment supporting virtual project teams involved in synchronous distributed design (Newton 1994; Newton and Crawford 1995; see also Chapter 8, Gul and Maher) and closing the information loop between the design office and the project site (Alexander *et al.* 1998; see also Chapter 17, Weippert and Kajewski).

Buildings – transitioning from buildings that satisfy a limited number of performance criteria centred principally on safety and minimum first cost solutions, to eco-efficient buildings whose life cycle performance is assessed virtually before construction – optimizing sustainability at building and precinct levels. Most chapters in this book are directed towards the achievement of this goal.

The architecture, engineering, construction and operations sector

Significance

As one of the largest sectors in the Australian economy, property, design, construction and facilities management accounts for 14 per cent of GDP (ISR 1999). In 2008, the cumulative value of site-based residential, non-residential and engineering construction was A$160 billion (Econtech 2008). The industry employs around 950,000 people through 250,000 firms, the vast majority of which are small to medium-sized enterprises (SMEs), and contributes significantly to the rest of the economy as an enabler. Construction is also a major industry globally (see Figure 1.2). The US far exceeds all other countries, but Australia is ranked equal fourteenth, making it a significant market, especially when considered in terms of construction expenditure per capita. The Australian Bureau of Statistics estimates that from an initial $1 million of extra output in construction, a possible $2.9 million in additional output would be generated in the economy as a whole. This would create nine jobs in the construction industry and 37 jobs in the economy as a whole (ACIF 2002). Productivity gains in the AECO sector have also been shown to have the most significant nationwide spill-over effects of any of the service sectors (Stoeckel and Quirke 1992).

Key challenges

Australia's construction sector operates against a background of industry fragmentation, intense competition, limited investment in research and development and new challenges including IT advancements; increasing public expectations in environmental protection and enhancement; increasing demand for packaged construction services; and moves towards private-sector funding of public infrastructure. Innovation and innovative behaviour are seen as key opportunities to raise the sector's performance and meet new challenges.

The sector's performance is further constrained by 'a focus on short-term business cycles and a project-to-project culture' (ISR 1999: 8). Construction contracting in Australia is regarded as a competitive and high-risk business (Uher 1994). This competitiveness is largely due to the fragmented nature

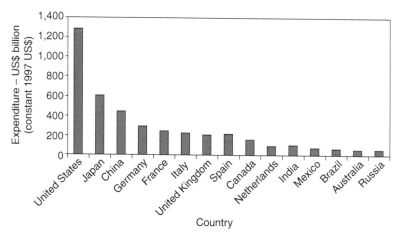

Figure 1.2 Total construction expenditure, 2008: top 15 nations (source: derived from data in *Engineering News-Record*, 254 (1): 12–13).

of the sector, with cost traditionally being the prime factor in the tender selection process (Hampson and Kwok 1997).

However, over the past few years, especially since the mid-1990s, there has been growing interest in developing a more robust, internationally competitive construction sector. Most major stakeholders across industry, government and research are involved in associated initiatives. There has been a significant improvement in the level and quality of communication and collaboration between stakeholders which is yielding initiatives that promise to lift future performance.

Three major industry policy initiatives have been prominent in developing this cultural shift in the AECO sector:

- the Building and Construction Industries Action Agenda;
- the Facilities Management Action Agenda;
- the Built Environment Design Professions Action Agenda.

Building and Construction Industries Action Agenda (1999)

In May 1999, the Australian government released the Building and Construction Industries Action Agenda, with a goal to enhance the construction sector's performance. The Agenda was the culmination of extensive discussions between major stakeholders over an 18-month period. A clear driver for this industry and government initiative was that, historically, Australian governments 'have not clearly articulated an industry development vision for the building and construction industry, and this is especially true for the non-residential segment' (ISR 1999: 35). Further, the key government department involved in developing programs to facilitate growth in the construction sector admitted that 'it is valid to argue that in the past the Commonwealth has not been particularly adept in coordinating its various policy arms in delivering policy to the industry' (ISR 1999: 39).

Key initiatives in the Building and Construction Industries Action Agenda strategy included:

- creating a more informed marketplace;
- maximizing global business opportunities;
- fostering technological innovation;
- creating economically and ecologically sustainable environments;
- creating a best practice regulatory environment.

The Agenda process resulted in the articulation of future policy directions to support sector growth. Despite public expenditures accounting for a very high proportion of total R&D expenditure in this industry in Australia, the relative level of public-sector commitment has been falling in recent years. The Australian government's *Building for Growth* report on the construction sector suggested that 'public investment in R&D is critical. The present level is probably too low for the size and importance of the building and construction industry' (ISR 1999: 20).

The innovation initiatives contained in the Building and Construction Industries Action Agenda promoted greater recognition of the value of enhanced collaboration between users and providers of R&D. They also encouraged a more active role in funding and participating in industry R&D. The government subsequently responded to calls from the construction sector and major R&D institutions for increased support to establish a national Cooperative Research Centre in construction in 2001 – the first CRC to be specifically centred in the sector. Australian government support for a CRC was viewed by all participants in the Australian construction sector as a positive public policy initiative to substantially enhance the level of medium- and long-term R&D in this critical national sector.

Facilities Management Action Agenda (2004)

The objective of the Facilities Management Action Agenda was to develop a strategic framework for the growth of a sustainable and internationally competitive Australian facilities management sector. Published in December 2004, it identified a total of 20 industry actions over five themes:

1   FM in the Australian economy
2   Maximizing innovation
3   Improving education and training
4   Addressing regulatory impediments
5   Towards a sustainable future.

This Agenda clearly acknowledged that growth in the facilities management industry would depend on promoting a culture of innovation and bringing innovations more rapidly into the marketplace. It recognized that, to date, much of the innovation within the industry had been ad hoc and iterative, building upon and adapting systems and services from other industry sectors. The Agenda also proposed actions to promote the benefits of innovation through greater industry collaboration and research and development, and to highlight the contribution that facilities management makes to workplace productivity.

Given the facilities management industry is in a strong position to influence business and government decisions to produce lower environmental impacts, the Action Agenda proposed that industry promote the role it can play in helping businesses respond to demands for sustainability. It also proposed that a web portal be established to disseminate information and promote industry awareness of the business benefits of improved environmental sustainability.

Built Environment Design Professions Action Agenda (2008)

In 2006, the Australian Council of Built Environment Design Professions (BEDP) – the peak body for architects, engineers, planners, quantity surveyors, lighting designers and landscape architects – was charged with the responsibility to work with the (then) Department of Industry, Tourism and Resources to deliver an Action Agenda for this important industry sector. In June 2008, the group completed a report, titled *Future Directions for the Australian Built Environment Design Professions*. This report was informed by studies conducted by a BEDP taskforce and four expert working groups, with support from the new Department of Innovation, Industry, Science and Research (DIISR).

The report responded to the imperative to develop an integrated and collaborative approach to management of key issues facing the built environment design professions in Australia. The following six issues were identified as being critical to the professions' future development:

1   Sustainability
2   Procurement
3   Innovation and technology
4   Industry capacity
5   Exports
6   Knowledge and training.

The report considered the impending changes that will shape our society over the next ten years and how they will impact on the built environment design professions. It identified key development opportunities for the professions, associated with innovation, sustainability and social inclusion, productivity and procurement, capacity development, global engagement and international competitiveness. The BEDP-led report provides clear recommendations on the actions that the built environment design professions, construction industries, government and community must take to realize the rich social dividends associated with a world-class built environment.

The key recommendations in the draft report (at July 2008) included the establishment of an Innovation Council for the Built Environment to link stakeholders in collaboration for innovation. In particular, the BEDP urged the Australian government to recognize the built environment design professions' strong support for funding of a cooperative research centre for the built environment in the 2008–09 grants round (given the impending closure of the CRC CI in June 2009).

It also provided some specific recommendations to establish indicators and encourage capital investment to fast-track reductions in greenhouse gases from the building industry. It sought to collaborate with industry suppliers and contractors to develop and introduce a single, national eco-efficiency rating tool to facilitate informed purchasing decisions. Finally, the report called for a workforce development strategy to encourage more effective recruitment of talent into the professions.

Key transitions

On 23 June 2004, the Minister for Science, together with the Minister for Industry, Tourism and Resources, launched the report *Construction 2020: A Vision for Australia's Property and Construction Industry* (Hampson and Brandon 2004). This was the culmination of an extensive process of industry engagement by the CRC CI, including a series of nationwide workshops in 2003 and 2004. The hundreds of attendees represented a broad spectrum from the public and private sectors, and included builders, contractors, architects, engineers and representation from industry associations. The initiative sought industry's views on how applied research and collaboration could best contribute to a robust, informed and strategic innovation agenda for Australia's property and construction industry.

The report identified eight key themes for the future of the industry. These *visions* described the major concerns of the industry and the improved future working environment favoured by its stakeholders:

1   Environmentally sustainable construction
2   Meeting client needs
3   Improved business environment
4   Welfare and improvement of the labour force
5   Information and communication technologies for construction
6   Virtual prototyping for design, manufacture and operation

7   Off-site manufacture
8   Improved process of manufacture of constructed products.

Figure 1.3 details these visions.

The first and clearest vision, agreed across the industry, was that of environmentally sustainable construction – the creation of buildings and

---

A summary of eight industry visions, together with suggested transition goals for achievement by the year 2020. For clarity, each one is presented separately, although in reality the visions are interdependent and the boundaries between them blurred.

Vision One

Environmentally sustainable construction – for industry to design, construct and maintain its buildings and infrastructure to minimize negative impacts on the natural environment, thereby preserving environmental choices for future generations. By 2020, the vision is for the industry to have comprehensive eco-efficiency evaluation tools for all stages of the construction life cycle.

Vision Two

Meeting client needs – for the design, construction and operation of facilities to better reflect the present and future needs of the project initiator, owners/tenants, and aspirations of stakeholders. This should take into account the need for improved quality and economic viability, as well as have the flexibility to adapt to future circumstances, technologies and the needs of society.

Vision Three

Improved business environment – for a regulatory, financial and procurement framework which encourages longer-term thinking and returns, a sharing of ideas and innovation between stakeholders, and a fair distribution of risk and returns. By 2020, the vision is for the industry to have a business environment achieving four types of dividends:

- Economic: with a fairer balance of risk and return to stakeholders;
- *Social:* providing equitable returns across the community;
- *Environmental:* striking a more sustainable balance between the built and natural environments;
- *Governance:* providing clarity of business responsibilities, leading to a more informed, transparent and honest marketplace.

This vision was considered the highest priority for an improved future for the industry and the most important future research topic.

Vision Four

Welfare and improvement of the labour force – for the industry workforce to be computer literate and highly skilled, showing mutual respect for each other through management and workers acting collaboratively, with improved health and safety conditions on site. A goal for 2020 is an ongoing supply of skilled workers to service this vital Australian industry. The fragmented set of occupational health and safety laws supports a call for a national code of construction safety management. The industry must also aim for a more internationally productive labour force operating in a less adversarial context. Almost 100 per cent of site respondents confirm that workplace-related issues should form a part of the future research agenda.

Vision Five

Information and communication technologies for construction – for communication and data transfer to be seamless and include mobile devices providing a commercially secure environment. These technologies will be embedded within both construction products and processes to improve efficiency and effectiveness. The knowledge economy will require property and construction to become more engaged in IT developments.

Vision Six

Virtual prototyping for design, manufacture and operation – for the opportunity to try before you buy – from inception to design, construction, demolition and rebuild. The prototype will be an electronic representation of the facility, from which relevant decisions can be made and from which the procurement processes can develop. Respondents considered that virtual prototyping would have the highest likelihood of becoming the basis for design, procurement and asset management in the next five to ten years

Vision Seven

Off-site manufacture – for a majority of construction products to be manufactured off site and brought to the site for assembly. This will enable better quality control, improved site processes including health and safety control, more environmentally friendly manufacture and possible reductions in cost. The goal is to establish the economic viability of off-site manufacture. Respondents considered off-site manufacture to have a very high likelihood of occurrence in the next five to 15 years

Vision Eight

Improved process of manufacture of constructed products – for developing new production processes, allowing the industry to work more efficiently.

The goal for 2020 is to re-engineer the supply chain to ensure that the property and construction process is as lean as possible. The industry will use IT to enhance the value of the product to the client and stakeholders through better quality control, organization and management of site activities. A substantial proportion of respondents reinforced the focus on the process of construction to achieve these improvements rather than the final constructed product or components.

Figure 1.3 The 2020 visions (source: CRC for Construction Innovation).

infrastructure that minimize their impact on the natural environment or are environmentally positive. Other significant areas of focus included the development of nationally uniform codes of practice, new tools to evaluate design and product performance, and comparisons with overseas industries, supported by a worldwide research network to ensure that Australian technology is at the cutting edge.

The overarching (or ninth) vision of achieving Australian leadership in research and innovation in delivering the 2020 vision was for industry to embrace the concept of industry, government and research working together through strategic applied research and innovation. A culture of self-improvement, mutual recognition, respect and support underpinned this vision. By 2020, the vision was for the industry to be taking more responsibility for leading and investing in research and innovation.

Innovation

Significance

Innovation is a process that leads to change – in a product, service, organization, industry sector or region – as a result of new ideas being developed into something of value. Three facets have been identified (Figure 1.4; Cutler 2008):

1   *Knowledge production* – the generation or adaptation of new knowledge, ideas, concepts
2   *Knowledge application* – the deployment of ideas in a real world context
3   *Knowledge diffusion and absorption* – the appropriation and adaptation of the knowledge by an individual or organization to provide new avenues to problem solving, creating new or large markets.

Innovation occurs once knowledge is productively incorporated into an entity's activities and outcomes (Productivity Commission 2007; refer again to the outer ring of Figure 1.1). The culture, receptiveness, stock of

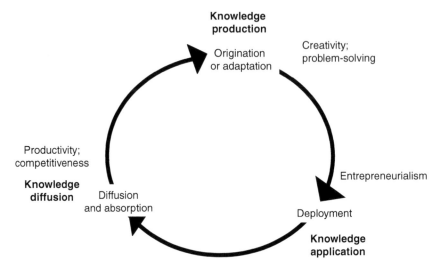

Figure 1.4 The innovation cycle (source: Cutler (2008), reproduced with permission).

human and technical capital and quality of information networks are all factors which differentiate organizations and nations in their ability to innovate.

In the most recent Innovation Capacity Index of OECD economies (Gans 2008: 19), Australia ranks thirteenth of 29 countries – a relatively stagnant position over the past decade. A similar ranking is achieved using industry expenditure on R&D (ABS 2006). How Australia performs in these and similar innovation and R&D rankings in future will depend on the effectiveness of its national innovation system. As illustrated in Figure 1.5, a national innovation system is an amalgamation of multiple and interdependent institutions and systems.

Cooperative Research Centres (CRCs) became a new feature of Australia's innovation system when they were established in 1990. The CRC program was introduced to improve the effectiveness of Australia's research effort by bringing together researchers in the public and private sectors with the end users. It links researchers with industry and government, with a focus towards research application. The close interaction between researchers and end users is the defining characteristic of the program. Moreover, it allows end users to help plan the direction of the research, as well as to monitor its progress.

There have been ten CRC selection rounds, resulting in the establishment of 168 CRCs across the Manufacturing, ICT, Mining and Energy, Agriculture and Rural-Based Manufacturing, Environment, and Medical Science and Technology sectors. The success of the program has been recognized not only within Australia but also internationally as it has been

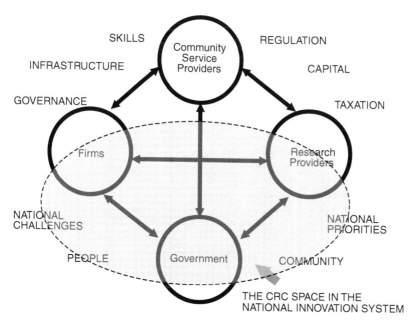

Figure 1.5 A national innovation system (source: Cutler (2008), adapted and reproduced with permission).

researched, emulated and even copied by a number of other nations. CRCs occupy a unique position and role within the national innovation system (refer again to Figure 1.5).

Key challenges

With a national contribution of approximately 2 per cent to global technical frontier knowledge creation and somewhat less within the AECO sector, Australian industry has the challenge of continuing to generate its share of new knowledge in order to capitalize directly in both a commercial and public-good sense from its application, but also to maintain a seat at the table in international forums of information exchange.

With shifts to networks of open innovation, globally networked operations, cyber infrastructure and demand-driven searches for applicable knowledge (Cutler 2008), there is an increasing need for capacity within government and industry for 'sudden catch-up' – representing significant improvements in products, processes and organizational arrangements. Figure 1.6 outlines three types of innovation. The first are novel innovations (segment A) that typically occur at the global technical frontier: *'This form of innovation is dynamically critical to economic growth and to social and environmental advances, since catch-up is premised on the*

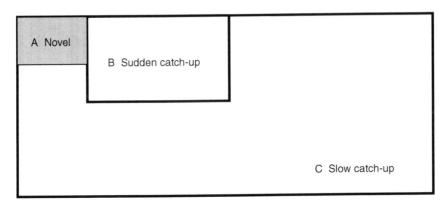

Figure 1.6 Innovation domains (source: Productivity Commission (2007), reproduced with permission).

existence of the original breakthroughs and revolutionary application'
(Productivity Commission 2007: 10). The level of frontier research in the
AECO sector in Australia and the associated rate of novel innovation
tend to lag other sectors, such as agriculture, mining, IT and medical,
which are more research-intensive. The CRC CI has focused its R&D in
segments A and B, with information diffusion programs addressing the
needs of 'slow catch-up' industry in segment C (e.g., Your Building,
www.yourbuilding.org/).

To be sustainable, the AECO sector needs to be able to appropriate
from a pipeline of innovative technologies, products, designs and
processes that can be substituted when existing ones begin to show signs
of obsolescence. Within this pipeline, three horizons of innovation have
been identified for delivering future sustainable built environments
(Figure 1.7; Newton 2007b).

Horizon 1 (H1) innovations are those that are commercially available
now and have a demonstrated level of performance which is clearly
superior to products or processes currently in the marketplace, and should
be widely substituted. Implementing H1 innovations primarily drives
efficiencies in existing systems. Their principal market is found among
organizations in segment C (Figure 1.6). Examples include energy- and
water-efficient appliances, building energy rating assessments, building
material eco-labelling schemes and the knowledge to be found on most
green building websites, e.g. <www.yourbuilding.org/>,and .

Horizon 2 (H2) and Horizon 3 (H3) innovations are those with capacity
for more fundamental transformation. H2 innovations are those where
there are some real-world examples in operation (i.e., early adopters of
novel innovation), but not widespread. They represent opportunities for

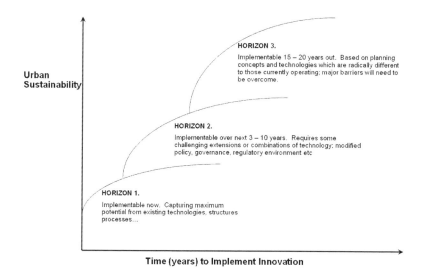

Figure 1.7 The three horizons of innovation (source: Newton (2007b), reproduced with permission).

sudden catch-up, given that evidence is becoming available on their performance-in-operation. Examples include water-sensitive urban design, distributed renewable energy and virtual building modelling.

Horizon 3 (H3) innovations are those which for the most part currently reside in research laboratories as prototypes, or are in the early stage of real-world trials, scale-up etc. If widely implemented, their impact would be transformational. They hold much of the promise for a transition to a sustainable built environment in the twenty-first century (Newton 2008b). Examples include solar-hydrogen energy systems, integrated urban water systems, embedded intelligence and cradle-to-cradle building materials manufacture. This transformation will not only be technical, as the impacts will have wide-ranging impacts across society. Transition to a hydrogen economy, for example, would require the development of new generation and storage capacity, distribution networks and infrastructures, together with the necessary training of the human capital to offer these services.

The speed with which innovations diffuse throughout an industry or community is difficult to predict. The logistic curve (Figure 1.8) is commonly used to depict the path of socio-technical innovation in both a *process* sense (i.e., from invention to full adoption: segments A through C, in Figure 1.6) and from a *stakeholder* perspective (i.e., innovators, early adopters, early majority, late majority and laggards).

In the AECO sector, it has been shown to be important for firms to be part of a consortium of leaders as they contemplate investing resources to move up the central steep part of this learning curve. The industry–research

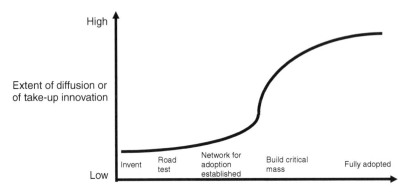

Figure 1.8 The innovation diffusion process.

collaboration (reducing the risks of first-mover disadvantages) is a central tenet of CRCs, reinforcing this key element of the innovation cycle. New innovation champions are also more likely to be encouraged into active participation in this environment, as more social (and financial) support is focused in the diffusion networks.

Key transitions

While providing a wider context of national and international applied research, this book embodies the results of six years of research by members of the CRC CI in a program designed to assist a transition to:

- a more innovative AECO sector; and
- a more eco-efficient built environment,

guided in part by *Construction 2020*, a priority-setting study of the Australian AECO sector.

All chapters in the book can be 'located' within the three-dimensional built environment performance framework developed by Peter Newton and Greg Foliente at CSIRO (see Figure 1.9; also Newton, Chapter 9). For ease of assembly within this book, chapters have been assigned to five substantive parts that recognize the significance of:

- *materials* as the building blocks of the built environment which are assembled into buildings and infrastructures that occupy spaces ranging in scale from the cadastral to the mega-metropolitan (Part II);
- *the life cycle* of built assets, and their key phases of increasingly *integrated* operation via BIM: *design*, *construction* and *facilities management* (Parts III to V);

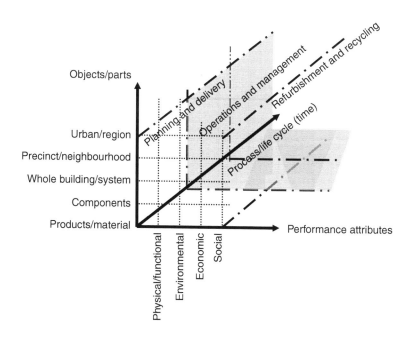

Figure 1.9 Framework for assessing sustainability performance in the built environment (source: Greg Foliente and Peter Newton (CSIRO), reproduced with permission).

- *the organizational industry context* of the AECO sector and its *capacity for innovation* across a wide spectrum of disciplines that include design science, engineering, material science, computer science, environmental science, management science and their *convergences*, and in the context of the challenges the sector faces in delivering a *high-performing built environment* (Part VI).

The chapters are not, however, 'islands' of new knowledge and insight, and the reader will find important links between chapters, as indicated in Figure 1.1. Some of the key messages and linkages are now outlined.

How built environment knowledge can be best represented to enable more sophisticated levels of sustainability performance assessment on individual physical assets, as well as entire urban systems, is a key reason for our focus on *building information models* (BIMs) and how they can be applied throughout the project/built environment life cycle. BIMs, as described by Kiviniemi (Chapter 6) and applied as a key element in integrated design – construction – operations analysis for AECO organizations (Drogemuller *et al.*, Chapters 7, 14; Fischer and Drogemuller, Chapter 16; Gul and Maher, Chapter 8; Reffat and Gero, Chapter 21) represent a *platform for virtual building*, a necessary precursor to sustainable building.

The capability for real-time assessment of environmental performance (Jones *et al.*, Chapter 3; Seo *et al.*, Chapter 10), cost estimation (Drogemuller *et al.*, Chapter 7), indoor air quality estimation (Brown *et al.*, Chapter 11), service life performance (Reffat and Gero, Chapter 21; Cole and Corrigan, Chapter 4) and deconstructability (Crowther, Chapter 12) prior to construction is now possible. They represent some of the first offerings in the toolkit for the design office of the future.

Future development will see the emergence of more extensive integrated design toolkits as a result of improved interoperability of software, improved capability of existing software, reinforced by improved collection and assembly of data and databases. Emergence of new tools within the framework of the International Alliance for Interoperability (<www.iai-international.org/>), including multi-scale models linking BIM and GIS, provides the required platform for broad-based energy-efficient and water-efficient design at both dwelling and neighbourhood scales (Ambrose, Chapter 13; Kenway and Tjandraatmadja, Chapter 15; Drogemuller *et al.*, Chapter 14).

The new integrated design platform will be (after Fox and Hietanen 2007):

- *automational* – increasing productivity through direct substitution of labour on routine tasks, e.g. real-time processing of cost and environmental estimates of buildings, structural calculations;
- *informational* – integrating and presenting a complex body of knowledge in real time, sufficient to making more effective decisions, e.g. computational fluid dynamics assessments of fire-spread or ventilation in buildings, or deposition of corrosivity agents on building and infra-structures (Cole and Corrigan, Chapter 4); LCADesign providing opportunity for an eco-efficiency performance assessment of a building in real time during the design process, compared to weeks or months for equivalent spreadsheet processes, consequently providing greater opportunity for design experimentation, consideration of options, innovation and value-adding (Seo *et al.*, Chapter 10); and software capable of incorporating climate change impacts into built environment program assessment (Drogemuller *et al.*, Chapter 14);
- *transformational* – enabling a major step change compared to traditional practices, e.g. restructuring inter- and intra-organizational relationships. Examples of transformational change emerge from the manner in which broadband internet-based systems can be developed to create virtual design/project teams (Gul and Maher, Chapter 8); closed loop information and communication systems between head office, design office and the project site, irrespective of location (Weippert and Kajewski, Chapter 17); and enabling clients to become designers through their involvement in virtual building processes (Fischer and Drogemuller, Chapter 16).

The ability to represent the built environment as a '*city of bits*' is now a reality, at least in the lab, but not widely implemented. Part of the reason for this lies in an incomplete assembly of knowledge pertaining to the performance of the building blocks of the built environment – the *material objects*; their attributes as manufactured and as they operate-in-use over their life cycle. In this book, we seek to present the current state of knowledge in relation to:

- methods for the *environmental performance assessment of materials* via life cycle analysis (Jones *et al.*, Chapter 3), including emissions from materials (Brown *et al.*, Chapter 11);
- progress towards international *product declarations* and *eco-labelling* systems that can be adopted for global materials supply, procurement and product stewardship (Jones *et al.*, Chapter 3);
- state-of-the-art developments in measuring the *service life performance and maintainability* of materials (Cole and Corrigan, Chapter 4; Reffat and Gero, Chapter 21);
- the level of *recycling and reuse* of building materials (Miller and Hardie, Chapter 5) and the ability to introduce *deconstruction* as part of the design process (Crowther, Chapter 12);
- future trends in *material science* (Turney, Chapter 2) that will radically influence facades, structures and interiors.

A key transition required in construction is the implementation of a *cradle-to-cradle process* analogous to that which exists in manufacturing (Kaebernick *et al.* 2008). Off-site manufacturing (Blismas and Wakefield, Chapter 19) and broader product stewardship on the part of building product manufacturers represent positive steps towards this goal – critical in a resource- and carbon-constrained world.

Virtual building is another step-transformation towards a sustainable city. This involves creation of building elements (from the design model) to an appropriate level of detail, definition of construction sequences for the assembly period, a connection of the activities and building elements to create the simulation, and an analysis of the simulation of the construction process to identify issues or conflicts for resolution. 4D CAD enables virtual construction before building in reality (Fischer and Drogemuller, Chapter 16). It is a core component of the future AECO platform, and enables other emerging processes such as virtual project teams and the off-site manufacturing of buildings. The application of 4D CAD should impact positively on project health, process planning and resource allocation (Weippert, Chapter 18).

Facilities management (FM) involves management of the built environment, for the benefit of both the building owner and the tenant. It should be acting as a major driver for technology change during procurement when 'as built' BIMs become a required deliverable supplied by the contractor together with the building. A BIM-centred FM operation will enable:

- spaces to be maintained to desired performance criteria (Ding *et al.*, Chapter 20);
- data-mining of asset management information to identify areas for possible savings in operations and learnings for future refurbishments and new designs (Reffat and Gero, Chapter 21);
- more effective commissioning of building services (e.g., Moller and Thomas, Chapter 22, in relation to right-sizing HVAC);
- better understanding of changes in indoor environment quality (Brown *et al.*, Chapter 11) and occupant productivity that may be linked to the design and layout of indoor space (Paevere, Chapter 25);
- more efficient re-lifing processes for buildings, ranging from the assessment of residual service life (Setunge and Kumar, Chapter 24) to material reuse (Crowther, Chapter 12) and overall functional performance (Boyd, Chapter 23).

How close these innovations can bring us to the 'construction enlightenment' envisioned by Brandon (Chapter 29) will be revealed over time as new technologies together with innovations in organizational management (see, for example, Walker, Chapter 26; Manley *et al.*, Chapter 28; Wasiluk and Horne, Chapter 27) are better integrated and linked with a new (more participatory) model of the urban development process – all of which will be required to deliver a twenty-first century transformation of the built environment.

Acknowledgements

The editors of this book are indebted to the contribution that David Hudson, editor at the Swinburne Institute for Social Research, has made to enhancing its readability, look and feel, without any apparent loss to his inimitable character. Also to Professor John Bell (Queensland University of Technology) and Professor John Wilson (Swinburne University of Technology) for their comments on a draft of this chapter.

Bibliography

Abbs, D. (2008) 'Flood', in P.W. Newton (ed.) *Transitions: Pathways Towards Sustainable Urban Development in Australia*, Melbourne: CSIRO and Dordrecht: Springer.
ABS (2006) *Research and Experimental Development, All Sector Summary, Australia 2004–05*, Cat. No. 8112.0, Canberra: Australian Bureau of Statistics.
ACIF (2002) *Innovation in the Australian Building and Construction Industry: Survey Report*, Canberra: Australian Construction Industry Forum for the Department of Industry, Tourism and Resources.
Alexander, J., Coble, R., Crawford, J., Drogemuller, R. and Newton, P.W. (1998) 'Information and communication in construction: closing the loop', in B.-C. Bjork and J. Adina (eds) *The Life Cycle of Construction IT Innovations: Technology Transfer from Research to Practice: Proceedings of the CIB Working Commission*

W78 *Information Technology in Construction Conference*, Stockholm: Royal Institute of Technology.

BTRE (2007) *Estimating Urban Traffic and Congestion Cost Trends for Australian Cities*, Working Paper 71, Canberra: Bureau of Transport and Regional Economics.

Centre for International Economics (2007) *Capitalising on the Building Sector's Potential to Lessen the Costs of a Broad Based GHG Emissions Cut*, Canberra: report prepared for ASBEC Climate Change Task Group. Online. Available at HTTP: <www.asbec.asn.au/files/Building-sector-potential_Sept13.pdf>.

Church, J., White, N., Hunter, J., McInnes, K., Cowell, P. and O'Farrell, S. (2008) 'Sea level rise', in P.W. Newton (ed.) *Transitions: Pathways Towards Sustainable Urban Development in Australia*, Melbourne: CSIRO and Dordrecht: Springer.

CSIRO, Maunsell Aust. Pty Ltd and Phillips Fox (2007) *Infrastructure and Climate Change Risk Assessment for Victoria*, Melbourne: CSIRO.

Cutler, T. (2008) *Review of the National Innovation System*, Canberra: Department of Innovation, Industry, Science and Research. Online. Available at HTTP: <www.innovation.gov.au/innovationreview> (accessed 8 July 2008).

Diaper, C., Sharma, A. and Tjandraatmadja, G. (2008) 'Decentralised water and wastewater systems', in P.W. Newton (ed.) *Transitions: Pathways Towards Sustainable Urban Development in Australia*, Melbourne: CSIRO and Dordrecht: Springer.

Dicks, A. and Rand, D. (2008) 'Hydrogen energy', in P.W. Newton (ed.) *Transitions: Pathways Towards Sustainable Urban Development in Australia*, Melbourne: CSIRO and Dordrecht: Springer.

Dodson, J. and Sipe, N. (2008) 'Energy security, oil vulnerability and cities', in P.W. Newton (ed.) *Transitions: Pathways Towards Sustainable Urban Development in Australia*, Melbourne: CSIRO and Dordrecht: Springer.

Econtech (2008) *Construction Forecasting Council 13th Forecast Presentation*, Brisbane. Online. Available at HTTP: <www.cfc.acif.com.au/>.

Fell, C., Hinkley, J., Imenes, A. and Stein, W. (2008) 'Solar energy', in P.W. Newton (ed.) *Transitions: Pathways Towards Sustainable Urban Development in Australia*, Melbourne: CSIRO and Dordrecht: Springer.

Fox, I. and Hietanen, J. (2007) 'Interorganizational use of building information models: potential for automational, informational and transformational effects', *Construction Management and Economics*, 25 (3): 289–96.

Gans, J. (2008) 'Advance Australia where?', *Innovation 08*, 5 May: 17–19. Online. Available at HTTP: <www.australianinnovation.net.au/display.php>.

Graham, P., Reedman, L. and Cheng, J. (2008) 'Energy futures', in P.W. Newton (ed.) *Transitions: Pathways towards Sustainable Urban Development in Australia*, Melbourne: CSIRO and Dordrecht: Springer.

Hampson, K. and Brandon, P. (2004) *Construction 2020: A Vision for Australia's Property and Construction Industry*, Brisbane: CRC for Construction Innovation.

Hampson, K. and Kwok, T. (1997) 'Strategic alliances in building construction: a tender evaluation tool for the public sector', *Journal of Construction Procurement*, 2 (1): 28–41.

Hampson, K. and Manley, K. (2001) 'Construction innovation and public policy in Australia', in *Innovation in Construction: An International Review of Public Policies*, Brussels: CIB TG35, Innovation Systems in Construction.

Hardwicke, L. (2008) 'Transitions to smart, sustainable infrastructure', in P.W. Newton (ed.) *Transitions: Pathways Towards Sustainable Urban Development in Australia*, Melbourne: CSIRO and Dordrecht: Springer.

Hennessy, K. (2008) 'Climate change', in P.W. Newton (ed.) *Transitions: Pathways Towards Sustainable Urban Development in Australia*, Melbourne: CSIRO and Dordrecht: Springer.

House of Representatives Standing Committee on Environment and Heritage (2005) *Sustainable Cities*, Canberra: Parliament of Australia.

—— (2007) *Sustainability for Survival: Creating a Climate for Change. Inquiry into a Sustainability Charter*, Canberra: Parliament of Australia.

Howden, M. and Crimp, S. (2008) 'Drought and high temperatures', in P.W. Newton (ed.) *Transitions: Pathways Towards Sustainable Urban Development in Australia*, Melbourne: CSIRO and Dordrecht: Springer.

ISR (1999) *Building for Growth: An Analysis of the Australian Building and Construction Industries*, Canberra: Department of Industry, Science and Resources.

Jones, T. (2008) 'Distributed energy systems', in P.W. Newton (ed.) *Transitions: Pathways Towards Sustainable Urban Development in Australia*, Melbourne: CSIRO and Dordrecht: Springer.

Kaebernick, H., Ibbotson, S. and Kara, S. (2008) 'Cradle-to-cradle manufacturing', in P.W. Newton (ed.) *Transitions: Pathways Towards Sustainable Urban Development in Australia*, Melbourne: CSIRO and Dordrecht: Springer.

Leicester, B. and Handmer, J. (2008) 'Bushfire', in P.W. Newton (ed.) *Transitions: Pathways Towards Sustainable Urban Development in Australia*, Melbourne: CSIRO and Dordrecht: Springer.

Meadows, D.H., Meadows, D.L., Randers, J. and Behrens, W.W. (1972) *The Limits to Growth*, London: Earth Island.

Mitchell, W.J. (1995) *City of Bits: Space, Place, and the Infobahn*, Boston, MA: MIT Press.

Newman, P. (2008) 'The oil transition and its implications for cities', in P.W. Newton (ed.) *Transitions: Pathways Towards Sustainable Urban Development in Australia*, Melbourne: CSIRO and Dordrecht: Springer.

Newton, P.W. (1994) 'Networking CAD', *Environment and Planning B: Planning and Design*, 21 (6): 731–47.

—— (1995) *Information Technology and Living Standards*, Canberra: Australian Institute of Health and Welfare.

—— (2006) *Australia State of the Environment 2006: Human Settlements Theme Commentary*, Canberra: Department of Environment and Heritage. Online. Available at HTTP: <www.environment.gov.au/soe/2006/publications/comment-aries/settlements/index.html>.

—— (2007a) '2006 Australia State of the Environment: Human Settlements', *Environment Design Guide*, Building Design Professionals and Royal Australian Planning Institute, February: 1–9.

—— (2007b) 'Horizon 3 planning: meshing liveability with sustainability', *Environment and Planning B: Planning and Design*, 34 (4): 571–5.

—— (2008a) 'Metropolitan evolution', in P.W. Newton (ed.) *Transitions: Pathways Towards Sustainable Urban Development in Australia*, Melbourne: CSIRO and Dordrect: Springer.

—— (ed.) (2008b) *Transitions: Pathways Towards Sustainable Urban Development in Australia*, Melbourne: CSIRO and Dordrecht: Springer.

Newton, P.W. and Crawford, J.R. (1995) 'Networking construction and the emergence of virtual project teams', in F. Williams, H. Brake and J. Nolan (eds) *Broadband Islands 95: Global Broadband and Beyond: Proceedings of the 4th International Conference, Dublin.*

Newton, P.W., Baum, S., Bhatia, K., Brown, S.K., Cameron, A.S., Foran, B., Grant, T., Mak, S.L., Memmott, P.C., Mitchell, V.G., Neate, K.L., Pears, A., Smith, N., Stimson, R.J., Tucker, S.N. and Yencken, D. (2001) *Australia's Human Settlements: State of Environment Report 2001–2005*, Canberra: Environment Australia. Online. Available at HTTP: <www.environment.gov.au/soe>.

Newton, P.W., Newman, P., Manins, P., Simpson, R. and Smith, N. (1997) *Re-Shaping Cities for a More Sustainable Future: Exploring the Link between Urban Form, Air Quality, Energy and Greenhouse Gas Emissions*, Research Monograph 6, Melbourne: Australian Housing and Urban Research Institute.

OECD (2008) *OECD Territorial Reviews: Competitive Cities in the Global Economy*, Paris: OECD.

Productivity Commission (2007) *Public Support for Science and Innovation*, Research Report, Melbourne: Productivity Commission.

Rees, W.E. and Roseland, M. (1991) 'Sustainable communities: planning for the 21st century', *Plan Canada*, 31 (3): 15–26.

Regan, M. (2008) 'Critical foundations: providing Australia's 21st century infrastructure', in P.W. Newton (ed.) *Transitions: Pathways Towards Sustainable Urban Development in Australia*, Melbourne: CSIRO and Dordrecht: Springer.

Stoeckel, A. and Quirke, D. (1992) *Services: Setting the Agenda, Report 2*, Canberra: Centre for International Economics, report to Department of Industry, Technology and Commerce.

Uher, T.E. (1994) 'What is partnering?', *Australian Construction Law Newsletter*, 34: 49–60.

United Nations (2008) *World Urbanization Prospects: The 2007 Revision Population Database*, New York. Online. Available at HTTP: <http://esa.un.org/unup/> (accessed 1 July 2008).

Part II
Materials

2 Future materials and performance

Terry Turney

Developments in materials and manufacturing processes are literally reshaping our urban environment. They are helping us adapt to the vagaries of life in more innovative ways – from the effects of earthquakes to the debilitation of Alzheimer's disease. The real prospect of intelligent, self-repairing, biomimetic structures, operating in a highly energy-efficient and waste-free manner, is enabling a fundamental shift in how we design our living environments. Ultimately, the cost and availability of energy and natural resources and the capacity of the Earth's ecosystem services to continue absorbing our activities will determine the nature of our urban infrastructure.

Within the complex tensions of natural and man-made forces, materials development is being driven by a series of major issues:

1   We are unable to maintain current consumptive growth at present global levels. This driver is reflected by increasing trends towards closed-loop materials usage (the three Rs – recycle, reclaim and remanufacture) and extended product lifetimes.
2   In the face of rapid climate changes and limitations in usable water and energy supplies, options regarding how we are going to sustain the utilization of materials resources are rapidly closing.
3   Rapid and massive urbanization in Asia and other regions is creating increased pressure for substitution of existing materials and resources.
4   A growing tension in resource equity between First World and other economies is also driving materials replacements. Maintaining economic measures of growth and generation of high-value manufactured products in developed economies is competing with the need to attend to basic human needs in developing and emerging economies.
5   The effects of commoditization: bringing down the unit cost of goods and services, including materials and processes.
6   The trend towards mass customization – flexible production of goods and services to meet individual customer's needs with near mass production efficiency demands new materials solutions (Kaplan and Haenlein 2006).

7   The increasing pervasiveness of globalized supply chains, markets and technology-enabled services ensures that materials advances are adopted everywhere.

Economic impacts of maintaining and growing our cities are not adequately addressed. In particular, the replacement costs of our ageing urban infrastructure (water, electricity, ports, rail and road systems) may soon exceed projected government revenues in many developed countries. Together with the increasing expenses of health care, greenhouse gas mitigation, food production, and land and water degradation, there are insufficient economic resources available to deliver solutions to all issues simultaneously. Improved materials design, production and performance can present opportunities for product differentiation and improved performance with stronger, lighter, cheaper, more intelligent, more energy efficient, etc., materials. The benefits of materials designed for their tenth or hundredth use, rather than for a single use, are not just in greater efficiency, but in the opportunity for businesses to provide services through manufacturing, rather than just manufacturing and selling the product. The potentially socially disruptive outcomes of global competition for economic, materials and energy resources may well result in a 'wicked problem' that is highly resistant to resolution, with multiple stakeholders each convinced that their version of the problem and their approach is correct (Briggs 2007).

Each of the above issues presents multiple opportunities for materials and processes to impact on building and construction innovation. The technological and scientific advances described in this chapter are dependent upon emerging technologies, such as nanotechnology and our understanding of molecular design, soft matter, biological systems and of surface science. The examples have been chosen to illustrate our first steps towards creating intelligent or self-repairing structures and towards biomimetic approaches in materials design and performance as we progress into the twenty-first century.

Clever constructions

Iron and other metal alloys are traditionally employed in structural applications for their tensile strength and tolerance to flaws. In contrast, ceramics and stone typically have high compressive strength but are brittle and do not tolerate defects, either in the bulk or on the surface. Polymers are often flaw-tolerant, but their poor strength and ductility at low applied stresses severely limits structural applications. Lignocellulose biomaterials, such as wood, straw and bamboo, have compressive and flexural advantages, but suffer from limited lifetime, flammability, and poor tensile strengths. As we reach the theoretical limits of traditional materials, it is becoming increasingly difficult to meet the stringent property demands now being sought. It is rare for the performance of a material to rely on one property alone. What is usually required is a combination of properties, which may well conflict with each other. A good example is a thermoelectric material, where one is seeking high electrical conductivity but poor thermal conductivity. Limitations in materials performance and correlations between different properties are well established (Ashby 2005). From that basic palate of materials, clever design at an electronic, molecular and nanostructural level allows construction of hybrid or composite materials and the deliberate engineering of surfaces to transform materials science and its applications.

Metals

Applications of metals are limited by their high density, high energy content, processing costs and corrosion issues. The understanding of the relationships between nano- and microstructure in metals and the resultant properties is now reaching a state where systems can be designed at a rational atomic level. Substantial progress in the development of alloys of the light elements, aluminium, magnesium and titanium, is changing their use (Polmear 2006). Similarly, development of low-density metal foams ('syntactic metals') with up to 80 per cent void fraction enables interesting lightweight structural applications, resulting in overall less materials usage. These foams also exhibit interesting functional properties, such as good impact resistance and acoustic insulation (Ashby *et al.* 2000; Banhart 2001). Improved understanding of the surface reactions is resulting in much better management of metal corrosion and the passivation of surfaces (Roberge 2006; Landolt 2007; Song and Atrens 2007) as well as inhibition of microbiologically influenced corrosion (Little *et al.* 2007). Novel and more economical metal processing methods, such as severe plastic deformation techniques to control nano- and microstructure, create stronger metals for use in extreme environments (Valiev *et al.* 2007).

Cement and concretes

Cementitious materials have traditionally been limited by low ductility, relatively high density and poor tensile strengths. However, through careful selection of cement precursors, the development of ultra-high performance concretes, with compressive strength of up to 250 MPa, has opened up many new applications. Examples include use of fine pozzolans (e.g., fumed silica), polymeric superplasticizers (often polyacrylates) to control rheology and lower the water:binder ratio, and careful process control (Rahman *et al.* 2005). A particularly effective way to high-performance or high added-value cementitious materials has been through composites containing fibre and even woven textile reinforcing. A striking example of the innovative use of such composites is the Concrete Canvas Shelter, consisting of a cement–canvas composite bonded onto an inflatable polyethylene liner. These shelters are rapidly deployable in less than one hour, simply by inflating and adding water, to create a robust, fire-proof, thin-walled structure that is optimized for compressive loading and ready for use within 12 hours (Concrete Canvas 2008).

Superplasticizers, together with ultrafine particles, such as fumed silica or kaolinite, create 'DSP cements' (densified with small particles), which have low porosity and high compressive strengths. The small particles occupy voids between the larger cement particles to afford both the strength and microstructure (Rahman *et al.* 2005; Shah and Weiss 1998). These design concepts have been extended in fibre-reinforced 'reactive powder concretes' (RPC) or 'engineered cementitious composites' (ECC), with compressive strengths of up to 230 MPa and flexural strengths of 30–50 MPa. The ductility of RPCs is remarkable for concrete, exhibiting good residual flexural and tensile strengths even after cracking (see Figure 2.1a).

A wide range of fibres, including steel and polymers, have been employed in these ECCs, which can withstand 3–7 per cent tensile strain without breaking or loss of strength. Their closed-pore structure also results in high durability, high resistance to attack by chlorides or sulphates and almost no carbonation when compared to standard Portland cement (Li 2003, 2006; Ahmed and Mihashi 2007). RPCs allow lightweight, long-life structures to be constructed, with substantial savings in materials construction costs and maintenance, and hence a lowered environmental impact. Importantly,

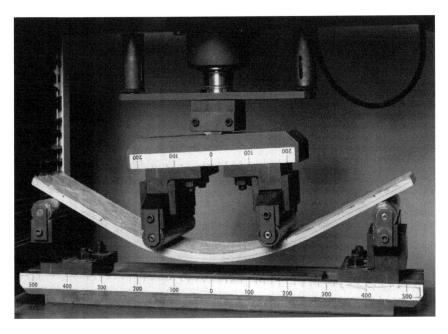

Figure 2.1a Bendable engineered cementitious composite (ECC) subject to flexural loading (source: photo courtesy of Kajima Corporation and Kuraray Co. Ltd).

Figure 2.1b Cable-stayed Mihara Bridge in Hokkaido, opened in 2005 with an ECC on
steel deck. The ECC stiffens the jointless deck and decreases stress, resulting in a 40 per cent weight reduction and a 50 per cent cost reduction
(source: photo courtesy of Kajima Corporation and Kuraray Co. Ltd).

ductile RPCs can be used in damage-tolerant structures, particularly for
improving the earthquake resistance of large buildings and civil engineering
structures (Meyer 2006). Recent examples of RPC use include the Mihara
Bridge in Hokkaido (see Figure 2.1b) as well as a 27-storey residential high
rise in Tokyo (Glorio Roppong in 2006) and the 41-storey Nabeaure Tower
in Yokohoma (2007), all of which use bendable ECC concrete in coupling
beams for seismic resistance in the building core.

Many aluminosilicate materials will react with very highly alkaline or
silicate solutions to produce cementitious inorganic polymers or 'geopolymers' (Davidovits 1991; Duxson *et al.* 2007). These materials can match
the performance of traditional cements, but allow use of otherwise waste
feedstocks, such as slag and fly-ash. Given the very high embodied energy
and CO_2 emissions associated with Portland cement manufacture, these
engineered inorganic polymers are certainly environmentally attractive
alternatives (Gartner 2004; Sofi *et al.* 2007).

Composite materials

It is often far easier to create new materials with demanding properties by
combining two or more existing materials into a composite structure.

Steel-reinforced concrete and carbon-fibre composites are man-made examples of how to stop otherwise brittle materials from failing under tension or flexure, and the tough nacre of abalone shells or the complex hierarchical collagen-hydroxyapatite structure of mammalian bone are natural examples. The attractiveness of composite materials is that their properties can generally be tuned to take advantage of their component properties (van Damme 2008). However, the exquisite hierarchy found in Nature's defect-tolerant structures still far exceeds man-made efforts (Fratzl and Weinkamer 2007; Meyers *et al.* 2008).

Polymer nanocomposites exhibit novel properties, partly arising from nano-particle, -fibre or -platelet being comparable in size to the polymer chain. Interactions at the high surface-area interface alter polymer chain motion and packing, and hence dominate composite properties, through a combination of enthalpic and entropic effects (Balzas *et al.* 2006; Crosby and Lee 2007). The importance of controlling structural defects in composites is illustrated by two recent reports of the careful layer-by-layer assembly of lamellar nanocomposites. In one case, montmorillonite clay platelets (approx. 1-nm thick) with polyvinyl alcohol layers produce a well-dispersed and low-defect nanocomposite which was transparent to light (fewer defects results in fewer light scattering centres) and exhibited an order of magnitude increase in tensile strength and stiffness when compared with similar but more disordered nanocomposites prepared by simple self-assembly methods (Podsiadlo *et al.* 2007). The other example involves 200-nm thick α-Al_2O_3 platelets dispersed in a chitosan matrix to afford a both strong and ductile nanocomposite, with a structure similar to nacre (Bonderer *et al.* 2008). Current composites in the building sector are far from reaching such sophistication in nanostructure control.

A stimulus-responsive, nanocomposite material that rapidly and reversibly alters its stiffness has been fabricated from a rubber copolymer and cellulose nanofibres (Capadona *et al.* 2008). Its overall stiffness, which increases by a factor of 40 simply through a change in solvent, is determined by transient interactions between adjacent nanofibres within the rubber matrix. This composite structure was deliberately designed to mimic the reversible change in stiffness found in Holothurians (sea cucumbers), which use it as a mechanism against predation. The material is remarkable in its behaviour, and opens the door to the design of a wide range of other stimulus-sensitive polymers.

Most applications for polymer nanocomposite, at least over the next ten years, will focus on relatively high added-value markets. Currently lightweight and impact-resistant automotive components and barrier food and beverage packaging predominate (BCC Research 2006). However, a range of other properties of relevance to the building sector are under active development, including UV resistance; combinations of optical, mechanical and electrical properties; electrostrictive or magnetostrictive effects; wear resistance; flame retardance; gas barrier or enhanced permeability; chemical inertness; and thermal stability. Future materials may well be designed to respond actively to mechanical, thermal, electrical, magnetic or chemical changes in their environment. Such materials will find many applications in the construction of truly 'intelligent' and adaptive buildings.

Functional facades

Building facades ought to be structural and decorative. In addition, multi-functional facades which combine energy management, condition reporting and self-maintenance are becoming a reality through rapid advances in 'smart' materials, coatings and energy devices. Such 'facades' are a characteristic of the natural world, where natural selection has evolved numerous mechanisms for keeping surfaces waterproof when necessary, relatively free of dirt, optimized for heat and light management, and generally self-repairing. The multifunctionality of human skin is a prime example. The past decade has seen a remarkable revolution in our understanding and mimicking of Nature's functional surfaces at a chemical, physical and atomic structural level. This understanding can now be applied to building facades.

Effective light management within facades requires control of both the intensity and wavelength of light entering a structure. One of the problems with using concrete in facades is that it is not transparent – and it is doubtful that it ever will be. A partial and innovative solution to opacity of concrete is the incorporation of large numbers of optical glass fibres to make a translucent light-transmitting material, Litracon™ (Losonczi 2007). The compressive strength of the concrete is not compromised by the relatively small volume fraction of fibres (*c.*4 per cent) needed. A similar translucent concrete product, Luccon™, has been made from fine-grained concrete and a fabric which is cast layer by layer in prefabricated moulds (Luccon Lichtbeton 2008). Although expensive compared with normal concrete, these products open new decorative and functional uses for an otherwise undifferentiated facade material.

Concepts from the cosmetics and personal-care products industries and from polymer film developments are being used to control the wavelength of light absorbed or transmitted by a surface. Films, containing highly dispersed nanoparticles such as ZnO or TiO_2 are effective UV absorbers, but still transparent to visible light (Tsuzuki 2008). At this stage, the chemical or photoreactivity of the nanoparticle additives limits film lifetimes to unacceptable levels. Some control of wavelength can be contained with museum glass, with UV-absorbing properties (Tru Vue 2004). However, its relatively high costs have prevented widespread uptake in the building sector. Self-cleaning glass products containing TiO_2 coatings, which are readily available, have the benefit of being highly UV-absorbing.

Switchable electrochromic glazing enables glass to change colour from clear to dark using an electrical current, which can be activated manually or programmed by sensors which respond to external light intensity. The production of electrochromic glass is a complex and costly undertaking, with five distinct layers (transparent conductor, electrochromic electrode, ion conductor, counter electrode, transparent conductor) being laid down by vacuum sputtering onto a glass sheet. The electrode assembly is reasonably fragile, and generally needs to be protected by double glazing (Manfra 2007). Long-lifetime electrochromic windows are now commercially available (e.g., Sage Electrochromics 2008; Lawrence Berkeley National Laboratory 2006). Such active glazing technologies have the ability to decrease peak cooling loads by up to 30 per cent in some commercial buildings; by darkening when necessary to reduce solar transmission into the building or brightening, they minimize the need for artificial lighting and cooling. They also allow designers to blur the distinction between walls and windows, with the possibility of variable light transmission depending on safety and privacy needs or the level and quality of light outdoors.

A related advance to the electrochromic facade is the photovoltaic facade, allowing the building to generate some of its energy requirements from the sun. Photovoltaic energy generation is one of the most intensively researched areas today. Low-cost substitutes for Si-based systems and flexible photovoltaic assemblies will soon be viable options. Although efficiencies comparable to Si (>20%) can currently be obtained in the laboratory, commercially produced cell assemblies rarely exceed 10 per cent. There are many companies currently competing to produce the most efficient, cost-effective thin-film solar cells (Konarka, Nanosolar, Global Solar, Ascent, DayStar, Miasole, First Solar, etc.). Their technologies vary in the manufacturing process used or in the photovoltaic materials (e.g., dye-sensitized solar cells or thin-film semiconductors containing fullerene, CdTe or chalcopyrite structures) (Snaith and Schmidt-Mende 2007; Hoth *et al.* 2007; Green *et al.* 2008). Production capacity of non-Si photovoltaics is growing very rapidly as costs decrease and efficiencies improve. Markets are being developed for building integrated photovoltaic (BIPV) products in commercial and residential sectors, and for centralized power generation. Flexible roll formats for BIPVs now allow rapid and low-cost system integration and retrofitting. A recent strategic assessment has indicated that the energy saving from replacing standard glazing with electrochromic glazing is comparable to the energy generated from photovoltaic facades in many cases, highlighting the importance of a multilateral approach to energy management (Mardaljevic and Nabil 2008).

Facades also play a key role in building heat-management. Controlled porosity coatings are now available, which substantially decrease heat transmission through surfaces. As an example, Industrial NanoTech Inc. produces a coating product, Nansulate®, which lowers thermal conduction through a controlled porosity inorganic oxide/hydroxide coating within a styrene-acrylic co-polymer matrix. The product appears to be quite effective for pipe insulation and tank insulation, and has the added benefits of being a mould and rust inhibitor. The very low thermal conductivity through the coating has been created by tailored sub-micron pore sizes, and the closed nature of the pore network (Industrial NanoTech 2008). More advanced methods of facade heat management involve IR reflective surface coatings. Partially transparent thin metallized surfaces are able to reflect up to 98 per cent of infra-red radiation and still achieve reasonable transmission in visible light conditions. However, traditional vacuum-coating technology for production of such films is too costly for widespread application to facades. As an alternative, a range of selectively emissive and reflective paints is becoming available in normal visible paint colours, but which give good IR reflectivity (Hyde and Brannon 2006; Ryan 2005). Recent laboratory developments in thermochromic VO_2 coatings could form the basis for thermally responsive, energy-efficient glazing (Vernardou *et al.* 2006). Vanadium dioxide undergoes a marked change in optical transmittance and reflectivity in the IR region, associated with an insulator-to-metal phase transition. In principle, such coatings would absorb IR radiation until they reach the transition temperature, and then become strongly reflecting.

The various heat management technologies outlined here permit interactive and real-time management of daily heating and cooling cycles within buildings. More widespread adoption of these materials would result in less heat transfer to and from buildings, reductions of 'heat island' effects within built-up areas, lower overall energy demands and an interactive mechanism for peak load levelling.

Of the many materials advances being made in multifunctional facades, progress in self-cleaning surfaces is having the greatest visual impact. Our understanding of surface wetting and reactivity is making the contamination-free surface a reality. It is worth examining this area in more detail, as it illustrates how scientific and technological understanding of a phenomenon translates into commercial products. Three alternative design strategies for self-cleaning surfaces are now emerging:

- easy physical removal of dirt through 'superhydrophilic' films of water which thoroughly wet the surface; or
- prevention of residual dirt adhesion with 'superhydrophobic' surfaces where water droplets readily roll off a surface, collecting dust particles on the way; or
- removal of organic or biological surface films by photocatalytic oxidation.

The physics of surface wetting has been well understood in terms of the surface tensions (γ) at the solid–liquid–gas interfaces as described by Young's equation ($\gamma_{SG} - \gamma_{SL} = \gamma_{LG}\cos\theta$, where θ is the contact angle at the interface; see Figure 2.2).

On strongly hydrophilic and smooth surfaces, a water droplet will spread out on the solid surface and the contact angle will be small. If it approaches 0°, the surface can be considered superhydrophilic. Naturally

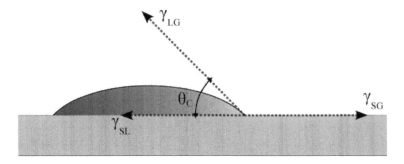

Figure 2.2 Surface tensions for a droplet resting on a smooth horizontal surface; the liquid/vapour interface meets the surface at a contact angle, $\theta°C$.

hydrophilic glass and other surfaces can be made to be hydrophobic (i.e. having static contact angles greater than 90°) by chemical surface treatment with silanes or fluoropolymers. In most cases, the static contact angle still does not exceed 130°. These treatments with long-chain alkyl silanes are the basis for the widely used damp-proofing of buildings. Fluoropolymer treatments have been developed, which have the added benefit of also being oil-repelling (lipo- or oleo-phobic). The release of Lumiflon®, a curable fluoropolymer coating by Asahi Glass in the 1980s, has been followed by a spate of hydrophobic coating products with diverse OEM and aftermarket applications, ranging from glass and concrete to textiles and paints (Nanogate 2008; Nanokote 2007; NanoSafeguard 2008; Nanotex 2008; Dow Corning 2008).

Very low surface energy, superhydrophobic and self-cleaning surfaces (with static contact angles greater than 150°) can be made by careful control of the roughness of hydrophobic surfaces at a micro- and nano-level (Blossey 2003; Feng and Jiang 2006; Roach *et al.* 2008; Lundgren *et al.* 2007). Water droplets simply rest on the superhydrophobic surfaces without actually wetting to any significant extent; they can exhibit contact angles approaching 180° and droplet run-off angles of less than 2° from horizontal. This phenomenon, often described as the 'Lotus Effect®', was inspired by the water-repellent properties of the lotus leaf (Barthlott and Neinhuis 1997, 2005). The detailed physics of wetting on rough surfaces is a subject of intense research, but two extreme cases have been well established. Cassie and Baxter found that increasing the roughness of a hydrophobic surface increases the contact angle of a water droplet. The energy penalty is typically too high for the water to follow the rough contours, resulting in the droplet receding and forming an even larger contact angle compared to the smooth surface (Parkin and Palgrave 2005; Wang and Jiang 2007). On a rough hydrophilic surface, the opposite effect is observed: the water droplets wetting the surface are pinned in place by the surface roughness, resulting in their inability to slide on the surface (generally called Wenzel wetting). The two extreme wetting states can be seen in Figure 2.3.

Nature's solution is even more elegant; the lotus leaf is an extreme state of Cassie wetting, showing a complex hierarchy of surface roughness (Li and Amirfazli 2008).

There are numerous commercial products which mimic the lotus effect, manufactured by hydrophobic modification of existing rough surfaces or by deliberate roughening of hydrophobic surfaces. As an example, BASF recently released Mincor TX TT, a polymer finishing material for water-proofing technical textiles, such as awnings, sunshades and sails, with the same self-cleaning effect as the lotus (BASF 2008a). The product creates an artificial surface roughness with nanoparticles, firmly embedded within a hydrophobic carrier matrix.

There are two approaches to superhydrophilic surfaces. It is possible to promote superwetting behaviour by introducing roughness at the right length scale (Bico *et al.* 2002). Textured surfaces with enhanced hydrophilic-ity have now been prepared by a variety of fabrication methods, including sol-gel coating, use of woven films, microlithography, modification of surface chemistry and creation of microporosity (Ogawa *et al.* 2003; Zhang *et al.* 2005; Cebeci *et al.* 2006). There are also several commercially avail-able superhydrophilic coating products, e.g. StoCoat™ Lotusan®, which rely solely on such nanotexturing or lotus effect (Sto Corp. 2007). In 2008, BASF AG released a new coating product, COL.9®, based on a dispersion of polyacrylate particles in which nanoparticles of SiO_2 have been incorpo-rated. In conjunction with Akzo Nobel it has launched a facade coating under the brand Herbol-Symbiotec (BASF 2008b). The system is claimed to work because of the combination of 'elastic' polymer particles and 'hard' nanoparticles, but may well be also an example of a superhydrophilic surface through roughness control.

A relatively simple method to produce superhydrophilic coatings employs photoactive semiconductor metal oxides, such as TiO_2, ZnO, WO_3, V_2O_5, or organic molecules, such as azobenzene or spiropyran. These films change their surface energies and become superwetting after exposure to UV or visible light, but often revert to being hydrophobic after a short period in the dark as the surface loses its charge (Wang *et al.* 1997; Wang

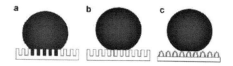

Figure 2.3 Different states of superhydrophobic surfaces: (a) Wenzel's state, (b) Cassie's superhydrophobic state, (c) the 'Lotus' state (a special case of Cassie's superhydrophobic state) (source: modified from Wang *et al.* (2007), © Wiley-VCH Verlag GmbH & Co. KGaA. Reproduced with permission).

and Jiang 2007). However, with suitable surface texturing at the nanoscale, relatively long superhydrophilic lifetimes can be achieved even with non-semiconducting metal oxides (Cebeci *et al.* 2006; Gu *et al.* 2004). Generally such superhydrophilic surfaces offer other advantages, such as anti-fogging or anti-reflection properties.

Superhydrophobic-to-superhydrophilic switching can be induced not only by UV light but also by mechanical changes, by electric or magnetic fields, by heating or with chemicals and solvents (Feng and Jiang 2006). Thus, it is possible to tune surfaces with nanostructured coatings to produce well-defined superhydrophobic, superhydrophilic, superoleophobic or superoleophilic domains all on the same substrate (Zimmermann *et al.* 2008). The prospect of such ambiphilic patterning of surface regions with varying affinity for water (or oils) may well lead to more sophisticated structures, allowing liquids to be transported away from areas of unacceptable condensation or, indeed, for water collection. Such passive structures mimic a natural strategy for water collection which has been found in a desert-dwelling beetle where fog, the sole source of moisture, is 'guided' to its mouthparts via amphiphilic channels (Parker and Lawrence 2001). These concepts could be extended to fabrication of tuneable coatings, for active transport of liquids on stimulus-responsive surfaces which were still macroscopically flat to the touch.

The photocatalytic effect of TiO_2 coatings has found numerous applications in 'green' building design. It has been incorporated into a wide range of paints, glazes and cements. Most major glass companies have developed products which exploit the photocatalytic effect of TiO_2 and its super-hydrophilic, anti-fogging and self-cleaning properties under irradiation. Examples are Pilkington Glass's Active®, which contains 15-nm photocatalytic TiO_2 particles together with a hydrophilic surface (Pilkington Group 2008). Similar products are available from other glass manufacturers, e.g. PPG Industries' SunClean® and Saint Gobain Glass' Aquaclean® product ranges. There are some outstanding architectural examples using photocatalytic technology. In particular, the Jubilee Church in Rome, also known as the Dives in Misericordia (see Figure 2.4), completed in 2003, comprises a self-cleaning concrete developed by the Italcementi Group under the brand name TX Active® (Italcementi Group 2006). TOTO in Japan has also been particularly active in applications, with licences to over 40 companies for TiO_2-based photocatalytic coatings for facades as well as for automotive and road materials (TOTO 2008).

Nanoparticulate TiO_2, particularly in its anatase form, is a potent photocatalyst under near-UV radiation. Its relatively low cost and chemical stability under both acidic and basic conditions and general chemical inertness make the material very attractive for control of airborne pollutants. One proposed use is to control volatile organics, NO_x and SO_x, over large urban areas. When Intalcementi's TX Active, TiO_2 was incorporated into paved surfaces in street tests in Segrate, near Milan, NO_x

Figure 2.4 The facade of Dives in Misericordia in Rome, built in 2003, consists of 256 precast, self-cleaning concrete sections assembled into 25-metre high curved white sails (source: photo courtesy of Claudio La Rosa, Rome (Flilckr – axez02)).

reductions of around 60 per cent from traffic air pollution were observed (Giussani 2006). The use of photocatalytic coatings for odour control in residential and commercial environments, such as bathrooms and food preparation areas, is an obvious future application, provided there is sufficient UV light intensity from ambient lighting. Special photoactive glazes for ceramic tiles were designed initially for hospital use, but are now readily available for domestic applications (Deutsche Steinzeug 2008). Not only are the photocatalytic coatings self-cleaning, they also show strong antibacterial effects.

Traditional fixing methods in building construction may be soon complemented by new products based on biological methods of adhesion. Of Nature's many methods of fixture, two have been subject to intensive recent research. The ability of many lizards and insects to climb on sheer surfaces appears to be a result of van der Waals forces, which operate only over atomic distances. The feet of these creatures have many fine flexible hairs or spatulae ordered in a hierarchical fashion into coarser hairs (setae) (Autumn *et al.* 2002). Collectively, they create a strong adhesive force which is compliant against both smooth and rough surfaces. Many research groups have now mimicked the biological architecture of these fibrillar surfaces to create synthetic adhesive structures, endearingly called 'Gecko Tape' (Chen *et al.* 2008). One notable example is the fabrication of micropatterned arrays of carbon nanotubes, with a hierarchical structure similar to that found on a gecko's foot, supported on a polymer tape. The tape has remarkable properties of adhesion, supporting a shear stress of $36\,N/cm^2$ – almost four times higher than that of the gecko's foot. This dry, reversibly adhesive, nanotube-based tape is also able to stick to a range of surfaces, including Teflon (Ge *et al.* 2007). Although under active development, a commercially viable 'Gecko Tape' is still a future prospect. There are, however, many other biomaterials which operate as more conventional 'wet' adhesives. Many organisms secrete strongly adhesive substances for attachment (e.g., the byssal threads of mussels) as well as in predation (e.g., secretions from ticks and velvet worms). One notable example, used as a defence mechanism against fish predation, is the adhesive structures (Cuvierian tubules) expelled by species of Holothurian (sea cucumbers). These proteinaceous materials are capable of strong adhesion in a matter of seconds (deMoor *et al.* 2003). Potential applications for a superglue that can rapidly set in water (even in seawater) are manifold.

So far, this section has illustrated materials solutions to heat and light management, energy generation, self-cleaning and adhesion. An intriguing challenge is creating a facade to diagnose internal or surface faults and to repair itself, either as a response to an external stimulus or even autonomously. Limited self-healing is not uncommon in metal alloys or at a nanoscale on surfaces (see, for example, Lumley 2007). Even concrete has some ability to self-repair cracks over time (Li and Yang 2007), and there is evidence that certain non-pathogenic bacteria could assist that healing process by promoting deposition of insoluble Ca-containing phases, making the facade of the future literally a truly living structure (Jonkers 2007). Certain classes of polymers, such as ionomers, with up to 20 mol% of ionic species within their structure, can undergo repeated self-healing. The materials, commercially available as DuPont's Surlyn® and Nucrel® polymers, will repair themselves, even after ballistic penetration, by recrystallization and then reordering of the physical cross-links between the ionomer components (Varley 2007). Unlike conventional cross-linked rubbers made of macromolecules, supramolecular elastomers have now been made that will self-heal by simply bringing together fractured or cut surfaces at room temperature (Cordier *et al.* 2008). There is a wide range of other potentially mendable polymers that have been developed, but at this stage none has been used commercially (Bergman and Wudl 2008).

A particularly exciting prospect for self-repairing structures lies in the 'vascularization' of materials. If a small hole or fracture appears in the structure, due to fatigue or external damage, a repairing compound or an adhesive would 'bleed' from embedded vessels near the point of damage to restore the integrity of the material, in much the same way as living organisms undergo self-repair. By embedding a three-dimensional microvascular network within an epoxy substrate, crack damage induced by bending the epoxy coating may be healed repeatedly through release of uncured epoxy in the presence of a curing catalyst (Andersson *et al.* 2007; Toohey *et al.* 2007). The concept of circulatory networks within materials provides a mechanism for delivery of healing agents for self-repair, but in future could also be employed to circulate sensors through a facade and impart additional functionality, such as self-diagnosis and temperature regulation, and variation of mechanical properties, such as flexural modulus.

Intelligent interiors

Of the many advances being made in materials for applications, two areas in particular will fundamentally change how buildings are designed and operated: the development of embedded intelligence through pervasive sensing, and improvements in energy generation, storage and use.

The rapidly decreasing size and cost of sensors and of computing capabilities is leading to a multitude of 'smart' consumer devices, from air conditioners and refrigerators to toothbrushes. However, the rapidity of these changes, extreme cost sensitivity and the risk-averse nature of the construction industry has resulted in minimal uptake of such technology in building design and construction. Most appliances and objects can now have embedded processing and communication capability, or be able to link online to data-processing capabilities. The current generation of radio-frequency identification (RFID) devices will soon be supplanted by more intelligent sensing devices linked to processing and possible actuation capabilities. Such technology, allowing building structures to be interrogated in real time, will be achieved through significant decreases in the cost of sensors and in mass customization through new inkjet printing technologies (Murata *et al.* 2005; Bidoki *et al.* 2007), flexible substrates for computing (see, for example, Kim *et al.* 2008) and in radically new sensor designs (e.g., NanoMarkets 2008; Israel *et al.* 2008). One can envisage integration of sensing devices for building condition monitoring and repair with other functions, such as care and supervision for people with physical or neurodegenerative disabilities, and with general health care.

Energy is becoming less available and more expensive than it has ever been. As this change will be permanent for the foreseeable future, it is important to consider all options for the production, storage and use of energy. Materials development has played a pivotal role in expanding our options for sustainable energy generation. Photovoltaic devices in relation to facades were discussed above. A variant, thermophotovoltaics (TPV), uses infrared and some of the visible part of the solar spectrum not captured by existing photovoltaic devices. TPV can convert high-grade heat into electricity via a radiation emitter and photocells. The main impediment to current uptake of the technology is a materials issue, related to mismatch of the radiation spectrum of the emitter to the quantum efficiency of the photocells (Yugami 2003; Chubb 2007). However, hybrid TPV generators with space heaters are available for commercial use (Fraas 2007). Alternatively, both low-grade and high-grade heat can be harvested by thermoelectric devices (Snyder and Toberer 2008). In addition, the use of piezoelectric devices to scavenge ambient vibrations within a building from passing traffic, the wind, etc., is also of significance and provides a means of powering ubiquitous nanosystems, essential for embedding intelligence into structures (Wang 2007). It is obvious that full utilization of available energy sources is far from being achieved in even the most advanced current building design.

Energy storage is an important issue in reducing the mismatch between supply and demand, in maintaining the performance and reliability of devices requiring power, and in energy conservation. Materials for energy storage devices have been dominated by improvements in battery design and performance, including flexible and paper-based batteries (Pushparaj *et al.* 2007), energy storage devices, such as supercapacitors, and heat storage devices, such as phase change materials for space cooling, air conditioning and solar energy storage (Zalba *et al.* 2003; Tyagi and Buddhi 2007; Kenisarin and Mahkamov 2007). Again, most of these storage options are not being taken up at present.

Light-emitting diode (LED) devices are playing an increasingly important role in the more efficient use of energy. Although the colour of objects can appear different under LED illumination than in sunlight or incandescent globes, materials and fabrication developments over the past decade have resulted in high-quality white LEDs, readily available at a low price and now giving better colour rendering than common fluorescent lamps. The current interest in LED technology arises from their far greater efficiency compared with other forms of common lighting. Internal quantum efficiencies of up to 80 per cent are attainable, although some of that light is still trapped by internal reflection and absorption within the device. However, recent improvements in light extraction from LEDs have been made by creating anti-reflection coatings by nanoimprint lithography on the surface of the device (see, for example, Kim *et al.* 2007). With the prospect of decreasing the national energy consumption by several per cent just through ubiquitous usage of LED lighting, it seems that the incandescent globe and fluorescent lighting will soon be of historical interest only. Organic light-emitting diodes (OLEDs) are also subject of intense research, as they do not require the complex fabrication methods that current LEDs need. The prospect of low-cost large OLED flexible displays is particularly attractive, to the extent that whole walls could not only be illuminated with a flexible display, but also act as an integrated audiovisual device. However, much work remains to be done in OLED materials development, as the current typical life expectancy of less than 1,000 hours is still far too short to be viable.

Conclusions

Remarkable advances in materials technologies are allowing the creation of multifunctional structures. These changes will flow through to how we interact with our working and living environments, and our expectations from them. The success of biomimetic and 'intelligent' structure will be measured in terms of sustainable use of energy and resources as much as amenity. Thus, the need for more sustainable generation and storage of energy will see distributed and more flexible sources of energy being exploited. Supply will be increasingly subject to local storage options and to the nature of the energy required (mechanical vs electrical vs high-grade heat vs low-grade heat). Such options are not available using our current centralized generation sources.

Second, the volume of materials flowing through the industrial and urban ecosystem into the human economy worldwide is now at roughly the same scale as the flow of materials occurring naturally through global bio-geochemical processes (Tibbs 2000). Unfortunately, the current manufacturing paradigm is one of increased materials consumption. It is important to replace single-use materials with better materials, not only fit for first use, but also more capable of reuse and remanufacture.

Finally, biomimetic materials and embedded intelligence will enable buildings to operate as 'organisms', with feedback loops, managing energy flows more efficiently and creating diverse functionality. The 'intelligent interior' will manage how buildings interact with their external environment, in terms of energy, light, heat and noise management. One can imagine a building, like an octopus or a chameleon that can change colour, being able rapidly to alter its state to take advantage of the availability of an external energy supply or of variations in weather conditions, or to mitigate dangers such as fires, storms or burglary. The same building will be able to sense its own condition and, at least in part, effect its own repairs and routine maintenance. If we wish, it will also be able to sense and care for the amenity, health and well-being of its occupants. Nature has operated a closed-loop, non-equilibrium system continuously for 3.5 billion years, finely optimized for purpose, purely through natural selection and the power of the sun ($<1\,kW/m^2$). By analogy, creating real-time interactions between various building elements, its users and external environmental factors would result in a building capable of maintaining and changing its fitness for purpose for extended periods.

The building of the near future may well be an integrated effort between materials science, biology, engineering and architecture, and operate akin to an organism, with multiple feedback loops controlling its 'behaviour'. Although the idea may seem far-fetched to some, developments in materials technology are for the first time making such intelligent structures a possibility. However, the introduction of new materials is as much constrained by our ability to use them effectively as by our inherent materials design capabilities. Use of these materials to build intelligent structures will lead to some obvious ethical debates regarding such issues as personal privacy and the equity of resource utilization, which fall outside the scope of this chapter. However, these issues need to be resolved, as history has taught us that when a scientific or technical advance becomes possible, it is generally adopted – for better or worse!

Bibliography

Ahmed, S.F.U. and Mihashi, H. (2007) 'A review on durability properties of strain hardening fibre reinforced cementitious composites', *Cement and Concrete Composites*, 29: 365–76.

Andersson, H.M., Keller, M.W., Moore, J.S., Sottos, N.R. and White, S.R. (2007) 'Self healing polymers and composites', in S. van der Zwaag (ed.) *Self Healing Materials: An Alternative Approach to 20 Centuries of Materials Science*, Dordrecht: Springer.

Ashby, M.F. (2005) *Materials Selection in Mechanical Design*, 3rd edn, Oxford: Butterworth-Heinemann.

Ashby, M.F., Evans, A., Fleck, N.A., Gibson, L.J., Hutchinson, J.W. and Wadley, H.N.G. (2000) *Metal Foams: A Design Guide*, Oxford: Butterworth-Heinemann.

Autumn, K., Sitti, M., Liang, Y.A., Peattie, A.M., Hansen, W.R., Sponberg, S., Kenny, T.W., Fearing, R., Israelachvili, J.N. and Full, R.J. (2002) 'Evidence for van der Waals adhesion in gecko setae', *Proceedings of the National Academy of Sciences USA*, 99: 12252–6.

Balzas, A.C., Emrick, T. and Russell, T.P. (2006) 'Nanoparticle polymer composites: where two small worlds meet', *Science*, 314: 1107–10.

Banhart, J. (2001) 'Manufacture, characterisation and application of cellular metals and metal foams', *Progress in Materials Science*, 46: 559–632.

Barthlott, W. and Neinhuis, C. (1997) 'Characterization and distribution of water-repellent, self-cleaning plant surfaces', *Planta*, 202: 1–8.

—— (2005) *Method for the Preparation of Self-Cleaning Removable Surfaces*, US Patent 2005136217, filed 23 June 2005.

BASF AG (2008a) Online. Available at HTTP: <www.corporate.basf.com/en/stories/wipo/mincor/story.htm?id=fWuLTCUFmbcp-mG> (accessed 23 June 2008).

—— (2008b) Online. Available at HTTP: <www.col9.de/portal/basf/ide/dt.jsp?setCursor=1_314695> (accessed 23 June 2008).

BCC Research (2006) *Nanotechnology: Nanocomposites, Nanoparticles, Nanoclays, and Nanotubes*, Report NAN021C, Norwalk, CT: BCC Research.

Bergman, S.D. and Wudl, F. (2008) 'Mendable polymers', *Journal of Materials Chemistry*, 18: 41–62.

Bico, J., Thiele, U. and Quéré, D. (2002) 'Wetting of textured surfaces', *Colloids and Surfaces A*, 206: 41–6.

Bidoki, S.M., Lewis, D.M., Clark, M., Vakorov, A., Millner, P.A. and McGorman, D. (2007) 'Ink-jet fabrication of electronic components', *Journal of Micromechanics and Microengineering*, 17: 967–74.

Blossey, R. (2003) 'Self-cleaning surfaces: virtual realities', *Nature Materials*, 2: 301–6.

Bonderer, L.J., Studart, A.R. and Gauckler, L.J. (2008) 'Bioinspired design and assembly of platelet reinforced polymer films', *Science*, 319: 1069–73.

Briggs, L. (2007) *Tackling Wicked Problems: A Public Policy Perspective*, Canberra: Australian Public Service Commission. Online. Available at HTTP: <www.apsc.gov.au/publications07/wickedproblems.pdf> (accessed 20 May 2008).

Capadona, J.R., Shanmuganathan, K., Tyler, D.J., Rowan, S.J. and Weder, C. (2008) 'Stimuli-responsive polymer nanocomposites inspired by the sea cucumber dermis', *Science*, 319: 1370–74.

Cebeci, F.C., Wu, Z., Zhai, L., Cohen, R.E. and Rubner, M.F. (2006) 'Nanoporosity-driven superhydrophilicity: a means to create multifunctional antifogging coatings', *Langmuir*, 22: 2856–62.

Chen, L., Glassmaker, N.J., Jagota, A. and Huia, C.-Y. (2008) 'Strongly enhanced static friction using a film-terminated fibrillar interface', *Soft Matter*, 4: 618–25.

Chubb, D.L. (2007) 'Thermophotovoltaic generation of electricity', in *AIP Conference Proceedings 890: Seventh World Conference on Thermophotovoltaic Generation of Electricity*, Melville, NY: American Institute of Physics.

Concrete Canvas (2008) Online. Available at HTTP: <www.concretecanvas.co.uk/index.html> (accessed 23 June 2008).

Cordier, P., Tournilhac, F., Ziakovic, A.S. and Leibler, L. (2008) 'Self-healing and thermoreversible rubber from supramolecular assembly', *Nature*, 451: 977–80.

Crosby, A.J. and Lee, J.-Y. (2007) 'Polymer nanocomposites: the "nano" effect on mechanical properties', *Polymer Reviews*, 47: 217–29.

Davidovits, J. (1991) 'Geopolymers: inorganic polymeric new materials', *Journal of Thermal Analysis*, 37: 1633–56.

DeMoor, S., Waite, H.J., Jangoux, M.J. and Flammang, P.J. (2003) 'Characterization of the adhesive from Cuvierian tubules of the sea cucumber *Holothuria forskali* (Echinodermata, Holothuroidea)', *Marine Biotechnology*, 5: 45–57.

Deutsche Steinzeug Cremer and Breuer AG (2008) Online. Available at HTTP: <www.deutsche-steinzeug.de/en/hydrotect_new/ the_solution.html?pe_id=122&pe_id=123> (accessed 23 June 2008).

Dow Corning (2008) Online. Available at HTTP: <www.dowcorning.com/content/construction/constructionprotect/6694_water_repellent.aspx?wt.svl=6694_launch_banner> (accessed 23 June 2008).

Duxson, P., Fernandez-Jimenez, A., Provis, J.L., Lukey, G.C., Palomo, A. and van Deventer, J.S.J. (2007) 'Geopolymer technology: the current state of the art', *Journal of Materials Science*, 42: 2917–33.

Feng, X. and Jiang, L. (2006) 'Design and creation of superwetting/antiwetting surfaces', *Advanced Materials*, 18: 3063–78.

Fraas, L.M. (2007) *JX Crystals*. Online. Available at HTTP: <www.jxcrystals.com/ThermoPV.htm> (accessed 28 June 2008).

Fratzl, P. and Weinkamer, R. (2007) 'Nature's hierarchical materials', *Progress in Materials Science*, 52 (8): 1263–334.

Gartner, E. (2004) 'Industrially interesting approaches to "low-CO_2" cements', *Cement and Concrete Research*, 34: 1489–98.

Ge, L., Sethi, S., Ci, L., Ajayan, P.M. and Dhinojwala, A. (2007) 'Carbon nanotube-based synthetic gecko tapes', *Proceedings of the National Academy of Sciences USA*, 104: 10792–5.

Giussani, B. (2006) 'A concrete step toward cleaner air', *Business Week*, 8 November: 108.

Green, M.A., Keith Emery, K., Hishikawa, Y. and Warta, W. (2008) 'Solar cell efficiency tables (version 31)', *Progress in Photovoltaics Research and Applications*, 16: 61–7.

Gu, Z.-Z., Fujishima, A. and Sato, O. (2004) 'Biomimetic titanium dioxide film with structural color and extremely stable hydrophilicity', *Applied Physics Letters*, 85: 5067–9.

Hoth, C.N., Choulis, S.A., Schilinsky, P. and Brabec, C.J. (2007) 'High photovoltaic performance of inkjet printed polymer-fullerene blends', *Advanced Materials*, 19: 3973–8.

Hyde, D.M. and Brannon, S.M. (2006) 'Investigation of infrared reflective pigmentation technologies for coatings and composite applications', in *Proceedings of the COMPOSITES 2006 Convention*, St Louis: American Composites Manufacturers Association. Online. Available at HTTP: <www.acmanet.org/resources/06papers/Hyde157.pdf>.

Industrial NanoTech (2008) Online. Available at HTTP: <www.nansulate.com/> (accessed 23 June 2008).

Israel, C., Mathur, N.D. and Scott, J.F. (2008) 'A one-cent room-temperature magnetoelectric sensor', *Nature Materials*, 7: 93–4.

Italcementi Group (2006) Online. Available at HTTP: <www.italcementigroup.com/ENG/Research+and+Innovation/Innovative+Products/> (accessed 23 June 2008).

Jonkers, H.M. (2007) 'Self healing concrete: a biological approach', in S. van der Zwaag (ed.) *Self Healing Materials: An Alternative Approach to 20 Centuries of Materials Science*, Dordrecht: Springer.

Kaplan, A.M and Haenlein, M. (2006) 'Toward a parsimonious definition of traditional and electronic mass customization', *Journal of Product Innovation Management*, 23: 168–82.

Kenisarin, M. and Mahkamov, K. (2007) 'Solar energy storage using phase change materials', *Renewable and Sustainable Energy Reviews*, 11: 1913–65.

Kim, D.-H., Ahn, J.-H., Choi, W.M., Kim, H.-S., Kim, T.-H., Song, J., Huang, Y.Y., Liu, Z., Lu, C. and Rogers, J.A. (2008) 'Stretchable and foldable silicon integrated circuits', *Science*, 320: 507–11.

Kim, S.H., Lee, K.D., Kim, J.K., Kwon, M.K. and Park, S.J. (2007) 'Fabrication of photonic crystal structures on light emitting diodes by nanoimprint lithography', *Nanotechnology*, 18: 055306–10.

Landolt, D. (2007) *Corrosion and Surface Chemistry of Metals*, 1st edn, Lausanne: EPFL Press.

Lawrence Berkeley National Laboratory (2006) Online. Available at HTTP: <http://windows.lbl.gov/comm_perf/Electrochromic/ec_reso.html> (accessed 23 June 2008).

Li, V.C. (2003) 'On engineered cementitious composites: a review of the material and its applications', *Journal of Advanced Concrete Technology*, 1: 215–30.

—— (2006) 'Bendable composites: ductile concrete for structures', *Structure Magazine*, July: 45–8.

Li, V.C. and Yang, E.-H. (2007) 'Self healing in concrete materials', in S. van der Zwaag (ed.) *Self Healing Materials: An Alternative Approach to 20 Centuries of Materials Science*, Dordrecht: Springer.

Li, W. and Amirfazli, A. (2008) 'Hierarchical structures for natural superhy-drophobic surfaces', *Soft Matter*, 4: 462–6.

Little, B., Lee, J. and Ray, R. (2007) 'A review of "green" strategies to prevent or mitigate microbiologically influenced corrosion', *Biofouling*, 23: 87–97.

Losonczi, A. (2007) Online. Available at HTTP: <www.litracon.hu/index.php> (accessed 27 June 2008).

Luccon Lichtbeton GmbH (2008) Online. Available at HTTP: <www.luccon.com/index2.htm> (accessed 23 June 2008).

Lumley, R. (2007) 'Self healing in aluminium alloys', in S. van der Zwaag (ed.) *Self Healing Materials: An Alternative Approach to 20 Centuries of Materials Science*, Dordrecht: Springer.

Lundgren, M., Allan, N.L. and Cosgrove, T. (2007) 'Modeling of wetting: a study of nanowetting at rough and heterogeneous surfaces', *Langmuir*, 23: 1187–94.

Manfra, L. (2007) 'A place in the shade', *Architect Magazine*, April: 15.

Mardaljevic, J. and Nabil, A. (2008) 'Electrochromic glazing and facade photo-voltaic panels: a strategic assessment of the potential energy benefits', *Lighting Research and Technology*, 40: 55–76.

Meyer, C. (2006) 'Opportunities and challenges of modern concrete technology in earthquake hazard mitigation', in S. Tanvir Wasti and G. Ozcebe (eds) *Advances in Earthquake Engineering for Urban Risk Reduction*, Dordrecht: Springer.

Meyers, M.A., Chen, P.-Y., Lin, A.Y.-M. and Seki, Y. (2008) 'Biological materials: structure and mechanical properties', *Progress in Materials Science*, 53: 1–206.

Murata, K., Matsumoto, J., Tezuka, A., Matsuba, Y. and Yokoyama, H. (2005) 'Super-fine ink-jet printing: toward the minimal manufacturing system', *Microsystems Technology*, 12: 2–7.

Nanogate AG (2008) Online. Available at HTTP: <www.nanogate.de/en/product-enhancement/functions/nanotension/water-repellant.php> (accessed 23 June 2008).

Nanokote (2007) Online. Available at HTTP: <http://nanokote.com.au/cms/> (accessed 23 June 2008).

NanoMarkets (2008) *Pushing Low-Cost Sensors to Market. A NanoMarkets White Paper*, Glen Allen, VA. Online. Available at HTTP: <www.nanomarkets.net/resources/summaries/NMLowCostSenors_08.pdf> (accessed 23 June 2008).

NanoSafeguard (2008) Online. Available at HTTP: <www.nanosafeguard.com/index.php?main_page=page&id=2&chapter=0> (accessed 23 June 2008).

Nanotex (2008) Online. Available at HTTP: www.nano-tex.com/applications/com-mercial_P1.html> (accessed 23 June 2008).

Ogawa, T., Murata, N. and Yamazaki, S. (2003) 'Development of anti-fogging mirror coated with SiO_2-ZrO_2-colloidal SiO_2 film by the sol-gel process', *Journal of Sol-Gel Science and Technology*, 27: 237–8.

Parker, A.R. and Lawrence, C.R. (2001) 'Water capture by a desert beetle', *Nature*, 414: 33–4.

Parkin, I.P. and Palgrave, R.G. (2005) 'Self-cleaning coatings', *Journal of Materials Chemistry*, 15: 1689–95.

Pilkington Group (2008) Online. Available at HTTP: <www.pilkingtonselfcleaningglass.co.uk/> (accessed 23 June 2008).

Podsiadlo, P., Kaushik, A.K., Arruda, E.M., Waas, A.M., Shim, B.S., Xu, J., Nandivada, H., Pumplin, B.G., Lahann, J., Ramamoorthy, A. and Kotov, N.A. (2007) 'Ultrastrong and stiff layered polymer nanocomposites', *Science*, 318: 80–3.

Polmear, I.J. (2006) *Light Alloys: From Traditional Alloys to Nanocrystals*, 4th edn, Oxford: Butterworth-Heinemann.

Pushparaj, V.L., Shaijumon, M.M., Kumar, A., Saravanababu Murugesan, S., Ci, L., Vajtai, R., Linhardt, R.J., Nalamasu, O. and Ajayan, P.M. (2007) 'Flexible energy storage devices based on nanocomposite paper', *Proceedings of the National Academy of Sciences USA*, 104: 13574–7.

Rahman, S., Molyneaux, T. and Patnaikuni, I. (2005) 'Ultra high performance concrete: recent applications and research', *Australian Journal of Civil Engineering*, 2: 13–20.

Roach, P., Shirtcliffe, N.J. and Newton, M.I. (2008) 'Progress in superhydrophobic surface development', *Soft Matter*, 4: 224–40.

Roberge, P.R. (2006) *Corrosion Basics: An Introduction*, 2nd edn, Houston, TX: NACE International.

Ryan, M. (2005) 'Introduction to IR-reflective pigments', *Paint and Coatings Industry Magazine*, August.

Sage Electrochromics (2008) Online. Available at HTTP: <www.sage-ec.com/> (accessed 23 June 2008).

Shah, S.P. and Weiss, W.J. (1998) 'Ultra high performance concrete: a look to the future', paper presented at Paul Zia International Symposium, Spring Convention, American Concrete Institute, Houston.

Snaith, H.J. and Schmidt-Mende, L. (2007) 'Advances in liquid-electrolyte and solid-state dye-sensitized solar cells', *Advanced Materials*, 19: 3187–3200.

Snyder, G.J. and Toberer, E.S. (2008) 'Complex thermoelectric materials', *Nature Materials*, 7: 105–14.

Sofi, M., van Deventer, J.S.J., Mendis, P.A. and Lukey, G.C. (2007) 'Engineering properties of inorganic polymer concretes (IPCs)', *Cement and Concrete Research*, 37: 251–7.

Song, G. and Atrens, A. (2007) 'Recent insights into the mechanism of magnesium corrosion and research suggestions', *Advanced Engineering Materials*, 9: 177–83.

Sto Corp. (2007) Online. Available at HTTP: <www.stocorp.com/allweb.nsf/lotusanpage> (accessed 23 June 2008).

Tibbs, H. (2000) 'Global scenarios for the millennium', in R. Slaughter (ed.) *Gone Today, Here Tomorrow: Millennium Previews*, Sydney: Prospect.

Toohey, K.S., White, S.R., Lewis, J.A., Moore, J.S. and Sottos, N.S. (2007) 'Self-healing materials with microvascular networks', *Nature Materials*, 6: 581–5.

TOTO (2008) Online. Available at HTTP: <www.toto.co.jp/docs/hyd_patent_en/case.htm> (accessed 23 June 2008).

Tru Vue (2004) Online. Available at HTTP: <www.tru-vue.com/Content.asp?pn=consumer/products> (accessed 23 June 2008).

Tsuzuki, T. (2008) 'Abnormal transmittance of refractive-index-modified ZnO/organic hybrid films', *Macromolecular Materials and Engineering*, 293: 109–13.

Tyagi, V.V. and Buddhi, D. (2007) 'PCM thermal storage in buildings: a state of art', *Renewable and Sustainable Energy Reviews*, 11: 1146–66.

Valiev, R.Z., Zehetbauer, M.J., Estrin, Y., Höppel, H.W., Ivanisenko, Y., Hahn, H., Wilde, G., Roven, H.J., Sauvage, X. and Langdon, T.G. (2007) 'The innovation potential of bulk nanostructured materials', *Advanced Engineering Materials*, 9: 527–33.

van Damme, H. (2008) 'Nanocomposites: the end of compromise', in C. Bréchignac, P. Houdy and M. Lahmani (eds) *Nanomaterials and Nanochemistry*, Berlin: Springer.

Varley, R. (2007) 'Ionomers as self healing polymers', in S. van der Zwaag (ed.) *Self Healing Materials: An Alternative Approach to 20 Centuries of Materials Science*, Dordrecht: Springer.

Vernardou, D., Pemble, M.E. and Sheel, D.W. (2006) 'The growth of thermochromic VO$_2$ films on glass by atmospheric-pressure CVD: a comparative study of precursors, CVD methodology, and substrates', *Chemical Vapour Deposition*, 12: 263–74.

Wang, R., Hashimoto, K., Fujishima, A., Chikuni, M., Kojima, E., Kitamura, A., Shimohigoshi, M. and Watanabe, T. (1997) 'Light-induced amphiphilic surfaces', *Nature*, 388: 431–3.

Wang, S. and Jiang, L. (2007) 'Definition of superhydrophobic states', *Advanced Materials*, 19: 3423–4.

Wang, S., Songa, Y. and Jiang, L. (2007) 'Photoresponsive surfaces with controllable wettability', *Journal of Photochemistry and Photobiology C*, 8: 18–29.

Wang, Z.L. (2007) 'Nanopiezotronics', *Advanced Materials*, 19: 889–92.

Yugami, H., Sasa, H. and Yamaguchi, M. (2003) 'Thermophotovoltaic systems for civilian and industrial applications in Japan', *Semiconductor Science and Technology*, 18: S239–46.

Zalba, B., Marín, J.M., Cabeza, L.F. and Mehling, H. (2003) 'Review on thermal energy storage with phase change: materials, heat transfer analysis and applications', *Applied Thermal Engineering*, 23: 251–83.

Zhang, J., Lu, X., Huang, W. and Han, Y. (2005) 'Reversible superhydrophobicity to superhydrophilicity transition by extending and unloading an elastic polyamide film', *Macromolecular Rapid Communications*, 26 (6): 477–80.

Zimmermann, J., Rabe, M., Artus, G.R.J. and Seeger, S. (2008) 'Patterned superfunctional surfaces based on a silicone nanofilament coating', *Soft Matter*, 4: 450–52.

3 Material environmental life cycle analysis

Delwyn Jones, Selwyn Tucker and Ambalavanar Tharumarajah

Worldwide environmental degradation arises from most building supply chain production sequences, according to reports of the United Nations Environment Programme (2003) as well as the Intergovernmental Panel on Climate Change (2007). Most industrial activities have some environmental impact, and many are only benign in some or a few aspects because the Earth's carrying capacity can renew their source of resource supply or assimilate their burdens in pollution sinks. Concerns about induced climate change, biodiversity and habitat loss, resource depletion and peak oil have shaped global business, industry, community and government responses, including those of:

- the World Business Council for Sustainable Development (2007);
- community organizations, for example the Worldwide Fund for Nature (2007);
- the Council of Australian Governments (1992);
- the Organization for Economic Co-operation and Development (2003);
- individuals such as Al Gore (2006) in delivering *An Inconvenient Truth*.

This chapter reviews the state of life cycle assessment (LCA) in the property and construction sector, including standard and novel conceptual and automated LCA globally and locally. Governments are incorporating Life Cycle Thinking (LCT) regarding environmental impacts into policy instruments covering development, procurement, construction, operation and disposition.

Standard approaches

LCA has evolved to become a global standard quality management environmental accounting method (International Organization for Standardization (ISO), 1998). It aims to foster systematic continuous improvement to enable practitioners to identify, evaluate and reduce operational impacts of manufacturing and assembly processes on Nature's carrying capacity so as to sustain community and economic wellbeing (Watson *et al.* 2004). The ISO LCA framework involves:

- definition of goals, scope, system boundary, functional unit, limits and assumptions;
- an inventory of operational inputs from and outputs to air, water and land;
- assessment of significant environmental burdens, damages and impacts;
- implementation of operational and policy improvement opportunities to mitigate impacts.

Scope and systems boundary definition

A full LCA cradle-to-cradle scope covers all phases of the product life cycle, from resource acquisition, refining, distribution, fabrication, use, repair, reuse and recovery to disposal over functional lifetimes. Operations are wide-ranging and include, among many others, drilling, refining, mining, smelting, forestry, farming, transport, digestion, fermentation, rolling, alloying, coating, shaping, assembly, use, repair, reuse, recycling, remanufacture, incineration and landfill.

The life cycle of a product begins in a 'cradle', acquiring raw materials and transforming them into delivered goods, used and disposed of to a range of fates, including an end-of-life 'grave'. Cradle-to-grave transformations require energy, water and intermediates, which are also on their own path of use and disposal. All material embodied in goods, co-products, waste and fuels eventually returns to air, water and land. So that a supply chain can continue to service a community's everyday needs, such cycles should involve a return to the cradle. Unless cycles are renewable or closed loop, at some stage they will physically exhaust the natural source of raw materials or overload the capacity of natural sinks to absorb the pollution generated.

LCA is a systematic study of environmental impacts of resource depletion and emissions generation in each and all operations throughout product life cycles. It can be limited to particular phases – for example, from extraction of bauxite to production of aluminium virgin metal ingots – or cover an entire life cycle to final disposal to land fill or refinishing, reprocessing, recycling and reuse. After defining the scope of work, the next steps involve compiling inventory and assessing impacts over the life cycle.

Life cycle inventory (LCI)

Compiling an inventory involves:

- identifying all operations involved in the study's scope and system boundary;
- tracking the sources of raw and intermediate materials and energy used throughout;
- quantifying how much raw and intermediate materials and energy is used throughout;

- identifying how many emissions are released to air, water and land throughout;
- tracking the fate of all emissions released to air, water and land throughout;
- determining how much of each emission is released to air, water and land throughout;
- comparing all outputs against inputs to check mass and energy flow is balanced.

Inventory results reveal profiles of resource use and pollution generation. A LCI can account for use of raw and recycled material resources, and operating and embodied energy and water use. It should account for generation of all emissions to air, together with their sequestrations.

Life cycle impact assessment (LCIA)

Environmental impacts arise from damages which can be represented as gross as well as detailed breakdowns. They are typically of four types:

- Human health impacts arise from, for example, carcinogens in emissions to air and water of chemicals such as arsenic, cadmium, nickel, vinyl chlorides and respiratory compounds of dust, carbon monoxide, ammonia, oxides of nitrogen and sulphur, aromatic hydrocarbons; ethylene and volatile organic compounds (VOCs).
- Climate change impacts arise from damages due to emissions of greenhouse gas and ozone layer depletion from CFC/HCFC emissions.
- Ecosystem quality impacts arise from emissions causing damages from acidification and eutrophication of water by ammonia, nitrogen and sulphur oxides, as well as ecotoxins in emissions to air of metals and water of ions of arsenic, cadmium, chromium, copper, nickel, mercury, lead and zinc.
- Resource depletion impacts arise from an increasing amount of resources expended (energy in particular), losses and damages in extracting fuels and minerals containing coal, crude oil, lignite, natural gas/condensates, copper, nickel, lead, zinc, uranium and others.

Inventory damage and impact assessment results are then used to identify hot spots in processes, supply chains, policy development, labelling and marketing. LCA studies extend to all stages of a product's value chain.

Challenges for regionalization in LCI and LCIA software tools

LCA software now on the market is customized for most national fuels and energy supply chains plus some state energy grids in OECD nations (Boustead 2007). Because of pioneering European work, some sound LCI data are available for typical energy and transport infrastructure, agriculture, chemicals, packaging, minerals and waste disposal – operations applicable to many nations. But pollution impacts of raw material extraction using different mining technologies, transportation and production processes can vary from country to country and region to region. Globally, many organizations have been working to overcome LCA limitations. Most seek sound information on resource acquisition and emissions to air, water and land for local and imported products.

Stakeholder needs and benefits

Because LCA considers a wide range of environmental burdens, it has become a vital analysis and reporting tool for a range of stakeholders in industry and government. Until recently, however, there remained serious and widespread concern about the high cost, time, data and skill requirements demanded by conventional LCA. Opinion has even suggested that LCA has evolved as an exclusively specialist endeavour, well beyond the reach of most practitioners needing to use it.

Scepticism about its relevance to real-world decisions facing today's business, governments and communities is also very common. Efforts to address such widespread concern, frustration and scepticism have stimulated new approaches to bring the power of LCA to bear on resolving global, national and local environmental issues (Jones *et al.* 2006). Stakeholders can now benefit from application of more practical user-friendly tools for LCT (Mitchell 2004; Watson 2004), streamlined LCA (Australian Greenhouse Office 2006) and automated LCA (Tucker *et al.* 2003).

Increasingly, stakeholders need streamlined approaches to provide for fast single impact assessments of specific publicly sensitive environmental elements, for example, ecotoxins, water use and embodied energy for carbon accounting, as well as for broader building ratings. Despite these being emergent methods, considered initially as having some limitations, many practitioners and clients are beginning to use them like a kit of complementary tools that together work better than one standard LCA tool.

General benefits

As a method of environmental accounting, LCA is a valuable decision-support tool for policy- and decision-makers to assess supply and procurement (Grant 2004; Cole *et al.* 2000). Drivers for its further development include:

- government accountability regulations for end-of-life landfill, reuse or take-back (European Topic Centre on Resource and Waste Management 2008);
- business participation in product stewardship initiatives, using it in continuous improvement processes (Queensland Government 2000);

- within-class, third-party accredited environmental performance labelling for product supply and procurement (Australian Environmental Labelling Association 2004);
- assessment to understand process or packaging burdens and improvement options;
- management to broaden the range of issues considered in regulation or policy;
- establishment of base-line data on emissions to drive improved practices in manufacture, use or disposal.

Avoiding end-of-pipe problems

LCA offers the capability to locate, reduce and then avoid end-of-pipe problems, such as waste and pollution, because it can assess the full life cycle of operations to provide infrastructure, urban and industry sector planners with an urban industrial ecology perspective capable of identifying problems before they arise or intensify (Martin and Verbeek 1998). Another benefit of LCA is its capacity to reveal unintended consequences of decisions and to avoid shifting environmental problems elsewhere, which is often the case when issues are dealt with in isolation and in ignorance of the natural and built systems that are affected (Wood and Jones 1996).

Business benefits

Industry is realizing that LCA has business value as a window to innovation, to revealing risk and new opportunities (Huysmans *et al.* 2007). Flow-on business benefits lie in enhanced product differentiation and marketability, satisfying tighter regulations and promoting corporate citizenship. Application of LCA can also assist business to identify:

- cost-effective options to improve designs, products and services;
- how to do more with less, avoid waste and reduce cost;
- costs and benefits of competing propositions, strategies and risk profiles (Wood and Jones 1996).

Corporate benefits

In the absence of a comprehensive LCA, life cycle thinking is one way to reap the benefits of using a life cycle approach. LCT conceptually considers the scope and systems boundary, including material flows and energy transformations. Even without numerical models, LCT can qualitatively map and consider what needs to occur in operational and decision-making phases. It can derive and define ecological outcomes required of policy, management, business, industrial, agricultural and natural systems. Rather than working with a narrow focus, LCT facilitates holistic deliberations that lead to more rational decision-making. Such approaches, as outlined in Table 3.1, for example, were applied in the project management of William McCormack Place in Cairns, Australia's first 5-star national building greenhouse rated office building that adopted both the (state) government Ecologically Sustainable Office Fitout Guideline (Queensland Government 2000) as well as the (federal) Australian Building Greenhouse Rating scheme (Jones *et al.* 2005). This project demonstrated how LCT can maximize benefits and minimize costs to the owner, economy, community, and local, national, state and global environments.

Streamlined LCA is useful for a company seeking to examine highest-impact operations with most potential for improvement. This is often an approach to taking the first step towards a comprehensive LCA (Mitchell 2004).

Table 3.1 LCT in property development

Phase	*Sustainable resource use*	*Environmental health*
Policy	Conserve water, fuel, minerals, heritage, habitat and biodiversity	Protect water, soil and air quality, safety and security
Invest to develop	Vision, mission and research into sustainable development	Vision, mission and research to improve environment health
Plan	Non-asset solutions; consistent heritage and cultural values	Equity of access, safety and security; built-environmental health quality
Design	In heritage and cultural context; end-of-life disassembly and reuse	Minimize ingress of traffic emissions; natural visual landscape and amenity
Procure	Local, renewable, reused content; avoid scarce-resource use	WH&S and EMS prequalification; avoid volatile hazardous materials
Construction	Conserve resources, water, soil; protect cultural heritage features	Avoid noise and dust emissions; avoid disrupting local habitats
Manage use	Maximize resource efficiency; reliance on renewable energy	Train HR in environmental health; improve pollution abatement
Maintain	Reliance on renewable supply; healthy soil and site biota	Environmental health; biodiversity of natural human habitat
Refurbish:	Uptake of local content and labour; reliance on renewable supply	Pre- and post-occupancy quality audits; reduced pollution to air, water, land
Disposal	Best-practice reuse and renewal; recovery of toxic soil and effluent	Best-practice prequalified contractors; minimize air, water, land emissions

Misuse of LCA

LCA was designed for continuous improvement of systems rather than for the competitive analysis that many LCA studies are used for, somewhat inappropriately (Watson *et al.* 2005). It is, for example, common practice to cite LCA evidence in promotional claims of superiority to rivals. But with so many parameters to consider and with trade-offs involving subjective assessment, LCA may not prove the overall superiority of a particular solution. Many such claims have been challenged and found simply to endorse their sponsor's subjectivity (Wood and Jones 1995).

One unfortunate legacy is that students and practitioners learn about and repeat such claims well after they have been debunked. This remains the case with early comparisons of residential-building operating and embodied energy, where results showed that operations strongly dominated. Some prominent studies ignored recurrent embodied energy in fitout and refurbishment, maintenance and cleaning, and replacement materials such as carpet over the building life, as well as other environmental impacts that have since been shown to be very significant (Huysmans *et al.* 2007). Such poor inheritance lingers, with many stakeholders believing that embodied impacts from product supply in new buildings are typically much less significant than operational impacts.

Global life cycle initiatives

A joint United Nations Environment Programme/Society for Environmental Toxicology and Chemistry (UNEP/SETAC) international life cycle partnership is promoting 'life cycle economy'. Its mission is to develop and disseminate practical tools for evaluating opportunities, risks and trade-offs associated with products and services over their entire life cycle. The goal is to enable comprehensive consideration of impacts of all life cycle phases in making informed decisions on production and consumption and in management policies and strategies (United Nations Environment Programme 2008). Uptake of LCT is to be facilitated by expanding the availability of better tools, data and indicators on a global scale. Regional networks are also being established to share experiences and data with trading partners (Curran 2006).

This life cycle partnership recognized the critical need to establish national databases to centralize LCI knowledge and data sources. This has now occurred worldwide, with many countries having initiated public access LCI databases (United Nations Environment Programme 2008). Providing accessible, transparent, quality LCI data are essential to pave the way for more sustainable practices in industry supply chains (Norris and Notten 2002).

Western European countries and Canada have been pioneers that have invested heavily in developing national LCI databases since 1980. Some are more advanced than others, as shown in Table 3.2, with the Boustead Model 5.0 from the United Kingdom (Boustead 2007), Athena LCI from Canada (Athena Institute 2008), ecoinvent from Switzerland (ecoinvent 2007), Spine from Sweden (SPINE@CPM 2006) and IVAM from the Netherlands (IVAM 2006) being among the most mature. Such databases also differ in quality, focus and accessibility. Some cover many economic sectors, while others cover only particular products; and while many countries have initiated national LCI databases, others focus on particular industries.

Australian initiatives

As noted previously, LCA alone is not a preferred tool to assess local damages and impacts because of the lack of accredited local, regional and national assessment methods. Australia, for example, despite having globally high value biodiversity and ecotourism, has no nationally accepted LCIA method to quantify biodiversity losses. The Australian Life Cycle Assessment Society (ALCAS) that established an LCI Group in 2002 has made several submissions to government to support such initiatives (Tharumarajah and Grant 2006). Providing a transparent, consistent, accessible and reliable national LCI database has many potential benefits to local industry, as well as governments and community stakeholders. A national all-sector Australian LCI database can become a single source of reliable local inventory data for undertaking life cycle impact studies on a wide range of products and services, as well as serving as a repository of information on best and worst practice and of case studies showing how to reduce environmental burdens in product supply chains (Tharumarajah and Grant 2006). AusLCI is to serve the needs of a wide spectrum of potential users, so data must:

- conform to ISO and Australian standards of LCA;
- meet transparency, consistency, quality and peer review criteria needs of the users;

Table 3.2 Summary of LCI database activities worldwide

LCI database activity	Countries
Coordinated data exchange	Italy, Switzerland, Australia, Canada, Taiwan, Japan, Korea, Sweden, USA
Significant without integration	Austria, France, Germany, UK, other Western European countries
For separate process chains	China, India, Argentina
Little LCI but use of LCA	Thailand, Malaysia, Vietnam, Eastern Europe, Brazil, Philippines, Indonesia, Singapore, Chile, Mexico, Taiwan

- be regionally specific, cover all core industry sectors, but reflect sector variations;
- be formatted for accessibility to maximize uptake and enable priority issues.

Many Australian groups have invested heavily in developing LCI for sectors such as building and construction, waste management, procurement labelling, minerals, metals, timber and packaging. The quality of this work ranges from preliminary to state of the art. The Forest and Woods Product Research and Development Corporation LCI project was particularly thorough in using the best intra- and extra-sectoral data and sensitivity analyses available (Forest and Woods Product Research and Development Corporation 2006). Despite world best practice LCI data acquisition and documentation processes, however, all such studies remain open to criticism for containing elements of unknown quality extra-sector data, simply because supply chains always rely on some input from operations outside their sector (Jones *et al.* 2003).

As with any emerging technology, such arguments are used to disparage competing claims by suppliers, consultants and software vendors alike; unfortunately, they also serve to undermine the value of LCA and delay its market uptake. Such delays also serve to frustrate industry, community, government and research efforts undertaken to meet the challenges of climate, resource depletion and environmental degradation.

The building supply chain LCA

Developing LCI databases for a supply chain takes decades of costly data-gathering activity, so it is important to learn first from what already exists for a particular sector. In Australia, industry stakeholders have been cooperating to develop LCA initiatives since the early 1990s. The first complete cradle-to-grave LCA of a public building was for Stadium Australia, by the NSW Department of Public Works and Services (NSWDPWS), initiated for the Sydney 2000 Green Olympics (Wood and Jones 1995, 1996). Full building LCAs have since been undertaken by:

- Broken Hill Pty Ltd (BHP) (Bluescope Steel 1995);
- CRC for Construction Innovation (Huysmans *et al.* 2007);
- Commonwealth Scientific and Industrial Research Organization (CSIRO) (Seo 2002).

These studies sourced information from building supply chain LCI databases, including:

- the ALCAS LCI Group established in 2002 (Tharumarajah and Grant 2006);

- the BHP-funded streamlined LISA building LCI, together with CHAPPY educational LCA software (LISA 2008);
- the CRC for Construction Innovation developed national building supply chain LCI databases for use with LCADesign, Automated Building LCA software (Jones *et al.* 2004);
- the CSIRO developed Australian Timber LCI public access databases (Forest and Woods Product Research and Development Corporation 2006).

Australian timber supply chain LCI

The first Australian public domain national forest and timber product LCI can now deliver rigorous representative quantitative information to enable stakeholder decisions and policy. It was created to enable stakeholders to better evaluate impacts of most common engineered timber systems, as shown in Table 3.3. Typical forests, mills and plants were visited for inspection, initial data collection, and identification and understanding of system processes. Collection, enhancement and verification of data provide industry with reliable information to improve production and procurement, considering the environmental bottom line. Benefits from this first national Australian LCI include:

- application of a common database for the wood industry;
- an objective quantitative basis for comparing competing wood products;
- data to compare environmental impacts of wood products from different manufacturing and materials processes;
- a database of wood products and building structures for use with LCA;
- the provision of credible Australian industry-based data for LCA of wood products;
- support to improve manufacturers' environmental performance and impact assessment;
- facilitation of communication of environmental information to customers and stakeholders;
- setting an industry best practice standard for handling and documenting LCA data.

This LCI has the potential to provide industry with understanding about prospective growth areas and value-adding in recycling and take-back schemes, and is a major advance in providing quality data on building products. Wide industry coverage also makes it representative of Australian wood products, and the ISO standard quality-assured procedures and documentation set a benchmark for other supply chain segment and national LCI initiatives.

Table 3.3 Coverage of products in Australian forest, timber and wood product LCI database

Class	Product	Sector (%)
Softwood log	Peeler, high-quality saw, low-quality saw and pulp logs, woodchip	50% plantation area
Hardwood log	Peeler, saw and pulp logs	22% forest regrowth
Sawmill product	Sawn green soft/hardwood and kiln-dried and planed timber, bark and woodchip products	40% soft, 30% hard wood
Veneer ply	3-mm, int/exterior and formply, tongue and groove flooring, structural	90% industrial plants
LVL	Laminated veneer lumber (LVL) (3 thicknesses)	60% industrial plants
Particleboard	Raw and decorated (3 thicknesses)	64% industrial plants
MDF	Raw and decorated medium-density fibreboard	92% industrial plants
Glulam	Glue laminated pine and hardwood lumber	55% industrial plants
I-beams	Oriented strand board, web/pine flanges, ply web, LVL flanges	65% industrial plants

LCI for Automated Building LCIA software

As yet Australia has no nationally accepted LCI database across all major industry sectors, and ongoing development is limited by a lack of industry capacity to deliver objective data (Jones *et al.* 2004). Recognition of these deficits emerged from CRC for Construction Innovation research involving compilation of LCI data to inform LCADesign. LCADesign is an automated environmental analysis software tool, using direct take-off from three-dimensional CAD models compliant with Industry Foundation Class (IFC) data transfer protocols (see Seo *et al.*, Chapter 10 in this volume).

LCADesign was developed to provide industry sector stakeholders benefits by facilitating users' direct analysis of building performance, without data re-entry. It employs repeatable evidence-based calculations aggregated up from each component to a whole building. The software tabulates the impacts of each object selected in design, and provides the following outputs to global standards:

- building information models (BIMs) used in modern design documentation practice;
- life cycle economic costing schedules considering service life;
- ISO 14000 EMS LCA methods for improvement assessment;
- LCI databases for building industry supply chain ecoprofiling in four nations;
- EcoIndicator-99 calculation of global, national and local damages and impacts.

LCADesign provides a range of inventory reports and globally accepted damage and impact assessments as well as a final point score, in real time, to identify hot spots of questionable environmental performance, and an ability to drill down on components and compare alternative materials and designs (Tucker *et al.* 2003).

Delivery of industry databases linkable to BIMs to generate aggregated and component specific environmental reports presents significant challenges (Jones *et al.* 2004). LCADesign's underlying LCI was developed on top of the Boustead Global 4 and 5 version models with input from the New South Wales Department of Commerce LCI model (Jones *et al.* 2004). This latter model was compiled with input from manufacturers supplying NSW government projects (Wood and Jones 1995) and also used to:

- conduct the first LCA analysis of an Australian public building (Stadium Australia);
- audit Sydney Olympic Games developments;
- inform development of BASIX and LCAid software;
- assess green performance clauses in NSW Supply recurrent contracts.

The following LCADesign pilot studies have been undertaken, using a customized LCI Database:

- California: quantitative analysis of Stanford University's Green Dorm. This study sought an optimum timber and steel composite rocking frame to mitigate earthquake damage potential, considering the site's proximity to the San Andreas Fault. Preliminary LCADesign results show highest human health and resource depletion damage from internal walls where most structural components arise (Tobias and Haymaker 2007).
- The Netherlands: analysis of KPMG's new 40,000-m^2 offices in Rotterdam compared to the Dutch developed GreenCalcs Tool, encompassing cradle to end of design life. LCADesign compared LCA results of substructure, structure and internal floors/walls. The building shell and interior structural elements from underground garage to level 14 were analysed by level to identify areas where enhanced environmental performance might be sought by substitution of materials and components (Huysmans *et al.* 2007).

Characterizing LCA impacts by product class

Mitchell *et al.* (2005) classified key commercial building supply chain LCI by typical supply chain and product characteristics as listed in Table 3.4.

The building supply chain material flows classed as infrastructure include supply and distribution of water, mineral, fuel, feedstock, energy, power, forestry, agriculture, and transport of commodities and services. Infrastructure operations claim the major share of fossil fuel use and related impacts on the environment. By virtue of reliance on land use it also offers the major share of opportunity for carbon sequestration, particularly in forestry and agriculture, to mitigate some impacts, such as the greenhouse effect. This class also has the largest impact on habitat loss, and hence the largest opportunity for enhanced flora and fauna conservation by provision of nature corridors, restoration of extractive sites of materials, etc.

The next class involves bulk products, including cement, glass, aggregate and structural steel (Table 3.5). These building commodities have high rates of local supply. By virtue of mass and volume, they form the major building share of resource and biodiversity depletion impacts as well as embodied energy-related impacts.

Table 3.6 provides examples of the class of shaped products which have relatively low levels of imports, high surface area and tensile strength, with price based on area or length and differentiated by finish. Operations are less resource and energy intensive than bulk operations per unit mass, but chemical finishing operations commonly involve emissions to air and water

Table 3.4 Classes of product types in the CRC for Construction Innovation database

Infrastructure	Bulk	Shapes	Items
Fuel, feedstock, power, water, transport, minerals, forestry, agriculture	Concrete, cement, sand; lime, plaster, stone, clay, masonry, metal, glass, structural steel, aluminium, grain, timber	Masonry, metals, cables, composites, ceramics, porcelain, polymers, fittings, furnishings	Paper, fibres and fabrics, paints, pigments, sealants, intermediates, glues, packaging

Table 3.5 Bulk class product lines

Base product	Components and lines
Concrete	Cement, mortar, crushed aggregate, sand; lime and plaster
Steels	Reinforcing and structural
Timber	Structural, formwork and laminated beams
Glass	Float, flat and coated
Clay, masonry	Tiles, bricks, blocks and pavers

Table 3.6 Shaped class product lines

Base material	Product lines
Board	Plasterboard, particleboard; ply; timber panelling, composite laminates
Panel and strip	Steel, aluminium, copper, polymer: PE, PP, PVC, PU and PA* and composites
Sheeting	Paper, aluminium; iron, copper and Cr/Al/Zn/Si/polymer coated steels
Coatings	Paint, sealants, finishes, pigments, lime-putty, plaster, render
Forms	Pipe, wire and extrusions: iron, aluminium, copper, steel and plaster
Cables	Copper, aluminium, glass, polymer and stainless steel composites
Fabric	Wool/cotton/hemp/PE/PP/PVC/PU//PA composites/carpet/underlay/linoleum
Wool/foil	Insulation batt/blanket: mineral, wool, polymer, aluminium, glass, resin, paper

Notes
*PE: polyester, PP: polypropylene, PVC: polyvinylchloride, PU: polyurethane, PA: polyacrylate.

that impact on human and ecosystem health. With shapes comprising the highest surface area per unit mass class, they contribute human health impacts from emissions connected with interior installation, cleaning and maintenance of surfaces.

As a class, itemized product lines, including glues, composites, connections and fittings (Table 3.7), form the smallest commercial building product mass flow. With the highest churn rate, however, the items comprise the major building fabric share of impacts related to solid waste to landfill and subsequent emissions to water and air. Potential for such impact reduction by adaptability, reuse, take-back and recyclability is considerable.

Looking to the future

If Australia is to achieve a life cycle economy appropriate for sustainable development in advanced nations, it will be essential to possess national LCA capability for evaluating opportunities, risks and trade-offs associated with the manufacture and assembly of products and associated services over their entire life cycles. Nationally integrated programs will be required to:

- promote the benefits and costs of adopting LCT, LCA and LCIA into routine practice;
- improve LCA tools and data as well as national LCIA impact and damage indicators;

Table 3.7 Itemized product lines

Type	Product lines
Composites	Glues, fillers, putties, adhesives, chemicals, solder, jointing tapes
Connections	Nuts, bolts, nails, screws, rods, tubes, hinges, flats and angles
Small shapes	Timber, ceramic, metal, glass and high density and low density PE, PVC, PS* polymers
Finished items	Timber, polymer, ceramic, metal, porcelain and glass
Fittings	Polymer, metal, timber, glass and ceramic components
Fabrications	Timber, paper, metal, polymer, ceramic, glass and laminations

Notes
*PE: polyethylene, PVC: polyvinylchloride, PS: polystyrene.

- ensure that LCA offers business value as a window to innovation, assessing risk and revealing new industrial ecology opportunities, with waste recognized as a potentially valuable resource;
- promote stakeholder recognition that the odds are against LCA providing a definitive universal offering in the near future because of the array of parameters that vary regionally and attract different 'importance weightings' from local, state and national governments;
- ensure that LCA is a continuous improvement tool;
- expose the history of subjective misuse of and flawed LCA to avoid their repetition;
- increase the focus on priority material human health and biodiversity loss impacts to reduce high risk;
- ensure maximum opportunity for linkage to emerging design assessment tools such as LCADesign.

Conclusions

This chapter has outlined the Australian and global state of the art in LCA. All nations have sustainable development issues straddling economic, community and environmental domains. These require quantitative assessment of risks, trade-offs and balancing acts. LCA is a powerful and vital environmental accounting method designed to meet such needs over a very wide range of performance criteria. One benefit is its capacity to reveal unintended consequences of decisions, to avoid the shifting of problems elsewhere (sectorally and/or geographically) that occurs due to the complex nature of material flows in a globalized industrial economy.

LCI databases are being developed internationally, and many Australian sectors are working to bring LCA to their stakeholders, who are pushing for more qualitative, streamlined, accredited and automated approaches. Among a range of Australian building industry firsts are the first public domain national Australian forest and timber product LCI to deliver quality assured information to enable stakeholder sustainability decisions. Another is LCADesign that has delivered real-time LCA of building architectural models in Australia, Germany, Holland and California. LCADesign opens the door now to automated ecoprofiling of any product derived from a digital CAD/CAM model, ranging from a building element to an entire building and an entire city.

Accessible high-quality national supply chain LCI data on material, energy and pollution flows are essential for better environmental practice across all supply chains, and the AusLCI initiative calls on support from all sectors of the economy. It is regrettable that, despite Australia's high-value biodiversity and threatened species, it has no nationally accepted LCIA method to quantify biodiversity losses from any supply chain. To operate as an advanced economy with sustainability goals, in particular relating to climate change mitigation efforts, energy transition and protection of biodiversity as well as human and ecosystem health (Newton 2008), Australia urgently needs two new national assets: a national LCI database and a LCIA method.

Bibliography

Athena Institute (2008) *Athena Life Cycle Inventory Product Databases*, Merrickville, OT. Online. Available at HTTP: <www.athenasmi.ca/tools/database/index.html> (accessed 11 April 2008).

Australian Environmental Labelling Association (2004) *State of Green Procurement in Australia*, Canberra. Online. Available at HTTP: <www.geca.org.au/green-procurement/home-2004sogp.htm> (accessed 12 December 2005).

Australian Greenhouse Office (2006) *A National System for Streamlined Greenhouse and Energy Reporting by Business: Draft Regulation Impact Statement*, Canberra: Department of the Environment and Heritage. Online. Available at HTTP: <www.climatechange.gov.au/reporting/pubs/ris.pdf> (accessed 12 April 2007).

Australian Life Cycle Assessment Society (2007) Online. Available at HTTP: <www.alcas.asn.au/alcas2/> (accessed January 2007).

Bluescope Steel (1995) *Case Study: LCA in the Building Industry*, Melbourne. Online. Available at HTTP: <www.bluescopesteel.com.au/go/brands/zincalume-steel/lifecycle-analysis> (accessed 1 May 2008).

Boustead, I. (2007) *Boustead Model 5.0*, Horsham, UK. Online. Available at HTTP: <www.boustead-consulting.co.uk/> (accessed 12 February 2008).

Cole, R.J., Lindsey, G. and Todd, J.A. (2000) 'Assessing life cycle: shifting from green to sustainable design', paper presented at the Sustainable Building 2000 conference, Maastricht, The Netherlands, 22–25 October.

Council of Australian Governments (1992) *National Strategy for Ecologically Sustainable Development*, Canberra: Department of the Environment, Sports and Territories.

Curran, M.A. (2006) 'Report on activity of task force 1 in the life cycle inventory programme: data registry–global life cycle inventory data resources', *International Journal of Life Cycle Assessment*, 4 (11): 284–9.

ecoinvent (2007) Online. Available at HTTP: <www.ecoinvent.ch/> (accessed 11 December 2007).

European Topic Centre on Resource and Waste Management (2008) *Life Cycle Thinking in Resource and Waste Management*. Online. Available at HTTP: <http://waste.eionet.europa.eu/themes/lca>.

Forest and Woods Product Research and Development Corporation (2006) 'Scenario planning: giving the timber industry a head start', *Leading Edge*, 4 (1): 1. Online. Available at HTTP: <www.fwpa.com.au/content/pdfs/Leading%20Edge/2006%20march.pdf> (accessed 1 May 2008).

Gore, A. (2006) *An Inconvenient Truth*, London: Bloomsbury.

Grant, T. (2004) *Eco-Design Centre: Review of Green Tools*, Melbourne: RMIT University.

Huysmans, M., Jones, D. and Slavenburg, S. (2007) 'ICT pilot of KPMG building sustainability assessment in design-build contracting', paper presented at the Second International Conference on Construction Project Management, TU Delft, The Netherlands, 24–26 October.

Intergovernmental Panel on Climate Change (2007) *Climate Change 2007: Impacts, Adaptation and Vulnerability*, Contribution of Working Group II to the Fourth Assessment Report of the IPCC, Cambridge: Cambridge University Press.

International Organization for Standardization (ISO) *ISO 14040 (1998) Environmental Management – LCA – Principles and Framework, Geneva; ISO 14040 Goal and Scope (1997); ISO 14041 LCI Analysis (1998); ISO 14042 LCIA (2000); ISO 14043 Life Cycle Interpretation (2000)*, Geneva.

IVAM (2006) Online. Available at HTTP: <www.ivam.uva.nl> (accessed 27 June 2006).

Jones, D., Mitchell, P. and Watson, P. (2004) *LCI Database for Australian Commercial Building Material*, Report 2001–006-B-15, Brisbane: CRC for Construction Innovation.

Jones, D., Messenger, G. and Lyon Reid, K. (2005) 'Sustainability at William McCormack Place', in K. Brown, K. Hampson and P. Brandon (eds) *Clients Driving Construction Innovation: Mapping the Terrain*, Brisbane: CRC for Construction Innovation.

Jones, D.G., Johnston, D.R. and Tucker, S.N. (2003) 'LCI for Australian building products', in *Proceedings of the International CIB Conference on the Smart and Sustainable Built Environment*, Brisbane.

Jones, D.G., Watson, P., Scuderi, P. and Mitchell, P. (2006) 'Client building product ecoprofiling needs', in K. Brown, K. Hampson and P. Brandon (eds) *Clients Driving Construction Innovation: Ideas into Practice*, Brisbane: CRC for Construction Innovation.

LISA (2008) *Case Studies*, LCA in Sustainable Architecture. Online. Available at HTTP: <www.lisa.au.com/caseStudies.html> (accessed 1 May 2008).

Martin, P. and Verbeek, M. (1998) *National Products Accounting Strategy: A Path to Competitive Advantage for Australian Industry*, Armidale, NSW: Profit Foundation.

Mitchell, P. (2004) 'LCT implementation: a new approach for "greening" industry and providing supply chain information: a plywood industry study', PhD thesis, Brisbane: School of Geography, Planning and Architecture, University of Queensland.

Mitchell, P., Jones, D., Watson, P., Johnson, D. and Seo, S. (2005) 'A national building products inventory', paper presented at the Fourth Australian Life Cycle Assessment Conference, Sydney, 23–25 February.

Newton, P.W. (ed.) (2008) *Transitions: Pathways Towards Sustainable Urban Development in Australia*, Melbourne: CSIRO and Dordrecht: Springer.

Norris, G. and Notten, P. (2002) *Current Availability of LCI Databases in the World*, Working Draft 2a, LCI Program of the Life Cycle Initiative, Boston, MA: Harvard University.

Organization for Economic Co-operation and Development (2003) *Environmentally Sustainable Buildings: Challenges and Policies*, Paris.

Queensland Government (2000) *Ecologically Sustainable Office Fitout Guideline*, Brisbane: Department of Public Works. Online. Available at HTTP: <www.build.qld.gov.au/aps/ApsDocs/ESDMasterDocument.pdf>.

Seo, S. (2002) *International Review of Environmental Assessment Tools and Databases*, Report 2001–006-B-02, Brisbane: CRC for Construction Innovation.

SPINE@CPM (2006) Online. Available at HTTP: <www.globalspine.com/> (accessed 12 January 2006).

Tharumarajah, A. and Grant, T. (2006) 'Australian national life cycle inventory database: moving forward', paper presented at the 5th ALCAS Conference, Melbourne, 22–24 November.

Tobias, J. and Haymaker, J. (2007) 'A model based LCA process on Stanford University's Green Dorm', in *InLCA/LCM Conference Proceedings*, Portland, OR.

Tucker, S., Ambrose, M., Johnston, D., Newton, P., Seo, S. and Jones, D. (2003) 'LCADesign: an integrated approach to automatic eco-efficiency assessment of commercial buildings', in *Proceedings of 20th CIB W078 Conference on Information Technology in Construction*, Auckland, New Zealand.

United Nations Environment Programme (2003) 'Sustainable building and construction: facts and figures', *Industry and Environment*, 26 (2/3): 5–8.

—— (2008) *The Life Cycle Initiative*. Online. Available at HTTP: <http://lcinitiative.unep.fr/> (accessed 17 March 2008).

Watson, P., Jones, D. and Mitchell, P. (2005) 'Temporal and physical life cycles', paper presented at the Fourth Australian Life Cycle Assessment Conference, Sydney, 23–25 February.

Watson, P., Mitchell, P. and Jones, D. (2004) *Environmental Assessment for Commercial Buildings: Stakeholder Requirements and Tool Characteristics*, Report 2001–006-B-01, Brisbane: CRC for Construction Innovation.

Watson, S. (2004) 'Improving the implementation of environmental strategies in the design of buildings', PhD thesis, Brisbane: School of Geography, Planning and Architecture, University of Queensland.

Wood, G. and Jones, D. (1995) 'LCA: how it works and practical applications', paper presented at the Ecologically Sustainable Development in Architecture and Building Conference, Sydney.

—— (1996) 'Using life cycle analysis to understand environmental impacts', paper presented at the International Conference on Design for the Environment, Sydney.

World Business Council for Sustainable Development (2007) *Builders Overestimate Cost of Going Green*. Online. Available at HTTP: <www.wbcsd. org/ plugins/DocSearch/result.asp?txtDocText=overestimate&txtDocTitle=overestimate> (accessed 11 August 2007).

Worldwide Fund for Nature (2007) *Climate Change*. Online. Available at HTTP: <http://wwf.org.au/ourwork/climatechange/>.

4 Service life prediction of building materials and components

Ivan Cole and Penny Corrigan

This chapter looks at critical problems encountered in the life prediction of building materials and components, and addresses how a database of life estimates can be compiled for the vast number of component/environment/usage combinations that may exist in a building. Further, it focuses on how an estimate of actual component life can be made for a component subjected to conditions that differ from the reference component in the constructed database. The chapter considers the different uses of life prediction data, and the varying levels of data complexity required for them. The possible methods of deriving life prediction data are analysed in terms of the above research challenges, as well as their applicability to the differing uses of life prediction data. For two methods – Delphi studies and process-based models – detailed case studies are presented.

The chapter will not attempt to review or summarize the body of research work on the service life prediction of building materials and components; rather, it will use specific examples to illustrate the particular challenges to life prediction. These challenges arise from the huge matrix of conditions for which prediction is required, and from the wide range of uses for life prediction data. The following simple calculation illustrates the enormity of the predictive task. A standard Australian dwelling, which may be erected in at least ten climatic sub-zones and a minimum of five pollutant zones, will comprise of more than 1,000 different components that may be situated in at least five different microclimates. In addition, there are up to five different material combinations per component, and a component will be built to three levels of quality, with three levels of workmanship in installation and three levels of maintenance. This equates to in excess of 15 million possible combinations.

There is a wide variety of possible uses for life prediction information, and users range from material and component manufacturers, to designers and facility managers of buildings, to managers of building portfolios. Common uses of service life data, in increasing degree of complexity, are:

1 *life prediction of known products in known environments* – this is important to manufacturers who provide product guarantees, and to designers for materials selection;

2   *comparative materials selection based on life estimates of known materials in known environments* – this has always been important to designers, but is increasing in importance as online life cycle assessment tools allow the selection of a variety of designs and material types;

3   *comparative life prediction in new environments* – as for (2) above;

4   *effect of design changes on component life* – as for (2) above;

5   *estimation of maintenance schedules for already built facilities* – this is critical for facility and portfolio managers;

6   *estimation of maintenance schedules from the design of facilities* – this is important to building designers, managers and owners;

7   *effect of workmanship and human factors on maintenance and component life* – this is critical for builders, owners and managers of dwellings;

8   *prediction of remaining life of inspected facilities* – this is critical for owners and managers of building portfolios.

The different uses require different levels of detail in the service life databases. For use 1, for example, in order to predict the life of a building component its materials of construction, its position in the building, the building location, the climate at that location and the estimated life would be required; for use 7, additional information is needed on workmanship errors and human usage patterns and their impacts on materials, and how damage progresses with time, not just the time to failure.

Thus, the generic problem of life prediction can be seen as how to effectively use a wide variety of information sources (historic data, survey information, modelling, expert opinion, etc.) to generate information on an enormous number of conditions. Buried within this overall question are a number of subsidiary questions, for example:

• What are the different classes of information available, and what are the different methodologies to be followed in their use?
• How can project service life from a known case be transferred to an unknown case where not all the conditions are the same?

Classes of information

Knowledge of service life can be derived from a variety of sources. One way to classify these sources is in terms of the degree of knowledge that already exists, the level of uncertainty associated with the data that can be derived, and the uncertainty in the phenomena that are being documented. The degree of knowledge can be classified into four levels of increasing uncertainty:

1   Well-known and documented situations
2   Known but not documented situations

3   The life and damage rate are not known but are predictable from damage causes (environmental and human), even if this prediction is complicated

4   The damage causes and damage progression are not predictable but can be measured.

The most reliable prediction can be made if a situation is well known and documented, which is the case where code books or databases have been derived from years of observation and measurement.

The next level down from this is when a situation is understood, but has not been documented. Knowledge may be gleaned by surveying expert opinion or through data-mining information buried in maintenance records, but these forms of knowledge are not often applicable to new situations.

Where the factors that cause damage are known and predictable, then modelling can be used to predict the progression of damage, especially when there is a strong relationship between damage and its cause. This class of knowledge would incorporate data derived from service life models.

If the known causes of damage cannot be well defined but are able to be measured, then sensor-based approaches can be used to predict damage progression. This method is particularly relevant when the onset of damage is controlled by human factors, such as building usage or workmanship. While it is very difficult to predict human errors, areas of risk can be monitored by sensing.

Life estimation can also be based on accelerated testing. In principle, this is an extension of the first knowledge class, with the particular issue of how a well-developed and documented measure of component life in one situation (accelerated testing) can be applied in another (real service conditions).

Individual prediction methods

Use of code books or databases

There is relatively little documented knowledge of component life available (at least in Australia), although it could be argued that the 'deemed to comply' durability provisions in Australian Standards do reflect industry experience and, to some extent, codify existing knowledge. Internationally, degradation versus time curves are used in asset management and maintenance programs (Kyle 2001). However, both the scientific rigour of such curves and the protocols for applying them to new buildings are not well established.

Surveys of the performance of actual buildings can be made as a means of generating databases of component life. Indeed, a number of extensive building condition surveys have been undertaken worldwide, with perhaps the benchmark being set by Vanier and Kyle (2001) in a survey of 600 roofs in seven climate zones across Canada. This survey used a specially designed data collection tool that facilitated the inclusion of so-called 'tombstone information' (location, roof type, area, roof and building age, CAD drawings, etc.), as well as data on condition, and distress type, severity and quantity for the different elements of the roof fabric. This tool dealt with all but one of the limitations of maintenance data discussed below. The survey defined the condition of a roof in terms of a 'state', where state 7 was a roof in excellent condition and state 1 was a roof requiring immediate replacement. The survey found relatively few roofs that were in an advanced state of disrepair (state 1 or 2).

The low level of data from roofs at an advanced stage of degradation does present a significant methodological problem in the use of survey data to derive life predictions: how can data that define condition states where damage has not reached the failure criteria be used to estimate the time to reach the failure criteria? In general, there are three ways of achieving this:

1 *Fitting methods.* A fitting method relies on forming a mathematical expression that treats damage condition as a numerical value, and then finds the best fit between this numerical damage value and the time to reach that value. The limitation of this approach is that, in general, damage condition is defined on the basis of what is readily observable, and thus increases in damage condition do not necessarily correspond to uniform increases in damage. For instance, in defining damage state for corrodible materials such as zinc or steel, it is common to define damage state as a function of rust coverage, e.g.:

   Damage state 1 = 0 per cent rust coverage
   Damage state 2 = 1 per cent rust coverage
   Damage state 3 = 5 per cent rust coverage.

   However, the time interval to progress from damage state 1 to 2 may be quite different to that from state 2 to 3.

2 *Transition probability methods.* In this type of method, the damage condition is not converted into a numerical value, as in a fitting method, but remains as a discrete condition, and then a transition probability from one state to another in a given time interval is calculated. Using the above example, transition probabilities from state 1 to state 2 and from state 2 to state 3 can be calculated. These probabilities can and have been derived from field surveys (Lounis *et al.* 1998).

   A major issue with this method is having sufficient data to determine transition probabilities, particularly for the high damage states where, as indicated by Vanier and Kyle (2001), it is difficult to obtain extensive data. If data cannot be obtained to define transition probabilities, then, in principle, these can be derived from modelling or accelerated testing (discussed later).

A major advantage of a transition probability method is that, in conjunction with condition monitoring, it is ideally suited for the prediction of remaining life. For example, in the Canadian case mentioned above, if condition monitoring were to assess a roof as being in state 5, transition probabilities could be readily used to assess the time required to reach state 1.

3   *Factor methods.* Once a database of service life is obtained, the question highlighted in the introduction arises: how can this information be transferred to real situations where all the conditions may not be the same as in the original data? The original database is referred to as 'reference service life' in ISO standard 1586–2:2001. Hypothetically, if a structure were built to the same design, with the same materials and to the same level of workmanship as the 'building' from which the reference service life was derived, there would be a one-to-one correlation between the actual service life and the reference service life. In fact, no two structures will be built in identical ways in identical environments; to deal with this issue, the building durability community has developed the factor method (ISO 2001), namely:

$$PSL = RSL \cdot f_A \cdot f_B \cdot f_C \cdot f_D \cdot f_E \cdot f_F \cdot f_G$$

where *PSL* is the predicted service life of the component of interest, *RSL* is the reference service life, and the f values are the factors (with values between 0 and 1) that account for variations in the quality of design, use, workmanship, etc., such that:

A = component
B = design
C = work execution
D = indoor environment
E = outdoor environment
F = in-use condition
G = maintenance.

This method may also be defined on a probabilistic basis (Arseth and Hovde 1999), so that *PSL* becomes *PSLDC* (predicted service life distribution of the component) and *RSL* becomes *RSLDC* (reference service life distribution of the component). While international efforts are underway to develop databases for *RSLDC* and the various other factors (Hans *et al.* 2008), there are a number of issues or limitations that currently affect the method. On a theoretical basis, there is no strong reason why the various factors should be treated independently.

For instance, the quality of work execution could easily impact on both the indoor and outdoor environment. Second, large databases need to be derived to define the factors, but the derivation of these databases is complicated by the fact that the factors are not truly independent.

Nevertheless, the factor method is currently the only internationally accepted system of estimating predicted service life from reference service life. It significantly extends the utility of databases on service life by allowing them to be applied outside the narrow domains from which the original reference service life data were collected. Thus, in theory, a combination of *RSL* databases and the factor method can be utilized for uses 1 to 7 (listed in the introduction), with perhaps the exception of use 5; in contrast, *RSL* databases would only be useful for uses 1 and 2.

Data-mining and expert opinion

Rather than surveying actual buildings, knowledge of the service life of buildings can be derived either by examining existing databases that inherently contain information on component life, such as the maintenance records of large collections of buildings, or by surveying the opinion of experts in the field.

The authors investigated the use of data-mining the maintenance records of public authorities in Australia, but found that these records were not generally of sufficient detail to extract meaningful data on service life (Cole *et al.* 2007). For example, while records would detail that 'interventions' had occurred on 'parts' of dwellings, they suffered from the following limitations:

- insufficient details of the nature of the intervention;
- in general, a 'part' was defined in a very broad sense ('roof', 'wall', etc.);
- no record was made of the type of material being maintained (e.g., a roof intervention would be recorded without reference to the type of roofing material);
- no detail was given regarding the reasons for maintenance or the degree of damage.

Of course, if more detailed maintenance information were to be kept in such records, then data-mining could provide valuable information of service life. Internationally, a number of systems have been established to ensure that the conditions of buildings and infrastructure can be monitored and recorded. Their use will lead to the development of meaningful information on the variations in building and component condition with time, and may enhance the accuracy of the 'building degradation' curves discussed previously. One such package is the Builder EMS package developed in Canada by Uzarski and Burley (1997).

However, the level of detail required in such condition assessment tools is far in excess of what is currently recorded by the maintenance units of public infrastructure owners. Once detailed maintenance data are collected, the same issues discussed in relation to using survey data to predict life will have to be addressed. That is, how can data that define condition at specific times and discontinuous times be used to estimate the final life of a building, and how can the variations of building and environmental conditions between the buildings in the databases and the building for which the prediction is required be taken into account? Respectively, transition probability and the factor method may be solutions to these two problems.

Further, some researchers indicate that case-based reasoning (CBR) is being used to match lifetimes in maintenance databases to those required for prediction (Kyle 2001). In CBR, the characteristics of the buildings in the maintenance database are searched to select those with the most similar characteristics to the building of interest, and then it is assumed that the lifetime of the building of interest is a weighted combination of the lifetime of the similar buildings (Kyle 2001). Given such enhancements, data-mining or CBR techniques could be used to derive life prediction for uses 1, 2, 5 and 8.

An accepted method of quantifying expert opinion is the use of Delphi surveys. An example of a Delphi survey will be given later in the chapter, and technical details will be discussed at that point. The advantage of expert surveys is that a large amount of information can be efficiently collected to allow the estimation of the life of components as a function of material type, geographic and climatic zone, and maintenance level, thus dealing to some extent with the enormous number of (for example) component/material/environment combinations that may require life predictions. The disadvantage is that the method cannot be used for new materials or new environments where practitioners have no experience, and it does not reveal any of the underlying processes that control degradation and so cannot be extended to take into account changes in conditions. Thus, while it is a good method for deriving life data for uses 1 and 2, it is not useful for the others.

Modelling

In developing models of degradation, it is vital that the appropriate degradation modes of the highest-risk components are modelled. Identification of such components and their degradation in buildings and infrastructure (with a vast variety of components, material types and microclimates) is a complex procedure, but there are a number of solutions to this problem – for example:

- The most thorough has evolved in the field of building pathology, where systematic approaches and standardized nomenclature have been developed to analyse which components may be at risk (Fatiguso and De Tomasi 2008).
- Many researchers analyse only a single building sub-system and, through in-depth knowledge (but often not defined knowledge), select only the component and failure mode that presents the greatest risk.

The methods used to model the degradation of components at risk include dose functions and damage indexes, neural networks (Pintos *et al.* 2000) and process models.

Dose functions and damage indexes

Dose functions define the degradation of a component or material as a function of the exterior environment, and may be derived from field exposures, laboratory tests (including accelerated tests), or the intrinsic properties of a material, or a combination of all three. For example, for atmospheric corrosion, a large number of dose functions have been derived that connect mass loss of exposed metal with the pollutant level and climatic parameters at exposure sites.

A typical dose function (from Tidblad *et al.* (1998)) is:

$$\text{Mass loss of zinc} = 1.35[SO_2]^{0.22} \exp\,[0.018\,\text{RH} - 0.021\,(T-10)]t^{0.85} + \text{Rain}[H^+]t$$

where mass loss is in gm^{-2}, T is temperature in °C, t is time in years, $[SO_2]$ is the atmospheric gaseous sulphur concentration in μgm^{-3}, RH is per cent relative humidity, Rain is the amount of rain in mm, and $[H^+]$ is the hydrogen ion concentration in rain in mgl^{-1}.

From the 1970s to the 1990s, many researchers derived dose functions for metallic components (Cole 2002), but no consensus was reached as to the correct formulation of the dependence of mass loss on climatic parameters. This is not surprising, as the formulations were generally derived from regression analysis of collected data sets, and thus were highly dependent on the variations that occurred in them. This is not in itself a problem, provided the dose functions are not used to predict mass loss in environmental conditions dramatically different from those in the data sets used to derive them. This definitely imposes geographic limits on the use of dose functions; however, more seriously, it may also impose temporal limits. For instance, the formulation of Tidblad *et al.* (1998) is highly dependent on the gaseous SO_2 concentration in the atmosphere and the $[H^+]$ concentration in rainwater, reflecting the significant effects of industrial gaseous pollution and the acidification of rain in northern Europe in the 1980s and 1990s. However, since that time industrial pollutant levels – and particularly [SO₂] – have dropped significantly in Europe, so it cannot be guaranteed that the formulation is still accurate and will remain accurate in the twenty-first century.

A second major issue with dose functions is how to correlate the performance of a simple specimen exposed in a relatively simple environment, to a building component within the local microclimates induced by a building. The complexity of this problem depends on the type of parameters used in the dose function formulation. If the parameters relate to the local condition of the material, such as moisture content for timber or deposited salt for metallic corrosion, then the problem is much reduced. This can be explained with reference to Figure 4.1, which is a simplified flow diagram of the influence of climate.

For a house in a given location, the climate in the vicinity of the dwelling is definable (this is not quite as simple as it sounds, but will be discussed later). While this climate will have an influence on the microclimate in the vicinity of a particular component, it may differ significantly from the component microclimate, as the house itself will perturb the climate in a way highly dependent on its geometry. In addition, the microclimate of components even facing the external environment will be highly influenced by the internal microclimate within the dwelling. Material responses (moisture content of timber, surface temperature of metal, etc.) will be strongly influenced by the local microclimate, and also by the exact geometry of components (for example, the moisture content in timber joints may be significantly different to the other timber members) and by usage and service issues. The material response (including salt and pollutant retention) of a component will determine component degradation.

Therefore, if a dose function is defined in terms of material response parameters, it can be applied to predict the degradation of a component within a building. However, a methodology to predict material response that takes into account geometry, usage and microclimate is required. If a dose function is defined in terms of external climate parameters (gaseous SO₂ in the case of Tidblad *et al.* (1998)), then it cannot be used to predict

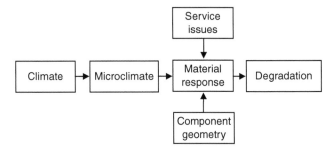

Figure 4.1 Flow diagram of the influence of climate, service issues and component geometry on component degradation.

component life, as the external and local microclimate variables may be significantly different. This is true for components facing the external environment, as well as those within the building fabric.

The degradation of metal cladding in a coastal environment may best illustrate this. Most dose functions for coastal environments contain a dependence of mass loss on airborne salinity. Of course, marine salts must be deposited and retained on a metal surface to promote corrosion. Applying a standard dose function to estimate the corrosion of cladding assumes that the retention of salt on an atmospheric test specimen (used to define a dose function) will be the same as that retained on external cladding. However, work by Cole and Paterson (2004) has shown this to be far from true, as deposition onto a building is controlled by airflow turbulence, and it tends to be up to three times greater at the edges of buildings than at the faces. There are also marked variations in deposition depending on the angle that a wall makes with the prevailing sea winds. Thus, direct application of dose functions in these circumstances would lead to very significant errors in prediction. However, if the dose function were expressed in terms of retained salt on a metallic surface, then it could be applied directly to predict component life. Therefore, appropriately defined dose functions could be used to define service life for uses 1 and 2 and, if combined with the factor methods, could be used for uses 3 and 7.

Damage indexes are very similar to dose functions in that they relate an index to environmental factors. They differ in that the index is normally defined as a risk index for degradation, not as a degradation rate, and, as such, damage indexes cannot be directly used for life prediction. However, it is commonly assumed that if the damage indexes of a material in two different applications are the same, then the degradation rate will be the same. One of the most widespread degradation indexes is Scheffer's (1971) climate risk index (CRI), which defines the rotting tendency of wood, and is given by:

$$CRI = \Sigma(T-2)(D-3)/17$$

where T is the average monthly temperature, D is days with precipitation above 0.01 inches, and the sum is over the months of the year. Scheffer's index has been used to define geographical zones with risks of timber rot across the USA. Haagenrud *et al.* (1998) have developed a local Scheffer index that can be applied directly to timber facades or components on buildings. Their index, which they define as WC-CRI, is given by:

$$WC\text{-}CRI = (t_{wet}/t_{total}) * 100$$

where t_{wet} is the time when both the timber surface is wet (defined as a current above a given value on a WETCORR unit, a sensor for measuring wetness) and the surface temperature is above a limit for fungus growth, and t_{total} is simply the total time of monitoring or analysis. The advantage of WC-CRI is that it does relate directly to material response parameters, and so can be used to compare the risk of different components.

Process-based models

Process-based models analyse the physical or chemical processes that promote degradation and, in principle, they avoid the various limitations highlighted for other approaches. For instance, as they are based on fundamental processes and not data sets, they should not be limited in geographic space or time. In practice, however, such models do contain some empiricism that requires adjustment when they are applied in different geographic zones.

The microclimatic conditions and degradation mechanisms at an actual component of interest can be modelled, so the issue of how to go from experimental sample to building component is sidestepped. Models for timber structures (Leicester *et al.* 1998) and for metallic components (Cole *et al.* 2003) have been developed that enable service life estimates that are applicable to uses 1 to 4, 6 and possibly 7.

Sensory systems

It is very difficult to predict exactly how workmanship or usage issues will impact material degradation, but it is possible to monitor the potential at-risk areas in a building or structure. Further, multisensory systems have been developed that incorporate sensors to monitor both damage and damage causes, which can be used for damage prognostics (Muster *et al.* 2005). Although in its infancy, the field of sensor-based prognostics for buildings will be an extremely valuable technique to provide data for uses 5, 7 and 8 in the future.

Accelerated tests

Accelerated tests have been used in national standards and guidance documents as either explicit or implicit guarantees of service life, whereby a product is deemed to have an appropriate lifetime if it passes a given accelerated test. In some cases there is an explicit connection made between performance in an accelerated test (see, for example, Cole *et al.* 1995), while in others there is no direct connection to service life. Rather, the justification for use of the standard is that materials and components that have shown acceptable lifetimes in service have passed the accelerated tests, and thus it is assumed that if new components or materials pass the test, then they will also show acceptable performance in service. ISO 15686–2:2001 sets out requirements for accelerated tests that are to be used to provide lifetime guarantees. The key amongst these is that the degradation modes in the accelerated tests should be the same as in service.

Nevertheless, despite the guidance of the standard, there are a large number of accelerated tests currently in use that do not have an established connection between performance in the tests and life in service. To overcome this issue, many workers are developing new accelerated tests that more closely replicate the microclimatic conditions and the degradation modes that occur in service. A major issue with these tests is how they can be accelerated while maintaining microclimate replication. Two approaches are commonly followed:

- to exclude those parts of the climate sequence where damage is unlikely to happen (Daniotti *et al.* 2008);
- to increase the level of damaging environmental agents (temperature, UV dose, salt concentration, etc.).

In practice, both techniques are often required to give sufficient degradation. If the former method is used to accelerate the test, then the time connection between the accelerated test and service duration is transparent (if only 20 per cent of an annual climate cycle needs to be used in the accelerated test, the acceleration factor is thus 5). Establishing an acceleration factor using the second approach is more problematic. The most common method is to determine the rate of increase of degradation on a standard material at the higher dose with regard to the lower dose, and then use this increase in degradation as the acceleration factor for the test. This, of course, assumes that the damage in subsequent materials will be accelerated in the same manner as damage in the standard material, which may not always be the case. In summary, there is a range of accelerated tests that acceptably duplicate the degradation modes in service. However, life prediction from accelerated tests remains problematic, especially for new materials.

Case studies

In this section, two example methods to estimate component life will be given in depth in order to expand on the issues highlighted above.

Expert opinion: Delphi survey

Surveying expert opinion is one method of acquiring a large amount of information in an efficient way. In particular, Delphi surveys offer an established protocol to refine the responses through feedback loops. A Delphi survey is a structured group-interaction process and is an established technique for obtaining consensus (Duffield 1993). The technique consists of a number of 'rounds' of opinion collection and feedback. A series of questionnaires is used for opinion collection, with the results from each round being used as the basis for the formulation of the questionnaire used in the next round.

A project applying the principles of a Delphi survey to collect expert opinion on the durability of building components has been carried out within the Cooperative Research Centre for Construction Innovation (Cole *et al.* 2004a). In addition to developing a database of estimated service life of components, the project aimed to assess whether the opinion of experts would be:

• sufficiently consistent to derive life estimates for components;
• internally consistent across different component and environmental types;
• consistent with both the lifetimes predicted by other methods and lifetimes that would be expected given the basic physics and chemistry of degradation.

Application of Delphi survey to building components

Material degradation is a complex process, and one of the strengths of the Delphi process in this context is the ability to gather information from experts with a wide range of backgrounds: professionals such as builders and architects will have a mix of practical experience and theoretical knowledge; building material suppliers will have intimate knowledge of their specific products; and academics and scientists working in material durability will understand the scientific principles relevant to the construction of a durability model.

In the example case study, 30 different building components – from nails and ducting through to roofing, window frames and door handles – were chosen as the basis of the survey. These were chosen to be representative of not only the much wider range of components within a building (>120 of the commonly used components in Australian domestic and small commercial construction), but also a range of possible materials, coatings and environments. The survey included service life (both with and without maintenance) and aesthetic life, and time to first maintenance, and covered both commercial and residential buildings in marine, industrial and benign environments.

For each component in a variety of situations and environments, the web-based questionnaire asked respondents to designate an estimated life from the ranges:

• <5 years
• 5–<10 years
• 10–<15 years
• 15–<20 years

- 20–<30 years
- 30–<50 years
- >50 years.

An example from the online questionnaire is shown in Figure 4.2.

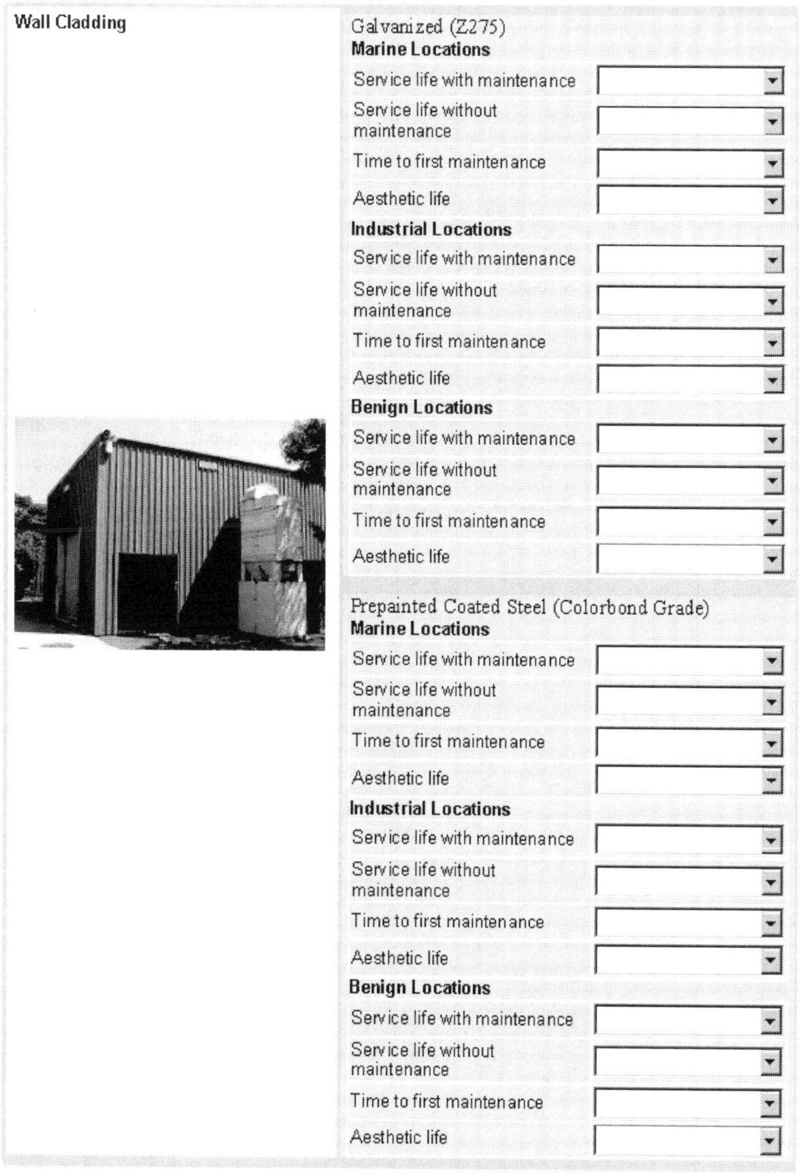

Wall Cladding	Galvanized (Z275)
	Marine Locations
	Service life with maintenance
	Service life without maintenance
	Time to first maintenance
	Aesthetic life
	Industrial Locations
	Service life with maintenance
	Service life without maintenance
	Time to first maintenance
	Aesthetic life
	Benign Locations
	Service life with maintenance
	Service life without maintenance
	Time to first maintenance
	Aesthetic life
	Prepainted Coated Steel (Colorbond Grade)
	Marine Locations
	Service life with maintenance
	Service life without maintenance
	Time to first maintenance
	Aesthetic life
	Industrial Locations
	Service life with maintenance
	Service life without maintenance
	Time to first maintenance
	Aesthetic life
	Benign Locations
	Service life with maintenance
	Service life without maintenance
	Time to first maintenance
	Aesthetic life

Figure 4.2 Example of online Delphi survey form.

The aggregated responses to individual questions were classified into four classes based on a simple rule relating to the consistency of the responses:

- Class 1 – one interval contained more than 50 per cent of responses;
- Class 2 – two adjacent intervals contained more than 50 per cent of responses;
- Class 3 – three adjacent intervals contained more than 50 per cent of responses;
- Class 4 – none of the above, or cases where there were two or more (non-adjacent) intervals with the same maximum number of occurrences.

An example of a class 2 response is shown in Figure 4.3.

The second round of the survey was used to gain additional information for categories that initially fell into class 3 or 4 due to lack of consistency in the responses. At the end of the second round, the information was assembled into a database and a table was generated for users to look up predicted lifetimes for metallic components in a comparable environment. In the table, the predicted life is presented in two forms, the mode and the mean, as well as a standard deviation for the mean value.

Returning to the methodological assessments discussed above:

- The Delphi survey did lead to consistent responses from the three groups of experts, with 78 per cent of the sets of responses for each component falling in either class 1 or 2 after the first round of the survey, and 95 per cent after the second;
- The Delphi survey was internally consistent. For example, it would be expected that responses on commercial and domestic roofing material would be very similar as, in this case, the use does not significantly change the life and, in all conditions for all materials, the difference between roof-sheet life and gutter life for the different

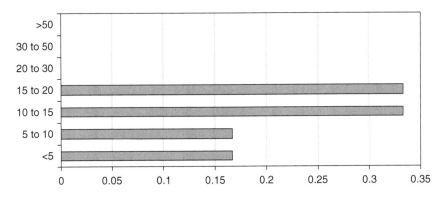

Figure 4.3 Summary of responses (class 2) for life expectancy of galvanized steel roof without maintenance in a marine environment.

types of buildings was never more than four years and generally less than three;

- The survey was consistent with basic physical and chemical principles, and with results of data derived from other sources. Again considering roofs, basic corrosion science indicates that roof life in a marine environment would be shorter than in an urban setting, as demonstrated by the survey. Roof life should also increase when the protection level of the component is increased – i.e., the life of Colorbond®> Zincalume>galvanized. This was in fact observed. For example, Table 4.1 shows a comparison of Delphi survey results with values derived from experimental studies (equating the mass loss of exposed metal plates with that of roof sheeting), process modelling (see next section) and maintenance records. It is apparent that, although there are variations, the Delphi survey was consistent with the other methods.

The Delphi survey therefore appears to be a reasonable system for obtaining a large amount of data covering a wide range of conditions. In terms of the possible uses of service data, the survey appears to provide reasonable estimates of component life in known environments, and thus could be used to select between known materials (uses 1 and 2). It cannot in itself account for variations in environment, workmanship or usage, but it could be combined with the factor method to extend its flexibility in this regard. It is not capable of assisting in the requirements of uses 3 to 8.

Process-based models

As discussed above, process-based models offer the possibility of direct prediction of component life and, in principle, they can take into account the local and usage factors that control the life of individual components. The issues and capabilities of such process-based models will be outlined with reference to the holistic model of corrosion developed in the Cooperative Research Centre for Construction Innovation (CRC-CI) Learning System for Life Prediction of Infrastructure project (Cole *et al.* 2007) and

Table 4.1 Comparison of survey predictions and database for roof sheeting

Data	Environment	Mode	Mean	SD
Delphi	Marine	10–20	12	6
Experimental	Marine	5–10	14	12
Process model	Marine	5–10	9	5
Maintenance	Marine		16	
Delphi	Benign	30–50	35	13
Experimental	Benign	>50	>50	
Process model	Benign	>50	>50	
Maintenance	Benign		41	4

in a series of journal papers (Cole *et al.* 2003, 2004b, 2004c, 2004d, 2004e; Cole and Paterson 2004, 2006).

The holistic model has to deal with two crucial issues:

1   referring to Figure 4.1, how to combine climatic data, usage patterns and component geometry to predict the material response of a component;
2   given the material response of a component, how to predict its degradation.

Figure 4.4 shows a model flow pattern for the analysis of downpipes, in which the basic structure is that climatic conditions and use conditions are combined to determine the local microclimate. The microclimate, material type and local material features are then used to calculate the damage rate and thus pipe life. However, particular usage cases may modify differing parts of the model. The set of downpipes is divided into a series of cases (six in all) that reflect different environments and usage patterns of the different sections of a downpipe (Table 4.2). The downpipe is first divided into the exterior surface or the interior surface, then the exterior surface is further broken into two: the section just below the roof eaves which is sheltered from rain, and the section at the lower part of the wall which will be cleaned by rain (though not by vertical rain). The interior surface of the downpipe is broken into two usage cases: maintained (equivalent to cleaned in this application) and not maintained. If the downpipe is not maintained, it may become blocked due to the accumulation of leaf litter and other debris, leading to three more classes: above, at and below the blockage.

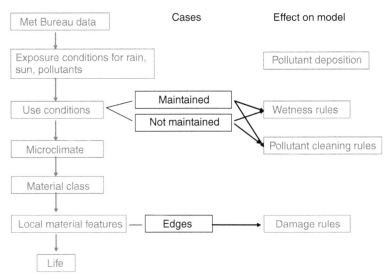

Figure 4.4 Model flow pattern for the analysis of downpipes.

Table 4.2 Sub-component and usage cases for downpipes

Case	Sub-component	Exposure	Usage	Position
1	Exterior	Sheltered from rain	All	
2	Exterior	Exposed to rain	All	
3	Interior	All	Maintained	
4	Interior	All	Not maintained	Above blockage
5	Interior	All	Not maintained	At blockage
6	Interior	All	Not maintained	Below blockage

As indicated in Figure 4.4, the usage cases will change different parts of the model. As discussed later, the model contains a module that calculates how rain will clean a metal surface of deposited marine salts. However, if the surface is covered in debris, it will trap salts and significantly decrease the efficiency of rain cleaning. Further, the accumulated debris will absorb moisture and prevent evaporation, thus increasing the time of wetness within the interior of a blocked downpipe. Each component that is modelled (roofs, gutters, etc.) is similarly broken into a series of cases.

A multiscale model is used to define the material response for each case of each component. The principle of the model is shown in Figure 4.5, in which the processes controlling atmospheric corrosion are presented in a range of scales: from macro through meso to local, micro and micron, and lastly electrochemical (Cole *et al.* 2003).

In Australia, corrosion is promoted by the effect of marine aerosols, so the model analyses their production, transportation and deposition. A major generator of marine aerosols is wind blowing over or across breaking waves (so-called whitecaps) in the open ocean, and the extent of aerosol pick-up is proportional to whitecap coverage (percentage or fraction of the ocean that is covered by breaking waves). Figure 4.6 shows a map of whitecap coverage as a fraction for the Australian region in January to February.

From such maps and from an analysis of aerosols produced by surf, the production of marine aerosol around the country is calculated. A computational fluid dynamics model is then used to calculate the transportation of aerosol across the country (Figure 4.7).

Aerosols are convected by wind, lifted by diffusion and dragged down by gravity. The effects of gravity depend on aerosol mass and thus diameter, with the wet particle diameter for salt produced by the sea being critically dependent on local RH. The aerosol production and transportation models are linked in a geographical information system (Cole *et al.* 2004b) that allows airborne salinity levels to be calculated in the vicinity of a building. The efficiency of aerosol deposition onto objects and dwellings will depend on aerosol size, air speed and air turbulence. This is illustrated in Figure 4.8, which graphs aerosol deposition efficiency as a function of

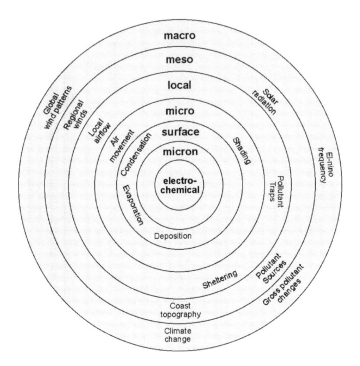

Figure 4.5 Schematic of multiscale model of corrosion.

size onto 'salt candles' (standardized airborne salinity measuring devices (ISO 9225)), and is given as:

$$\eta = 100 * D/C\bar{U}A_s$$

where D is the deposition rate, C is the atmospheric concentration of the aerosol, $\bar{U}$ is the mean air speed and A_s is the area of the surface that the aerosol impacts on. The fact that deposition efficiency depends on air turbulence leads to marked variations across a building, with heightened deposition onto building edges where air is more turbulent. The practical effect of this in the multiscale model is that pollutant deposition onto gutters and downpipes is heightened with respect to general deposition onto walls and roofs, leading to correspondingly faster degradation rates of gutters and downpipes.

Having calculated the deposition rate onto a component, the next stage is to calculate any cleaning effects of the natural environment. Experimental studies indicate that wind is not in general effective at removing aerosols from surfaces, but rain can be, provided it exceeds a critical level in any one rain event (Cole and Paterson 2007). The efficiency of salt removal from metal surfaces can be approximated by:

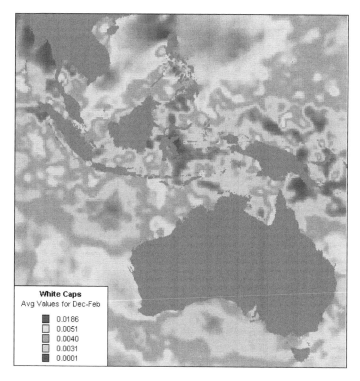

Figure 4.6 Map of whitecap activity in Australasia and South East Asia.

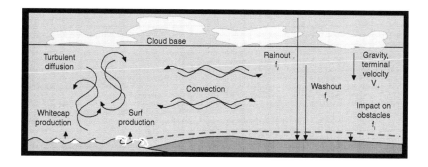

Figure 4.7 Schematic of CFD model of aerosol transportation.

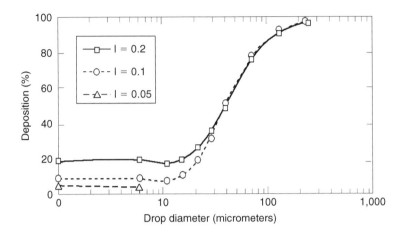

Figure 4.8 Deposition efficiency of aerosol onto a salt candle.

$$S_f = S_i * e^{-\alpha R} \text{ if } R - R_i > 0; \text{ or}$$
$$S_f = S_i \text{ if } R - R_i < 0$$

where S_i and S_f are the salt content before and after a rain event respectively, R is the rainfall that impacts on the surface during the rain event, R_i is the minimum rain required to clean the surface and α is a constant. The values of R_i and α depend on the nature of the rain-impacted surface, and while the values are moderately constant for metals, they may vary considerably between materials (R_i is much higher for concrete, which will absorb moisture, than metal which will not). The model thus implies that no cleaning will occur until the rain in a given event exceeds R_i, so that cleaning of porous materials such as concrete and unglazed ceramics may be limited. The rainfall parameter R represents the rain impacting on the surface, which may differ from the meteorological measure of rainfall, particularly for walls protected by eaves, which can only be impacted by rain at a significant angle to the vertical.

The net impact of rain cleaning is to further differentiate microclimates on the exterior components of a structure, depending on their exposure to rain and the nature of the material. The model thus predicts enhanced corrosion in areas sheltered from rain, such as under eaves or the underside of gutters, relative to those fully exposed to rain.

The local ambient temperature and RH in the vicinity of a dwelling are calculated using standard geographic interpolation techniques applied to data from meteorological stations in the region (Cole *et al.* 2004b). Rainfall and wind data are taken from the closest meteorological stations (with some constraints to ensure that coastal locations are matched to coastal stations). The surface temperature of metal surfaces is calculated to take into account surface heating during the day and cooling at night, and surface RH is then calculated from the ambient RH and surface temperature (Cole and Paterson 2006).

Given knowledge of pollutant levels on surfaces and local climatic parameters, the state of a surface is calculated on a three-hourly basis with:

- State 1 = a dry surface;
- State 2 = wet from rain;
- State 3 = wet from the wetting of hygroscopic salts.

A surface is said to be wet if the surface RH exceeds the deliquescent RH of any contaminating salts, and the possible contaminating salts depend on the type of environment, as detailed in Cole *et al.* (2004c).

The damage that occurs over a three-hour period in each state has been derived from laboratory tests. In contrast to the accelerated tests discussed previously, these were conducted while controlling the local conditions at the surface, and so can be directly used to define the damage under each state.

Thus, the model uses a wide range of factors, including usage factors, local geometry, and climatic and pollutant factors, across a wide range of scales to calculate the degradation of individual components. Further, individual components are classified into a number of cases and each is analysed separately.

The model can readily predict the life of products in known and new environments. It can also be used to compare the life of components made from different materials, and the effect of maintenance if it can be defined in terms of changes to material response or local microclimate (as in the example of the effect of maintenance on downpipes). The model can also be used to determine the maintenance requirements from designed facilities. In principle, the model can be used to calculate the impacts of design, workmanship and human factors, again provided these changes can be translated to changes to microclimate and material response. While this last task is complex, it can be completed if the multiscale model approach has been integrated with that of building pathology. The estimation of maintenance of already built facilities and the prediction of remaining life are best accomplished with other techniques.

Summary

This chapter has outlined the different types of information that can be assembled to define the service life of components. As indicated, depending on the source, information will contain different levels of detail and thus be suitable for different uses. The appropriateness of each technique to the different applications of component life predictions is outlined in Table 4.3.

As indicated in the introduction, some applications – such as prediction

Table 4.3 Applicability of different prediction methods to usage requirements

Prediction method	Data use*							
	1	2	3	4	5	6	7	8
Degradation curves	Yes	Yes						
Condition surveys + data fitting					Yes		Yes	
Condition surveys + transition probabilities					Yes		Yes	Yes
Data mining maintenance records	Yes	Yes				Yes		
Expert opinion	Yes	Yes				Yes		
Dose function/damage indexes	Yes	Yes	Yes			Yes		
Process models	Yes	Yes	Yes	Yes		Yes		
Sensing					Yes		Yes	Yes

Note
*See introduction.

of the effect of design on service life – will require much information, with service life being developed as a function of component geometry, service issues and microclimate, and so will require a rich database such as that developed with process-based models. Other uses – such as knowledge of known products in known environments – will require relatively few data, which can be derived from simple databases, such as those formed by degradation curves or expert opinion.

There is a second dimension to this problem: some of the uses require information which has a low predictability, such as the effect of workmanship or human factors on life, and thus their effect is better assessed by techniques that permit direct assessment of the built facilities, such as condition surveys or sensing.

The chapter has highlighted (but not resolved) two contradictory requirements in life prediction. If all components in a building are analysed under all the variations in conditions that may prevail, then the number of cases is literally in the millions. On the other hand, if component life is to be analysed accurately to give the depth of dependency required to address design and maintenance conditions, each component may need to be broken into a number of sub-cases that can be analysed by refined and time-intensive techniques such as process-based modelling. Clearly it would not be possible to analyse the millions of possible scenarios by a process-based model. Therefore, a combination of techniques will have to be developed where the majority of service life data is obtained by efficient techniques such as expert opinion, while the components at high risk of failure are analysed by more precise techniques such as process-based modelling.

Bibliography

Arseth, L.I. and Hovde, P.J. (1999) 'A stochastic approach to the factor method for estimating service life', in *Proceedings of the Eighth International Conference on Durability of Building Materials and Components, Vancouver, Canada, 30 May to 3 June 1999*, vol. 2, Ottawa: National Research Council of Canada.

Cole, I.S. (2002) 'Recent progress in modelling atmospheric corrosion', *Corrosion Reviews*, 20 (4–5): 317–37.

Cole, I.S. and Paterson, D.A. (2004) 'Holistic model for atmospheric corrosion: Part 5: Factors controlling deposition of salt aerosol on candles, plates and buildings', *Corrosion Engineering, Science and Technology*, 39 (2): 125–30.

—— (2006) 'Mathematical models of the dependence of surface temperatures of exposed metal plates on environmental parameters', *Corrosion Engineering, Science and Technology*, 41 (1): 67–76.

—— (2007) 'Holistic model for atmospheric corrosion: Part 7: Cleaning of salt from metal surfaces', *Corrosion Engineering, Science and Technology*, 42 (2): 106–11.

Cole, I.S., Ball, M., Bradbury, A., Chan, W.Y., Corrigan, P.A., Egan, S., Ganther, W.D., Ge, E., Hope, P., Martin, A., Muster, T., Nayak, R., Paterson, D., Sherman, N., Trinidad, G. and Vanderstaay, L. (2007) *Learning System for Life Prediction of Infrastructure: Final Report*, CRC-CI Project 2005–003-B, Brisbane: CRC for Construction Innovation.

Cole, I.S., Bradbury, A., Chen, S.-E., Gilbert, D., MacKee, J., McFallen, S., Shutt, G. and Trinidad, G. (2004a) *Final Report of Delphi Study*, CRC-CI Project 2002–010-B, Brisbane: CRC for Construction Innovation.

Cole, I.S., Chan, W.Y., Trinidad, G.S. and Paterson, D.A. (2004b) 'Holistic model for atmospheric corrosion: Part 4: Geographic information system for predicting airborne salinity', *Corrosion Engineering, Science and Technology*, 39 (1): 89–96.

Cole, I.S., Ganther, W.D., Sinclair, J.O., Lau, D. and Paterson, D.A. (2004c) 'A study of the wetting of metal surfaces in order to understand the processes controlling atmospheric corrosion', *Journal of the Electrochemical Society*, 151 (12): B627–35.

Cole, I.S., Lau, D., Chan, F. and Paterson, D.A. (2004d) 'Experimental studies of salts removal from metal surfaces by wind and rain', *Corrosion Engineering, Science and Technology*, 39 (4): 333–8.

Cole, I.S., Lau, D. and Paterson, D.A. (2004e) 'Holistic model for atmospheric corrosion: Part 6: From wet aerosol to salt deposit', *Corrosion Engineering, Science and Technology*, 39 (3): 209–18.

Cole, I.S., Linardakis, A. and Ganther, W. (1995) 'Controlled humidity/salt dose tests for the estimation of the durability of masonry ties', *Masonry International*, 9 (1): 11–15.

Cole, I.S., Paterson, D.A. and Ganther, W.D. (2003) 'Holistic model for atmospheric corrosion: Part 1: Theoretical framework for production, transportation and deposition of marine salts', *Corrosion Engineering, Science and Technology*, 38 (2): 129–34.

Daniotti, B., Spagnolo, S.L. and Paolini, R. (2008) 'Climate data analysis to define accelerated ageing for reference service life evaluation', in A. Nil Turkeri and O. Sengul (eds) *Proceedings of the Eleventh International Conference on Durability of Building Materials and Components, Istanbul, Turkey, 11–14 May 2008*.

Duffield, C. (1993) 'The Delphi technique: a comparison of results obtained using two expert panels', *International Journal of Nursing Studies*, 30: 227–37.

Fatiguso, F. and De Tomasi, G. (2008) 'Technological features and decay processes of a new building type at the beginning of the XX century', in A. Nil Turkeri and O. Sengul (eds) *Proceedings of the Eleventh International Conference on Durability of Building Materials and Components, Istanbul, Turkey, 11–14 May 2008.*

Haagenrud, S., Veit, J., Eriksson, B. and Henriksen, J. (1998) *Final Report: EU Project ENV4-CT95–0110 Wood-Assess*, NILU Publication OR40/98, Kjeller: Norwegian Institute for Air Research.

Hans, J., Chorier, J., Chevalier, J.-L. and Lupica, S. (2008) 'French national service life information platform', in A. Nil Turkeri and O. Sengul (eds) *Proceedings of the Eleventh International Conference on Durability of Materials and Components, Istanbul, Turkey, 11–14 May 2008.*

International Organization for Standardization (2001) *Buildings and Constructed Assets – Service Life Planning: Part 2 – Service Life Prediction Procedures*, ISO 15686–2:2001, Geneva: ISO.

Kyle, B.R. (2001) 'Toward effective decision making for building management', paper presented at the APWA International Public Works Congress, Philadelphia, 9–12 September.

Leicester, R.H., Cole, I.S., Foliente, G.C. and Mackenzie, C. (1998) 'Prediction models for durability of timber construction', paper presented at the World Conference on Timber Engineering, Lausanne, 17–20 August.

Lounis, Z., Lacasse, M.A., Vanier, D.J. and Kyle, B.R. (1998) 'Towards standardization of service life prediction of roofing membranes', in T.J. Wallace and W.J. Rossiter Jr (eds) *Roofing Research and Standards Development: Fourth Volume*, West Conshohocken, Pa.: American Society for Testing and Materials.

Muster, T.H., Cole, I.S., Ganther, W.D., Paterson, D., Corrigan, P.A. and Price, D. (2005) 'Establishing a physical basis for the in-situ monitoring of airframe corrosion using intelligent sensor networks', in *Proceedings of the NACE Tri-Services Conference (TSCC05), Orlando, Florida, 14–18 November 2005*, Houston: NACE International.

Pintos, S., Queipo, N.V., Rincon, O.T. and Morcillo, M. (2000) 'Artificial neural network modelling of atmospheric corrosion in the MICAT project', *Corrosion Science*, 42: 33–52.

Scheffer, T.C. (1971) 'A climate index for estimating potential for decay in wood structures above ground', *Forest Products Journal*, 21: 10–25.

Tidblad, J., Mikaliov, A.A. and Kucera, V. (1998) 'Unified dose-response functions after 8 years of exposure', in *Proceedings of the UN/ECE Workshop on Quantification of Effects of Air Pollutants on Materials, Berlin, Germany, 24–27 May 1998.*

Uzarski, D.R. and Burley, L.A. (1997) 'Assessing building condition by the use of condition indexes', in *Proceedings of the Infrastructure Condition Assessment: Art, Science and Practice Conference, Boston, August 1997*, Washington: American Society of Civil Engineers.

Vanier, D.J. and Kyle, B.R. (2001) *Canadian Survey of Low Slope Roofs: Presentation of BELCAM Data Set*, NRCC-44979, Ottawa: National Research Council Canada.

5 Minimizing waste in commercial building refurbishment projects

Graham Miller and Mary Hardie

Within the general construction sector, renovation and refurbishment of commercial buildings is a high-volume generator of waste material destined for landfill. With major refurbishments carried out at an average of 20-year intervals, there is considerable potential for reuse and recycling strategies to help minimize waste generation in this sector. Retail refurbishments occur at much more regular intervals.

Waste minimization in refurbishment projects can involve a variety of approaches. Design strategies such as using long-lasting materials, planning for deconstruction, and specifying standard units, quantities or modules of different components can have a significant impact in minimizing waste. Management strategies that ensure the accurate ordering of materials and components, good storage and site control and appropriate education and induction of site personnel can also contribute significantly to the economic as well as the environmental performance of a construction project. Accurate condition assessment of the existing building components, coupled with a strategy of 'repair in place' when appropriate, is likely to produce further savings, while close monitoring of all waste generated and recording of waste destinations will enable benchmarks to be established and continual improvement to be encouraged. All of these strategies need to be addressed if refurbishment is to become more environmentally sustainable.

Background issues and extent of the problem

In the following sections, 'reuse' refers to a second life for a building material or component without significant alteration or transformation. 'Recycling' refers to the use of salvaged material as feedstock for new material, and involves significant transformation and reprocessing.

The built fabric of cities is constantly being altered and upgraded in response to commercial and functional imperatives. Often, little attention is paid to the material waste resulting from this constant churn of renewal and renovation. There is a growing realization that if building construction is to become sustainable in any real sense, the issue of waste generation must be addressed. To this end, greater reuse of built fabric components as well as recycling of all significant building materials are of critical importance.

It is increasingly proving to be either financially beneficial or cost neutral to expend effort on reusing building materials and components, or recycling bulk waste such as concrete or high intrinsic value material such as metal. This, coupled with the increasing cost of dumping waste and the decreasing availability of landfill sites, is encouraging building contractors to re-examine their practices with regard to waste management. Improvement in such practices in recent years is having a snowball effect, resulting in economies of scale and encouraging more recycling and increasing the economic return from more efficient use of limited resources.

Most recent statistics (Australian Bureau of Statistics 2007) indicate that generation of solid waste from all sources in Australia increased from 22.7 megatonnes in 1996–97 to 32.4 megatonnes in 2002–03. Construction and demolition waste comprises 42 per cent of total waste generated, and of this 57 per cent is recycled. Many studies have shown that a considerable amount of the material currently destined for landfill from construction projects is, however, potentially recyclable (Anderson and Mills 2002; BRE Centre for Resource Management 2003; Faniran and Caban 1998; Fatta *et al.* 2003; Formoso *et al.* 2002; Hobbs and Kay 2000; Lawson *et al.* 2001; McGrath 2001; Poon *et al.* 2004; Wong and Yip 2004). It has been shown that dumping material in landfill is, as well as being economically unsound, also a component of the environmental problems leading to climate change (Ackerman 2000). If this issue is to be addressed, the construction sector's high usage of extracted natural materials (estimated at 40 per cent of total in the United States, for example) needs to be modified (Kibert *et al.* 2000). It is argued that the aim should be to achieve 'the metabolic behaviour of natural systems' (Kibert *et al.* 2000) and achieve a balance so that waste does not outstrip the ability to regenerate. Waste needs to be seen as a resource (Newton 2006, 2008), and therefore the consequences of waste generation should be part of the eco-efficiency evaluation process for any proposed construction works.

Increasing the percentage of material that is successfully recycled can reduce rates of resource depletion as well as the production of greenhouse gases generated, while at the same time improving the long-term profitability of commercial refurbishments. Construction waste minimization can reduce the production of methane and other gases generated in landfill, reduce the consumption of raw material resources, and reduce the energy expenditures associated with transportation of bulk waste.

It has been reported that integration of the design and construction process is crucial to successful waste management and resource recovery (Bell 1998; te Dorsthorst and Kowalczyk 2002). To date, the phase after demolition has largely been ignored in design considerations, but this is likely to change as some studies claim that it is more important for overall sustainability to design a building for recycling rather than to use low-energy materials (Thormark 2000). Adding weight to this, Schultmann and Rentz (2002) report that emphasis on the environmental aspects of maintaining and improving the building stock is likely to increase. The potential gains that can be made through the process of waste minimization represent a significant aspect of this increasing environmental emphasis on redevelopment. The natural world is characterized by the constant cycling of energy and materials. All waste products are a resource to be utilized and, according to Kibert *et al.* (2000), the built environment needs to mimic this cyclical regeneration in nature if sustainable building is to become a reality.

Emerging views and attitudes

Sustainability in the built environment need not necessarily be in conflict with commercial priorities. Bottom line savings of approximately 50 per cent of the budgeted amount for waste removal have been achieved by Australian projects incorporating waste minimization strategies (Andrews 1998; Australian Bureau of Statistics 2003). Data on the recycled *content* of building materials are available from several sources, including from manufacturers of building components, and these feed into the published assessments of life cycle evaluations of building materials. However, this front-end or supply-side information is not well matched with verifiable quantities of material from actual projects where recycling stock is generated. There is, in fact, very little available information about the ultimate destination of material removed in the course of a commercial refurbishment project, or the factors which encourage recycling. Hard data on key areas such as recycling *rates* for materials are therefore scarce. The indications are that recycling rates quoted in waste management plans compiled at the approval stage of a project may have little correlation with the actual rates that are achieved in practice. Consequently, there are no validated benchmarks or targets which can give guidance as to what can be achieved, or indicate areas of underperformance.

In order to address the paucity of hard data for critical aspects of waste minimization and management, a consultation with 15 industry experts involved in commercial refurbishment in Australia was undertaken to ascertain best practice strategies for waste minimization at the different phases of a project along with achievable target rates for recycling and reuse of different materials and components (Miller *et al.* 2006). The breakdown of the respondents was nine industry practitioners, five consultants and one waste contractor. In each case, the average of the responses for components is given. The categories are building fabric, fittings, finishes and services; reuse on site, reuse off site, recycling on site and recycling off site are also identified. The results of the expert consultations regarding best practice strategies are summarized in Figure 5.1, and are discussed in the following sections of this chapter.

Planning phase	Strip-out phase	Fit-out phase	Occupation phase
• Careful pre-refurbishment audit	• Early removal of hazardous materials	• Accurate estimating and ordering	• Client and end-user satisfaction achieved
• Identification of hazardous materials	• Sequential deconstruction	• Use of modular components and preferred quantities	• Cost-efficient project delivery
• Identification of components suitable for reuse	• Clear lines of responsibility	• Minimal design changes and rework	• Recognition via green ratings
• Establishment of benchmark targets for recycling practice	• Recording of quantity and destination of all waste leaving site	• Packaging covenants	• Triple bottom-line benefits
• Provision of appropriate staff training	• Equitable sharing of the benefits of recycling	• Monitoring waste generated against targets	• Validation of recycling benchmarks

Figure 5.1 Features of good waste minimization practice in commercial refurbishment projects.

Project phases

Although the stages of a refurbishment project do not always occur in a linear fashion and may be undertaken simultaneously, from an organizational point of view it is useful to divide the process into the conceptual or planning phase, the demolition or strip-out phase, the construction or fitout phase, and the commissioning and occupation phase.

Conceptual/planning phase

Before a renovation project is commenced, it is critical to carry out a pre-refurbishment audit in order to produce a clear picture of the condition of the existing structure, fabric, finishes and fittings. Condition assessment reports along with adequate ongoing maintenance regimes can reduce the scope of refurbishment required (Alani *et al.* 2002). It may be possible to avoid significant amounts of new construction with a 'repair in place' strategy. It is sometimes forgotten that making simple repairs as part of a refurbishment project can often be the most effective option from an environmental and cost perspective. Indeed, repair may be the appropriate strategy for situations as diverse as deterioration of finishes up to structural cracking in the building fabric. If this is so, it is best identified early in the process. Refurbishment projects particularly benefit from evaluation against the pyramidal structure of waste avoidance, as shown in Figure 5.2.

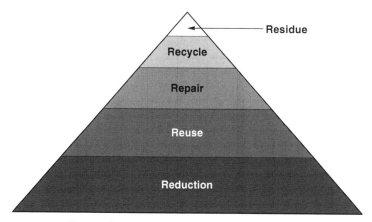

The small top triangle represents the residual waste
fraction for disposal in a well-managed project

Figure 5.2 Waste avoidance pyramid (source: Miller *et al.* (2006)).

The presence of any hazardous materials in the building to be refurbished should be identified in the pre-refurbishment audit. The most significant of these in recent years has been asbestos fibre. Office buildings constructed between the 1950s and 1970s commonly have some asbestos-based products which were formerly used for insulation and fire protection purposes. If left undisturbed this material is unlikely to be a hazard, but when airborne fibres are released by renovation work they represent a significant risk to human health. As a result, asbestos removal is covered by strict regulation and remediation protocols in most developed countries, including Australia (Australian National Occupational Health and Safety Commission 1988). The presence of asbestos in a renovation project was nominated as a factor that severely restricts potential recycling from refurbishments by most experts surveyed. Some reported that mere proximity to small quantities of hazardous materials such as asbestos can render otherwise recyclable materials contaminated. One waste contractor reported that the suspicion of asbestos being present in the source material could rule out the crushing of concrete for road base. This highlights the importance of careful assessment of the preconditions for any refurbishment project. Other materials and components commonly found in older buildings and whose presence needs to be identified and allowed for include lead (in roofing and flashings), polychlorinated biphenyls or PCBs (formerly used in coolants, stabilizers, flame retardants and sealants) and mercury (in mercury vapour lamps). Glass coatings and composite materials also present a barrier to higher value recycling.

An analysis of the options in relation to waste management at the outset of the project is fundamental for project planners to allow for inclusion of waste minimization practices. This should include a comparison of costs of no recycling versus recycling options. It is possible to identify some strategies that are likely to produce a positive return, some that will probably be cost neutral, and some that may result in future benefits but involve an initial cost. In this regard, it is helpful to engage as many stakeholders as possible at the outset. The building owner, for example, may see longer-term financial benefits of designing for disassembly compared to making major structural changes to meet new requirements. Similarly, commercial tenants may be persuaded of the benefit of a modular design if, at the expiry of the lease, they are required to return the space to its prior layout.

During the refurbishment planning phase, those building components that are suitable for reuse should be flagged. This may involve reuse in place, movement to another location on site, or reuse off site. The reuse of building components in new construction is becoming an increasingly economical practice. This is greatly aided by forethought at the initial design stage to make future disassembly possible.

Recycling targets should be set for each major building material waste stream generated by the refurbishment. Some guide to the percentages that can be achieved is given by the data collected from the surveyed experts and shown in Figures 5.3 to 5.6. However, site location, access, sorting space and building age are all likely to have some effect on what can be achieved. It is important to set realistic benchmarks and to seek continual improvement over time with each new project.

The final critical factor during the planning phase is to ensure that adequate training in waste minimization and disposal is provided for all staff who will be involved. Induction programs and on-site training in waste minimization for all construction workers should be provided. A change of culture and attitude may be required, and this needs to be specifically directed by regulatory requirements, market forces and environmental issues. Subcontractors will generally follow the lead set by the head contractor in this matter. Head contractors will either have their own in-house training system, or be ready to implement one at the request of the project owner or client who declares this matter to be a priority. General consciousness-raising on the issues of recycling and waste management is likely to have considerable benefits both for the environment and for the contractors' bottom line.

Demolition/strip-out phase

The strip-out phase of a refurbishment project generates the great bulk of the waste sent to landfill (Schultmann and Sunke 2007). The quality of material recovered at this stage is determined by the management of the demolition process, the time allocated for deconstruction and the space available for sorting of materials. Expert consultation has revealed that the first critical issue is the early removal of any hazardous substances. In particular, if asbestos fibre is present, it is important for full remediation to take place before any other renovation activity, otherwise all the waste generated will have to be regarded as contaminated. In addition, any activity in the building, including inspection and condition assessment, would have to be undertaken in full personal protection gear. Only licensed and experienced contractors should be permitted to deal with hazardous wastes. Responsibility for the overseeing of the waste removal process should be clearly identified and addressed before construction commences, a site coordinator appointed, and comprehensive records kept of the eventual destinations of all material removed. It should be clear that a risk assessment needs to be undertaken should hazardous materials be detected at any time during the construction phase, and on-site practices revised to minimize contamination. Potentially hazardous materials should be kept in secure bunded areas. Unavoidable waste must be disposed of in a safe and timely manner. As an example, the site manager should ensure that bagged hazardous materials are removed from the site before the bag is damaged from surrounding construction activity.

Sequential deconstruction is the term used to describe the careful dismantling of existing built fabric in order to maximize the quantity and quality of recyclable material recovered. If this becomes common practice, procedures for material recovery will become established and recycling rates will increase. This strategy relies on construction scheduling which allows time for disassembly. The cost is later recovered through the salvage value of the recyclable and reused product, as well as through the savings in disposal fees and waste transportation costs. Allowing for future deconstruction makes possible multiple reuse of building components, and so justifies the use of high-quality materials in the initial fabrication. Such 'design for disassembly' has the added benefit of being an aid to 'buildability' or efficient construction. Components such as partitions, ceiling panels, windows, doors, cupboards and light fittings are all readily reusable if they can be taken out of a building to be refurbished in a non-destructive manner. The establishment of secondary markets for such materials is already taking place, and should be encouraged by state and local government authorities.

A clear chain of responsibility for handling and sorting waste is essential for efficient management of refurbishment projects. Multiple movement of sorted waste materials is inefficient and likely to discourage future salvage efforts. Recording of the quantity and destination of all waste leaving the site is regarded as best practice among those experts committed to waste minimization. This is also important if valid benchmark targets are to be established and actual recycling rates achieved are to be verified. Equitable sharing of the benefits of recycling is considered to be important in creating an industry and site culture favourable to recycling. In some cases this can be achieved by incentive schemes and bonuses for teams who achieve or exceed their benchmark target (Tam and Tam 2008).

Fitout/new construction phase

Once the bulk of the demolition or strip out has occurred, construction fit out can commence for the refurbishment. Accurate estimating and ordering is critical to waste minimization at this point. 'Shallow estimating' or the wholesale rounding up of quantities 'just to be sure' results in significant waste unless there is an agreement that the supplier will take back unused quantities. Such agreements are becoming increasingly common. Some concrete suppliers even take back unused partly set concrete and use it for pulverized road base. This, however, is still a less desirable outcome than accurate ordering of quantities needed.

Aspirations for systems of modular components have a long history in construction. Successful systems in which modular components are regularly reused are, however, not at all common. The need to be able to modify or customize components for a particular situation tends to work against large-scale interoperability of building components. Buildings are significantly more complex artefacts than manufactured consumer products, and thus they require a higher level of architectural engineering to maintain their long-term value. Nevertheless, there is considerable potential to develop the 'disassemble and reuse' strand of waste minimization in commercial building construction.

Late design changes resulting in rework have been identified as an important cost for construction, and such changes also lessen recycling rates in refurbishment jobs (Love and Li 2000). They do this by creating pressure on the project schedule and thereby leaving no time for sorting and separation of off-cuts and alteration materials.

Packaging covenants – where, for example, return of unused materials or components is agreed – require 'product stewardship' throughout the supply chain, and have helped achieve significant reductions in the domestic waste stream, and could have similar results for construction waste. Group rather than individual packaging for items such as light fittings, door furniture and fixings which are used in multiple applications can reduce waste. Suppliers can also be asked to provide their product with minimal packaging and to take back packaging from the construction site. Containers can be returnable and refillable. Reusable protective coverings can be used in transporting goods, in preference to disposable packaging. Waste products such as old carpet can find a new life for this purpose. Reducing packaging not only minimizes waste but also provides savings on site by lessening the time spent removing and disposing of packaging.

Ongoing monitoring of waste generated against targets set at the beginning of the project is essential in order to verify that improvements in recycling practice are more than lip service. Tight materials control and material audit-ing should be required of all construction contractors and subcontractors. Unloading procedures, all-weather storage and handling of materials should be controlled to avoid wastage due to on-site materials damage.

Commissioning and occupation phase

Client and end-user satisfaction is a crucial goal for all those involved in the delivery of building projects. Repeat clientele provides stability and continuity for any construction business, and this can only result from high levels of client satisfaction with completed projects. Building owners will always require cost-efficient project delivery, but increasingly many non-cost criteria affect the choice of construction contractor. The commer-cial and government markets are starting to appreciate and accept the value of green rating schemes that assess the performance of buildings (including their embodied energy component). Not all green rating schemes include a recycling component in their requirements, but leading schemes do. Achiev-ing building recognition and status via green ratings is becoming increas-ingly important as a marketplace differentiator as community awareness of environmental issues gains momentum. Triple bottom line benefits are likely to accrue for companies that take reuse and recycling targets seri-ously in respect of their own buildings. Validation of recycling benchmarks can be achieved as the bank of delivered projects with good data records increases and it becomes possible to make rigorous comparisons between different levels of waste minimization effort.

Areas for future improvement

Establishing best practice guidelines and benchmark percentage rates

Although research outcomes have stressed the importance of tailoring waste minimization targets to the particular project circumstances, the building and construction industry will need to establish best practice guidelines if there is to be significant ongoing improvement. Figures 5.3 to 5.6 show the estimated achievable targets currently possible, and are based on data collected from experts in the field.

Some general results can be gleaned for the four component categories of building fabric, fittings, finishes and services. First, the building fabric removed in a commercial refurbishment project receives a significant level of recycling, almost all of which happens off site. Aluminium, structural steel and steel reinforcing are reportedly recycled at the rate of 86, 79 and 84 per cent respectively. Heavy masonry materials like bricks, blocks and concrete are also commonly recycled at rates of over 70 per cent for each element.

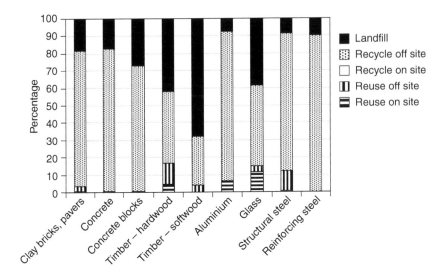

Figure 5.3 Current rates for building fabric (source: Miller *et al.* (2006), reproduced with permission).

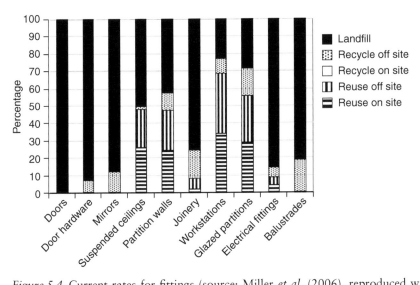

Figure 5.4 Current rates for fittings (source: Miller *et al.* (2006), reproduced with permission).

Second, landfill is the principal destination for most fittings removed from refurbishments except for suspended ceilings, partition walls, workstations and glazed partitions. Workstations were commonly reused both on and off site (35 per cent in each category). Very little recycling was reported for fittings.

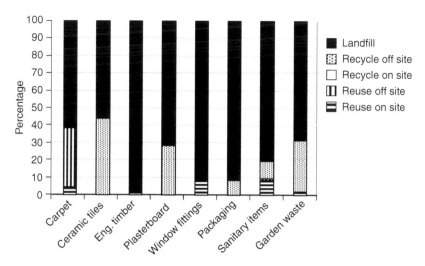

Figure 5.5 Current rates for finishes (source: Miller *et al.* (2006), reproduced with permission).

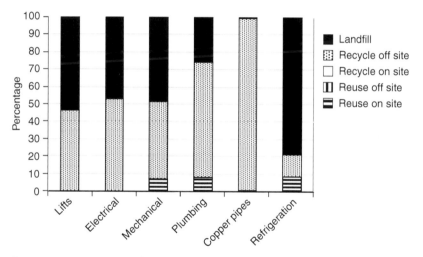

Figure 5.6 Current rates for services (source: Miller *et al.* (2006), reproduced with permission).

Third, the majority of all finishes removed during refurbishments end up in landfill and no recycling on site was reported. Reuse for carpet is reportedly a growing area. Plasterboard recycling was an area of considerable disagreement among the experts. While several reported that no recycling occurred, a few were able to report high levels of recycling. The differences appear to be location based, with Victorian recycling facilities being widely available, while very little plasterboard recycling occurred in other states.

Finally, high levels of recycling off site occur with most service components, but there was very little reuse reported. Refrigeration components appear to lag other service components in having recycling facilities available.

Problem areas

There are several commonly reported impediments to waste minimization in general building construction projects. These include lack of available space along with time restrictions which have been shown to limit on-site sorting of the waste stream (Poon *et al.* 2004; Shen *et al.* 2004; Kartam *et al.* 2004; Formoso *et al.* 2002; Touart 1998; Gavilan and Bernold 1994), work practices and attitudes that may militate against reuse and recycling (Teo and Loosemore 2001), inadequate management skills and knowledge (Egbu 1997), and small quantities of recyclable material that can be uneconomic to sort and transport to a recycling facility (Seydel *et al.* 2002).

Opportunities

Several areas of opportunity have been identified. Accessible warehousing of secondhand materials was seen as a commercial opportunity not yet fully realized on a sufficiently large scale. It is frequently lack of storage space that results in recyclable or reusable material going to landfill. Internet-based materials exchanges, perhaps established by local or regional governments, could be an initiative that would make much more use of the residual value in building waste. In many major Australian cities there is already a healthy secondhand market for office partitions, commercial carpet and office equipment. This could be extended to other materials areas with suitable storage facilities and good materials inventory systems in place.

Particular materials were identified as possible commercial recycling opportunities for the right investor or entrepreneur. In New Zealand, for example, there is a company recycling left-over paint. They divide it into light, medium or dark colours, and into acrylic or oil. Competitively priced, this recycled product finds a ready market in commercial projects as an undercoat or primer. Window or architectural glass cannot be included in the same recycling process as bottle glass from kerbside collections because its additives give it a different melt temperature. However, the quantities of such glass available from demolition and refurbishment mean that it can be recycled separately, and at least one company is using this as the raw material for glass reflectors for road lane marking. Used plasterboard can be pulverized and the gypsum content used to replace virgin gypsum in some industrial processes. Mostly, however, the collection and processing loop is not yet in place. Plastics are difficult to recycle because of the variety of kinds in use and because they are frequently bonded to other materials (composite and hybrid materials). Nevertheless, waste contractors report that they are now getting paid for some recovered plastic, and this is likely to increase. Once recovery facilities and systems are in place, they can accelerate the level of recycling achieved simply through ready availability.

Conclusions

It is clear that a growing number of designers, contractors and clients are aware of the possibilities for waste minimization and good waste management practices in helping to improve our environment and the economic viability of construction projects. Waste management practices in Australian commercial refurbishment projects have improved over recent years. As a result, markets are developing in the reuse and recycling of various materials, which it is hoped will encourage further growth and improvements in this critical area.

However, despite increased awareness and some degree of improvement, the construction industry generally continues to be a high generator of solid waste products, and refurbishment projects are a significant part of this waste stream. Waste minimization strategies in office building refurbishment can potentially make a significant contribution to the sustainability of the built environment as a whole. The refurbishment process is part of the loop of resource consumption. Refurbishments extend the useful life of a building thereby allowing continued use of the resources initially expended in its construction. If extended life cycles are allowed for, by means of design for deconstruction and disassembly, then the materials and energy savings generated by refurbishments can be ongoing, and something approaching the cyclic processes of systems in the natural world may eventually be achieved. This can certainly be aimed for as a worthwhile goal, as has been the case in the manufacturing sector (Kaebernick *et al.* 2008). More information is required on the specific benefits, especially cost and environmental benefits, of minimizing waste, and the next major step is for industry and regulators in partnership to develop a more systematic approach to benchmarking and the dissemination of best practice ideas in construction waste management.

Bibliography

Ackerman, F. (2000) 'Waste management and climate change', *Local Environment*, 5 (2): 223–9.

Alani, A.M., Tattersall, R.P. and Okoroh, M.I. (2002) 'Quantitative models for building repair and maintenance: a comparative case-study', *Facilities*, 20 (5): 176–89.

Anderson, J. and Mills, K. (2002) *Refurbishment or Redevelopment of Office Buildings? Sustainability Comparisons*, London: BRE.

Andrews, S. (1998) *WasteWise Construction Program Review: A Report to ANZECC*, Canberra: Department of the Environment.

Australian Bureau of Statistics (2003) 'The WasteWise construction program', *Yearbook Australia, 2003*, Cat. no. 1301.0, Canberra.

—— (2007) 'Household waste', *Australian Social Trends*, Cat. no. 4102.0, Canberra.

Australian National Occupational Health and Safety Commission (1988) *Guide to the Control of Asbestos Hazards in Buildings and Structures*, Canberra. Online. Available at HTTP: <www.workcover.act.gov.au/pdfs/guides_cop/AsbestosGuide.pdf> (accessed 15 December 2007).

Bell, N. (1998) *Waste Minimisation & Resource Recovery: Some New Strategies*, Canberra: Royal Australian Institute of Architects.

BRE Centre for Resource Management (2003) *Construction and Demolition Waste: Part 1*, BRE Good Building Guide 57, London: BRE.

Egbu, C.O. (1997) 'Refurbishment management: challenges and opportunities', *Building Research & Information*, 25 (6): 338–47.

Faniran, O.O. and Caban, G. (1998) 'Minimizing waste on construction project sites', *Engineering, Construction and Architectural Management*, 5 (2): 182–8.

Fatta, D., Papadopoulos, A., Avramikos, E., Sgourou, E., Moustakas, K., Kourmoussis, F., Mentzis, A. and Loizidou, M. (2003) 'Generation and management of construction and demolition waste in Greece: an existing challenge', *Resources, Conservation and Recycling*, 40 (1): 81–91.

Formoso, C.T., Soibelman, L., De Cesare, C. and Isatto, E.L. (2002) 'Material waste in building industry: main causes and prevention', *Journal of Construction Engineering and Management*, 128 (4): 316–25.

Gavilan, R.M. and Bernold, L.E. (1994) 'Source evaluation of solid waste in building construction', *Journal of Construction Engineering and Management*, 120 (3): 536–52.

Hobbs, G. and Kay, T. (2000) *Reclamation and Recycling of Building Materials: Industry Position Report*, London: BRE.

Kaebernick, H., Ibbotson, S. and Kara, S. (2008) 'Cradle to cradle manufacturing', in P.W. Newton (ed.) *Transitions: Pathways Towards More Sustainable Urban Development in Australia*, Melbourne: CSIRO and Dordrecht: Springer.

Kartam, N., Al-Mutairi, N., Al-Ghusain, B. and Al-Humoud, J. (2004) 'Environmental management of construction and demolition waste in Kuwait', *Waste Management*, 24 (10): 1049–59.

Kibert, C.J., Sendzimir, J. and Guy, B. (2000) 'Construction ecology and metabolism: natural system analogues for a sustainable built environment', *Construction Management and Economics*, 18 (8): 903–16.

Lawson, N., Douglas, I., Garvin, S., McGrath, C., Manning, D. and Vetterlein, J. (2001) 'Recycling construction and demolition wastes: a UK perspective', *Environmental Management and Health*, 12 (2): 146–57.

Love, P.E.D. and Li, H. (2000) 'Quantifying the causes and costs of rework in construction', *Construction Management and Economics*, 18 (4): 479–90.

McGrath, C. (2001) 'Waste minimisation in practice', *Resources, Conservation and Recycling*, 32 (3–4): 227–38.

Miller, G., Khan, S., Hardie, M. and O'Donnell, A. (2006) *Report on the Findings from Expert Surveys on Reuse and Recycling Rates in Commercial Refurbishment Projects*, Report no. 2003–028-B-T4b-01, Brisbane: CRC for Construction Innovation.

Newton, P.W. (2006) *Australia State of the Environment: Human Settlements*, Canberra: Department of Environment and Heritage. Online. Available at

HTTP: <www.environment.gov.au/soe/2006/publications/report/human-settle-ments.html> (accessed 29 February 2008).

—— (ed.) (2008) *Transitions: Pathways Towards More Sustainable Urban Development in Australia*, Melbourne: CSIRO and Dordrecht: Springer.

Poon, C.S., Yu, A.T.W. and Jaillon, L. (2004) 'Reducing building waste at construction sites in Hong Kong', *Construction Management and Economics*, 22 (5): 461–70.

Schultmann, F. and Rentz, O. (2002) 'Scheduling of deconstruction projects under resource constraints', *Construction Management and Economics*, 20 (5): 391–401.

Schultmann, F. and Sunke, N. (2007) 'Energy-oriented deconstruction and recovery planning', *Building Research & Information*, 35 (6): 602–15.

Seydel, A., Wilson, O.D. and Skitmore, M. (2002) 'Financial evaluation of waste management methods: a case study', *Journal of Construction Research*, 3 (1): 167–79.

Shen, L.Y., Tam, V.W.Y., Tam, C.M. and Drew, D. (2004) 'Mapping approach for examining waste management on construction sites', *Construction Engineering and Management*, 130 (4): 472–81.

Tam, V.W.Y. and Tam, C.M. (2008) 'Waste reduction through incentives: a case study', *Building Research & Information*, 36 (1): 37–43.

te Dorsthorst, B.J.H. and Kowalczyk, T. (2002) 'Design for recycling', *Design for Deconstruction and Materials Reuse*, Karlsruhe, Germany: CIB.

Teo, M.M.M. and Loosemore, M. (2001) 'A theory of waste behaviour in the construction industry', *Construction Management and Economics*, 19 (7): 741–51.

Thormark, C. (2000) 'Including recycling potential in energy use into the life cycle of buildings', *Building Research & Information*, 28 (3): 176–83.

Touart, A. (1998) 'Recycling at construction sites', *BioCycle*, 39 (2): 53–5.

Wong, E.O.W. and Yip, R.C.P. (2004) 'Promoting sustainable construction waste management in Hong Kong', *Construction Management and Economics*, 22 (6): 563–6.

Part III
Design

6 Building information models
Future roadmap

Arto Kiviniemi

The state-of-the-art and selected roadmap issues presented in this chapter are built on four major sources: *Review of the Development and Implementation of IFC Compatible BIM* (Kiviniemi *et al.* 2008), *Strategic Roadmaps and Implementation Actions for ICT in Construction* (Kazi *et al.* 2007a), *International Workshop on Global Roadmap and Strategic Actions for ICT in Construction* (Kazi *et al.* 2007b) and *Integrated Design Solutions: Scoping Paper* approved by the CIB board in April 2008 (Kiviniemi 2008).

Definition and background

The term 'building information model' (BIM) has several definitions. The one used here is: 'BIM is an object-oriented, AECO (Architecture, Engineering, Construction and Operations)-specific model; a digital representation of a building to facilitate exchange and interoperability of information in digital format based on open standards.' This definition emphasizes the external use (i.e., the use of BIM as a communication media between several stakeholders and between tasks during the life cycle of a building or even a building portfolio of the owner), instead of the use of BIM as a tool in an internal task or process, such as architectural design or structural engineering. The internal use is naturally also a valid viewpoint for BIM, but this chapter concentrates on the use of BIM as an enabler of data exchange, sharing and communication.

The implementation of the BIM concept for any internal use is significantly easier than for the external use, since it has to deal with a limited set of information and can be handled typically within one software application. Its implementation for external use, as a communication platform, is a challenging task. The International Alliance for Interoperability (IAI) has developed a data specification, Industry Foundation Classes (IFC), for BIM since 1994 and, despite several published versions of this, the use of the IFC-compliant BIM in real projects has so far been very limited (Kiviniemi *et al.* 2008: 21, 44, 79). The slow adoption of the BIM in the industry has been caused by several technical and human barriers, presented in the next section.

General barriers for BIM

The barriers in adopting BIM can be categorized into internal and external barriers in the same way as the use of BIM.

In the internal use of BIM, the main barriers are costs and human issues, mainly the learning of new tools and processes. The learning process is significantly more expensive than the actual costs of hardware and software, especially if the productivity losses during the learning period are considered. However, high investment costs and the constant need to upgrade hardware and software are seen as the two main obstacles for companies (Kiviniemi *et al.* 2008: 109–11) (Figure 6.1). Also, the unclear balance between the benefits and costs and the fear that the actual benefits go to other participants in projects are significant obstacles; the sufficient business drivers for the use of BIM are still often missing. Another internal barrier is the fear of lacking features and flexibility of the modelling tools, partly based on experiences of the early BIM tools, but also on lack of knowledge and on prejudices (Kiviniemi *et al.* 2008: 50).

Legal issues, responsibilities, copyrights and potential loss of intellectual property (IP) when sharing BIM data are obstacles which slow down the adoption of BIM as a project platform. These obstacles are closely related to the need to redefine the workflow, roles and responsibilities in the BIM-based processes. In general, the industry lacks agreements and common practices concerning how to use integrated BIM, although in Nordic countries the willingness to share BIM data seems to be higher than elsewhere. An interesting issue is the fear of increased transparency of the process, which is seen as a threat by some and as an advantage by others (Kiviniemi *et al.* 2008: 50).

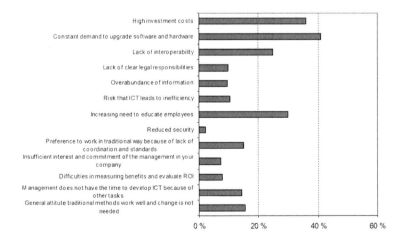

Figure 6.1 Problems and obstacles in increased use of ICT (source: Kiviniemi *et al.* (2008: 109), reproduced with permission).

Technical issues – mainly lack of sufficient and reliable interoperability between software applications – are significant obstacles, although perhaps not fully recognized by the industry yet, since most companies have no experience of the use of shared BIM (Kiviniemi *et al.* 2008: 109–11) (Figure 6.1). Those which have tried to use IFC-compliant BIM in real projects are aware of the problems, and are trying to find solutions for reliable data exchange. Unfortunately, the current IFC certification process does not guarantee sufficient implementation quality and reliable IFC data exchange (Kiviniemi *et al.* 2008: 30–33).

General drivers for BIM

One of the most efficient drivers for BIM is the requirement from the building owner, since it can be a categorical selection criterion for the designers. This has recently been the main driver at the global level. Similarly, construction companies can set the BIM requirement as a mandatory condition of collaboration for their suppliers, and at some levels this is already happening (Kiviniemi *et al.* 2008: 49).

National technology programs and other collective initiatives can promote the use of BIM, which may significantly accelerate its adoption. One example is that the utilization of shared BIM is higher in Finland than in the other Nordic countries, which may be related to its continuous support and promotion by Tekes, the funding agency for technology and innovation. Also, professional organizations can contribute to the deployment of BIM by promoting the idea and opportunities. Some professional organizations in the USA and Nordic countries are already encouraging their member companies to adopt the concept (Kiviniemi *et al.* 2008: 49).

According to a Nordic survey, only a few companies see the possibility of developing a new business as a reason for investing in ICT, which indicates that the business opportunities are not strong drivers yet. However, improved competitiveness and efficiency of technical work are the main drivers for ICT investments (Kiviniemi *et al.* 2008: 106–9) (Figure 6.2). According to a US survey (Khemlani 2007), the most important criterion for the use of BIM is the ability to produce final construction documents within the BIM tool itself, while integration is not seen as an important issue. This indicates that AECO practitioners are able to see the efficiency of the tools in the existing processes but are not able to identify the business potential of new processes. Despite this, investment in IFC-compliant BIM received the highest score regarding the areas for future investment in two recent Nordic surveys (Kiviniemi *et al.* 2008: 50, 114–15).

State-of-the-art in the deployment of BIM

The adoption of and drivers for BIM differ from country to country. In the USA the National BIM Standard (NBIMS 2007) and General Services

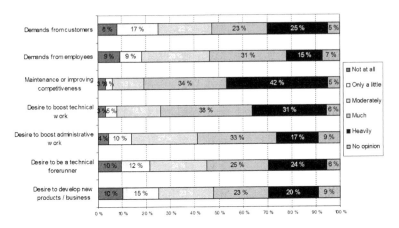

Figure 6.2 Motivation for ICT investments (all respondents) (source: Kiviniemi *et al.* (2008: 72), reproduced with permission).

Administration's (GSA) BIM requirements (GSA 2007) created a strong interest in the use of BIM, but the actual use of BIM in real projects is still very limited (American Institute of Architects 2006: 13) (Figure 6.3). In Denmark, the mandatory BIM requirements from the Danish state clients since January 2007 have strongly influenced the AECO market (Digitale Byggeri 2007). In Finland, the continuous support from Tekes, new technology strategy and projects of the Confederation of Finnish Construction Companies, and BIM requirements of Senate Properties have encouraged the industry to adopt the BIM concept (Senate Properties 2007). In Norway, Statsbygg and Norwegian Homebuilders Association have influenced the use of BIM in recent years (Statsbygg 2007). Several contractors have invested and implemented BIM systems in order to have integrated BIM support for their production of apartments and houses. In the Netherlands, the use of BIM may have been influenced by the limited willingness to share digital data between companies, but it has been used to gain benefits on the company level. In Sweden, the major contractors are playing an important role in the construction sector and have most likely influenced the use of BIM in that country (Kiviniemi *et al.* 2008: 49).

Despite the technical and human problems and limitations, adoption of BIM has started to accelerate around the world since late 2006. As documented above, the main driver for the adoption of integrated BIM has been the requirements of large public building owners. Since the public owners are bound by regulations requiring open competition, it is not possible to demand the use of a defined software product and thus the mandatory BIM requirements must be based on an open standard, such as IFC.

In some countries, the large construction companies have also started to adopt BIM as part of their processes, especially in their own production, and

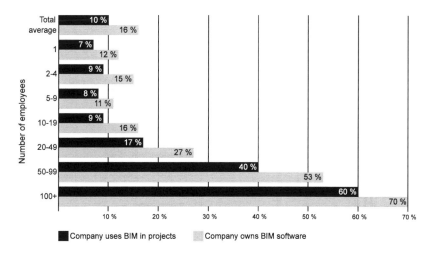

Figure 6.3 Architects using BIM in the USA, categorized by company size (sources: data: American Institute of Architects (2006). Image: VTT, Kiviniemi (2008), reproduced with permission).

to require BIM from their designers and suppliers, but they often operate within one software application and the use of IFC-compliant BIM is still very limited (Kiviniemi *et al.* 2008: 21).

The BIM requirements of the public owners have led to a peculiar situation where the software tools are not able to support the business needs of some major clients. One can claim that the advanced clients are ahead of their suppliers, which naturally creates pressures to solve the current conflict between demand and supply. One effort to identify the necessary actions from this specific viewpoint was the ERAbuild BIM study (Kiviniemi *et al.* 2008), and a wider viewpoint towards RECC industry's ICT R&D strategic needs was the Strat-CON project (Kazi *et al.* 2007a, 2007b).

Market demand and business drivers

The basic problem in the deployment of integrated BIM can be described as a 'wicked circle'. If adequate software support is missing, AECO projects cannot use integrated BIM → if the projects do not use integrated BIM, it is impossible to measure its benefits → if the evidence of benefits is missing, the end-users have no reason to demand integrated BIM tools → if the end-users do not demand integrated BIM tools, the software vendors have no motivation to invest in the development of such tools → which leads back to the start of the loop (Figure 6.4).

This phenomenon is very common for systemic innovation – that is, changes which affect the processes of several players in a complex supply

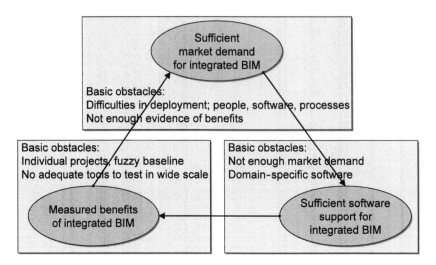

Figure 6.4 The 'wicked circle' of BIM implementation and deployment (source: VTT, Kiviniemi (2008), reproduced with permission).

chain (Taylor and Levitt 2004: 2). In such situations, most companies either wait to see what will happen or try to prevent the change as a potential threat. Only a few see it as an opportunity and want to drive the change. Because these companies are a minority which still have to work within the existing market, any major change takes a long time unless a significant player changes the market balance – as some major owners have recently done in the AECO industry by starting to demand BIM. However, deployment of a systemic innovation based on technology, such as BIM, requires also that at least reasonable technical possibilities exist.

This means that the best solution to speeding up a change process in the early phases of an innovation is to create demand and supply in a balanced fashion, so that there is a sufficient market for the early adopters on all levels in the value network, including both the software vendors and AECO companies. It seems that an efficient pathway is one that involves a significant national effort capable of creating the critical mass in a relatively short period of time (Kiviniemi *et al.* 2008: 49).

As described in the ERAbuild BIM report (Kiviniemi *et al.* 2008: 57–9):

the development, implementation and deployment of integrated BIM are complex issues depending on the co-existing demand and supply. There are practically an infinite number of factors influencing the path and speed on how this development could happen; will there be any key players who want to actively promote the change, which role they have, which parts of the development they see as a creation of a competitive advantage, do they want to participate in the creation of a 'new infrastructure' – collaboration platform for AECO – is public funding available or is the development based on private funding, etc.

The ERAbuild BIM report was initiated by several national funding agencies to help develop their future funding policy. It discussed two alternative scenarios based on either public or private interest and funding, trying to identify some key actions at the macro level.

The development of the early ICT tools started from automation of the manual processes, rather than from rethinking the processes (Figure 6.5). This is logical because otherwise only very few potential users could understand the usability of the tools, and it would be a less favourable scenario for software vendors. After some experience, the users can gradually start to see the new possibilities and to move through an informational phase towards transformational change.

This basic model applies also to the development and deployment of typical design software. The first step in the mid-1980s was 2D CAD – automation of drafting, which still dominates in the AECO industry. The first BIM software products inherited the technological basis from the drafting tools and are commonly used as a repository from which the designer can generate drawings, that is, still on the automational level of the development. This is exactly the situation which Khemlani (2007) identified as the main driver for BIM in the USA.

The implementation of an IFC interface into the BIM software represents a step towards the informational level by enabling information management among the shareholders.

The third step is when companies start to develop new business processes based on the potential of BIM. Changes in such processes will

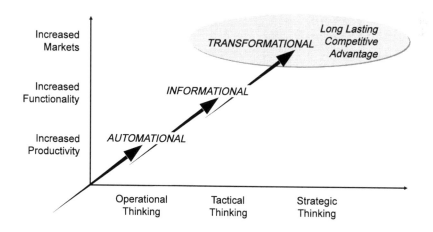

Figure 6.5 Adoption levels of new technology (source: VTT, VBE II/Stephen Fox (2006), reproduced with permission).

affect the roles and responsibilities of the whole network, thus there will be a need to change contractual and business models as well. This can be problematic, since shareholders in the network can have different, and sometimes conflicting, business drivers. Besides technical issues and lack of skills, this has been one of the reasons for the slow adoption of integrated BIM. In the current process all project participants try to sub-optimize their own benefits, and in the new processes the benefits do not necessarily come to the same party that has to invest in the change. Typically, the added value of improved information quality is created at the beginning of the supply chain and the benefits flow downwards, from the architect to other designers and further to the construction company and finally to the owner and end-users of the building. This means that there is a need for new contractual models which reward and motivate all shareholders in a fair way. Sufficient business drivers are imperative for extensive deployment of BIM.

Current interoperable BIM technology and quality problems

The current IFC specification covers much larger content than any of its implementations, except some model servers and software toolboxes. This follows from the domain-specific nature of software applications; they do not contain internal objects or applicable structures for all IFC objects. Thus, the IFC specification is not the bottleneck for the deployment of integrated BIM; the problems are in the implementation. However, there is no need to make applications which could cover the whole IFC specification; the problem is in the lack of relevant use-case definitions – for example, what is the data set needed to exchange between task A done in software type X and task B done in software type Y? The current IFC certifications are based on a much wider data set, a so-called coordination view, which is neither well defined nor well documented. This makes the implementation of IFC support unnecessarily complex, and still does not ensure sufficient quality for the data exchange. In many cases, a small subset from the upstream application would be sufficient for the downstream application. This situation is one reason for the slow progress of IFC implementation. If it requires significant effort and there is no sufficient market demand, the implementation is not lucrative for software vendors.

The current IFC certification process does not guarantee that the certified products can be successfully used in real projects. In addition, as stated above, the content of the certification is unclear and the end-users cannot know what certified software should support. It is practically impossible for average end-users to find out what information will be exported or imported (Kiviniemi 2007: 12). A prerequisite for a reliable certification process is also that it is managed by an impartial party – that is, not by the IAI and software vendors.

Necessary future steps towards integrated BIM

Some earlier roadmaps towards integrated BIM processes

Integrated BIM is just a small part of comprehensive technology roadmaps for the AECO industry. However, it relates to several areas in many of the existing roadmaps; for example, in the Strat-CON roadmap (Kazi *et al.* 2007a) it relates in some extent to at least five of the eight areas: digital models, interoperability, collaboration support, knowledge sharing and ICT-enabled business models (Figures 6.6 and 6.7). Likewise, the existing roadmaps have parallel areas. Figure 6.6 includes one example of the mapping between two roadmaps, FIATECH Capital Projects and Strat-CON.

For practical reasons, these comprehensive roadmaps must be divided further into themes and topics. A good example is the Start-CON Roadmap, which is divided into eight main themes and eight main topics (Figure 6.7). In the Strat-CON methodology, each theme is processed documenting the current state, drivers, short-, medium- and long-term steps, and the final goal (an example of the interoperability theme is shown in Figure 6.8). The end result is a very comprehensive, well-documented development path to an envisioned future, with cross-linking between the items in the theme-based roadmaps. The Strat-CON project was initiated to align the earlier ROADCON roadmaps with the main thematic areas addressed by the European Construction Technology Platform's (ECTP) focus area

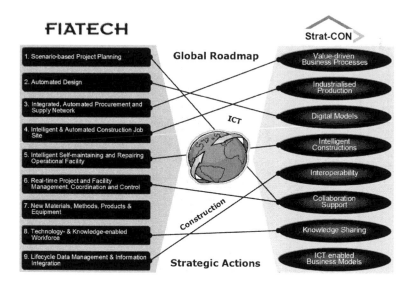

Figure 6.6 Congruence between FIATECH Capital Projects and Strat-CON roadmaps (source: VTT, Kazi *et al.* (2007b), reproduced with permission).

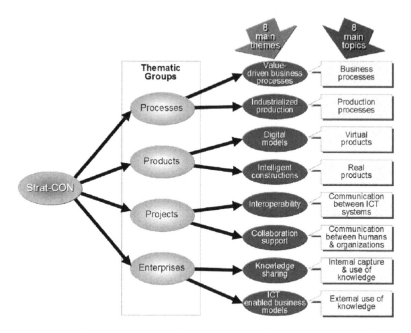

Figure 6.7 Thematic areas and main topics in the Strat-CON roadmap (source: VTT, Kazi *et al.* (2007a), reproduced with permission).

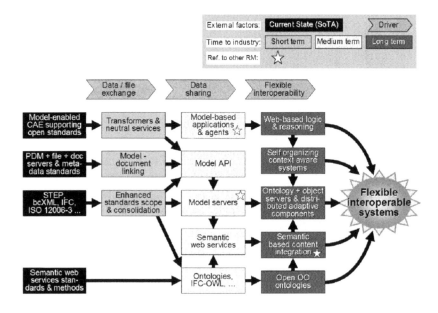

Figure 6.8 Strat-CON roadmap for interoperability theme (source: VTT, Kazi *et al.* (2007a), reproduced with permission).

on processes and ICT. Strat-CON identified and developed a set of strategic actions at the European Union level for realizing the vision of ICT in construction. The same methodology has also been successfully applied on smaller-scale roadmaps – even on technology roadmaps for individual companies.

Another and much simpler approach for an integrated BIM roadmap was used in Koivu's (2002) project 'Roadmap to Intelligent Product Model'. It identified two major drivers for the development, technology and business models, and took as its starting point two business scenarios ('minimizing the costs' and 'adding value to the customer') and two technology scenarios ('proprietary data') and ('open standards'). This solution space created four main scenarios (Figure 6.9).

Scenario 1: Only the cost-effective survive on the market. Business models are based on minimizing the costs and proprietary data formats. Companies must build their information systems based on products of one or very few software vendors, since efficient use of information is only possible inside one application. This limits the degree of freedom to choose or change software, and ties the customers strongly to the selected vendor. There is no incentive for data-sharing among project shareholders or during the life cycle of the building. It would also be technically challenging or impossible, since there are no interoperable applications.

Scenario 2: Optimizing 'islands of automation'. Business models are based on minimizing the costs and open standards. Data-sharing between applications is technically possible, but there is limited or no incentive for participants to do so, since they cannot benefit from the added value. The

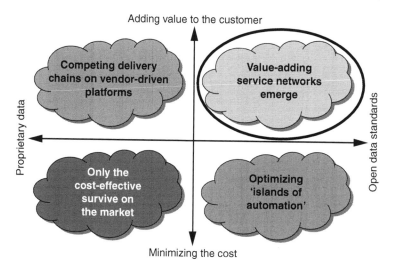

Figure 6.9 Four scenarios of the future (source: Koivu (2002)).

optimization happens at the company level and concentrates on the development of internal systems and their efficiency. The degree of freedom to select applications is significantly higher than in Scenario 1.

Scenario 3: Competing delivery chains on vendor-driven platforms. Business models are based on added value and proprietary data formats. Data-sharing is possible only within the same system or limited set of applications having point-to-point solutions. This limits the degree of freedom to choose or change the applications, and ties the customers strongly to the selected vendor. This scenario leads to relatively stable supply chains, since a prerequisite for efficient information management is the use of compatible ICT systems. The utilization of life cycle information is limited, since the probability that production systems of the project team and maintenance systems of the future clients would be compatible is relatively small.

Scenario 4: Value-adding service networks. Business models are based on added value and open standards. Data-sharing between applications is easy, and does not limit the selection or changes of applications of the participants. This scenario leads to open business networks, and new business and contractual models where the shareholders must agree how the added value will be compensated and shared in the project team. The utilization of the life cycle information is preferred, since the use of data between systems is technically possible and, business-wise, feasible.

Koivu (2002) studied the above scenarios using the Delphi method. The results indicated that the preferred future among the research and industry experts was Scenario 4, with the expected timeframe to reach the goal varying from four to eight years. Looking at the situation now, it seems that the outlook in the study was correct, although somewhat optimistic about the schedule. Scenario 4 is also the starting point of the following selected roadmap topics.

Integrated Design Solutions (IDS) is a new priority theme for the International Council for Research and Innovation in Building and Construction (CIB). The CIB board accepted the IDS scoping paper in April 2008 (Kiviniemi 2008). At this stage, the CIB identified four main sub-themes in IDS: information and communication technology, simulation and analysis tools, integrated work processes, and education. This structure is used below for each of the roadmap topics.

Information and communication technology

The integrated BIM technology must be developed to a platform which can efficiently support the new processes and communication between the project shareholders, including also the surrounding community. The key development themes are information creation (software tools, product libraries) and interoperable platforms (data standards, ontologies, interfaces, model servers).

Data creation

In general, the BIM tools in the design and construction process can be divided into two main categories: upstream applications which mainly create information, and downstream applications which utilize the information from the upstream applications. Naturally, this is a simplified image, since all applications need some input and create some output. However, it is useful when thinking about the emphasis in the development needs; the proportion of potentially automatic input from other applications is crucial for the use of software which traditionally needs a lot of manual work in data transformation, such as quantity take-off or thermal simulation. The upstream applications should be developed so that the information needed in the downstream processes is usable without extensive manual intervention. This is significantly different from the traditional development of domain-specific tools.

A significant part of the information creation in AECO is done by the manufacturing industries in their product development and production processes. This information is crucial for the as-built and maintenance models, and is also needed in the design and construction processes. Prerequisites for the use of this information in the integrated BIM environment are standardized product libraries in an interoperable format. Several research projects have investigated product libraries, with the latest and most complete project being the development of the International Framework for Dictionaries (2008) in Norway. However, there has not been any serious effort to produce content into these product libraries for the AECO industry. Additional research in this area is still needed, but the main question is the real commercial demand for compliant product information. This process requires a market – that is, extensive use of integrated BIM in design, construction and maintenance activities.

Interoperable platforms

The current BIM interoperability in the AECO industry is based on the IFC standard and file exchange. Although the IFC interfaces are a short-term issue because of the previously documented quality problems, they cannot solve the limitations which inherently come from the file-based exchange: large files; no partial exchange; no robust version, owner or access management; loss of information which is not supported by different applications, etc. There is a clear need to develop more advanced sharing technologies, such as model servers which can store and manage workflows and all the information produced and needed by different parties.

Another topic related to the interoperable platforms concerns flexibility of information structures. Mapping between ontologies and/or data formats will be a crucial issue when integrating the ICT systems of business areas (such as design, production, maintenance, procurement, financial administration) within and between companies.

Simulation and analysis tools

Integrated BIM provides a basis for the effective and efficient use of simulation and analysis tools which currently demand significant manual work for input. The extensive effort in the set-up has been a significant obstacle to the use of existing tools, such as thermal, lighting, acoustical, fire, 4D, layout, simulations, environmental and life cycle assessment; automated quantity take-off and cost estimation; visualization techniques; and data-mining. Integrated BIM can increase the commercial potential of such tools, and also create possibilities to add functionalities to the existing applications and create totally new tools.

The extensive use of simulation and analysis tools will open new possibilities for collaboration and communication, thus improving the shared understanding of issues among project shareholders and providing new possibilities for advanced decision support.

Energy and environmental issues especially are becoming increasingly important for all shareholders in AECO and the wider society. Integrated BIM can be used as the platform to develop software tools for analysing environmental aspects of project alternatives already in the early stages of design. One possibility is to link design models to environmental and life cycle databases as well as to cost databases. Commercial implementation of compliant environmental databases faces the same challenges as product libraries: critical mass of BIM users is needed to create the interest to create necessary information.

Another rapidly growing area is safety and security analysis of buildings and the built environment. Here, integrated BIM can provide not only a platform for analysis but also advanced user interfaces for emergency situations.

Integrated work processes

Process development for integrated BIM

Current BIM implementations are based on document-based processes, and are not optimal for integrated BIM processes. The complete process re-engineering will require extensive research, but already the current tools allow practical process development at the company or project level.

Contractual models

Moving towards integrated BIM processes will raise complex legal and contractual issues, such as IP, task definitions, legal questions of validity if documents and BIM are in conflict, use and ownership of the BIM during the building life cycle, and third-party libraries as a part of a BIM. These will have to be resolved before extensive deployment of BIM in the industry is possible.

Procurement and logistics in integrated BIM environment

Several research projects have studied the use of BIM in the procurement process and logistics. The main conclusions are that the technology is applicable, but the business implications must be identified and quantified with reasonable accuracy before it can be deployed on a wide scale.

Knowledge management in integrated BIM environment

Integrated BIM will enable significant possibilities in knowledge sharing within industry, companies and project teams, and enhanced performance in meeting cost, schedule, quality, safety and sustainability objectives. However, overcoming gaps concerning the implementation of usable knowledge systems will require increased understanding of related economic, contractual and motivational factors. A requirement for effective knowledge sharing and collaboration support is ubiquitous access to the correct information and relevant knowledge. This will require efficient knowledge-capturing and representation systems.

Education

Integrated BIM will fundamentally change the way the AECO industry will work. However, most current education is based on the traditional document-based processes, even if it includes the use of modelling software. There is an urgent need to change the curricula for architects and engineers. This will be a significant challenge for universities all around the world in the near future, since only few have started these developments and there is a lack of people with the necessary knowledge and skills to plan the new education system. The sharing of educational ideas, information and material among interested institutes is a crucial element for speeding up the development of curricula which can meet the future requirements of the industry.

Conclusions: necessary future steps to move towards the deployment of integrated BIM-based processes

Deploying new technologies, such as integrated BIM, is a challenging task, especially when it is done on the industry cluster level, such as AECO. One strategy is just to rely on market forces, trusting that feasible solutions will eventually succeed. However, passive waiting means slow and unpredictable progress. If the industry wants to actively influence the development of its processes and technologies, it is important to identify and respond to the key obstacles and drivers.

On the human side, the main obstacles are old processes, business models and contracts. In the short term, a necessity is to develop and deploy standard processes and contractual models for the use of integrated BIM, in the same way as AECO developed standard procedures for the document-based environment. Another important obstacle is the lack of competent people, and thus another necessary action is to introduce efficient BIM education to both the new and the existing workforce.

The main technical obstacle to the deployment of integrated BIM in the short term is the insufficient quality of software support for data exchange. This can be improved only by creating sufficient demand for high-quality applications, which means that the industry has to start using the existing possibilities and understand and overcome effects of the current limitations.

In the longer term, it is necessary to develop extensive university education for integrated BIM processes, more efficient collaboration platforms, and improved analysis and simulation tools. Although there is currently a lot of unused potential in the IFC specification, the need for improvements based on the lessons learned will become relevant as the deployment progresses. Facilitating the implementation of IFC support and improving the performance of the data exchange by reducing the complexity of the specification are important parts of the next-generation IFC specifications.

Bibliography

American Institute of Architects (2006) *The Business of Architecture: 2006 AIA Firm Survey*, Washington, DC: American Institute of Architects.

Digitale Byggeri (2007) *Digital Construction*. Online. Available at HTTP: <http://detdigitalebyggeri.dk/>.

GSA (2007) *BIM Requirements*, Washington, DC: General Services Administration. Online. Available at HTTP: <www.gsa.gov/bim/>.

IFD (2008) *The International Framework for Dictionaries*. Online. Available at HTTP: <www.ifd-library.org/>.

Kazi, A.S., Froese, T., Vanegas, J., Tatum, C.B., Zarli, A., Amor, R., van Tellingen, H., Moltke, I. and Testa, N. (2007b) *International Workshop on Global Roadmap and Strategic Actions for ICT in Construction*, Strat-CON project. Online. Available at HTTP: <http://cic.vtt.fi/projects/stratcon/const_IT_ws_22_24082007_report.pdf>.

Kazi, A.S., Hannus, M., Zarli, A. and Martens, B. (2007a) *Strategic Roadmaps and Implementation Actions for ICT in Construction*, Strat-CON project. Online. Available at HTTP: <http://cic.vtt.fi/projects/stratcon/stratcon_final_report.pdf>.

Khemlani, L. (2007) *Top Criteria for BIM Solutions: AECbytes Survey Results*. Online. Available at HTTP: <www.aecbytes.com/feature/2007/BIM Survey Report.html>.

Kiviniemi, A. (2007) *Support for Building Elements in the IFC 2×3 Implementations Based on 3rd Certification Workshop Results*, Finland: VTT (original discussion paper published in IAI in October 2007, and updated version in November 2007).

—— (2008) *Integrated Design Solutions: Scoping Paper*, CIB.

Kiviniemi, A., Tarandi, V., Karlshøj, J., Bell, H. and Karud, O.J. (2008) *Review of the Development and Implementation of IFC Compatible BIM*, ERABUILD funding organizations in Denmark, Finland, the Netherlands, Norway and Sweden. Online. Available at HTTP: <http://cic.vtt.fi/buildingsmart/index.php?option=com_docman&task=doc_download&gid=64&Itemid=85>.

Koivu, T. (2002) 'Future of product modeling and knowledge sharing in the FM/AEC industry', *ITcon*, 7 (Special Issue: ICT for Knowledge Management in Construction): 139–56. Online. Available at HTTP: <www.itcon.org/2002/9>.

NBIMS (2007) *National BIM Standard*, Washington, DC: National Institute of Building Sciences. Online. Available at HTTP: <www.nibs.org/nbims.html>.

Senate Properties (2007) *BIM Requirements*, Helsinki. Online. Available at HTTP: <www.senaatti.fi/document.asp?siteID=2&docID=517>.

Statsbygg (2007) *BIM Requirements*, Oslo. Online. Available at HTTP: <www.statsbygg.no/Aktuelt/Nyheter/Statsbygg-goes-for-BIM/>.

Taylor, J. and Levitt, R. (2004) *A New Model for Systemic Innovation Diffusion in Project-Based Industries*, CIFE Working Paper 86, Stanford, CA: Center for Integrated Facility Engineering, Stanford University. Online. Available at HTTP: <http://cife.stanford.edu/online.publications/WP086.pdf>.

7   Integrated design platform

Robin Drogemuller, Stephen Egan and Kevin McDonald

This chapter describes a suite of integrated design tools developed for the Cooperative Research Centre for Construction Innovation. This suite of software is indicative of the type of software platform that will be standard for the design firm of the future. It has been built around the concept of 'interoperability', where each separate computer program supports one type of activity. The intention is that the information produced in one program will then be available for other programs to access through a publicly available information exchange specification. The Industry Foundation Classes (IFCs) are the information exchange specification that was used throughout these projects. The IFCs were developed by the International Alliance for Interoperability, also known as BuildingSMART (IAI 2007).

This software platform is the largest group of IFC-based software that has been developed by one group to date. While the individual pieces of software are of interest and will be briefly discussed, the lessons learned in developing the computer programs over the period 2001–08 are more significant.

The software-based projects were undertaken in stages. The first stage was a proof of concept that showed that the research goal was achievable. The methods of user interaction were defined and the necessary types of information identified. The software deliverable from this stage was then used to gauge interest by Construction Innovation industry partners and other industry representatives on the potential for commercialization. Only those projects with a significant interest from industry went on to the next stage, which was the development of a full working prototype. A number of the computer programs described in this chapter are near commercialization.

At the time that Construction Innovation started working on the software described here, there was no significant understanding of the development issues of IFC-based software for AECO (architecture/engineering/construction/operations) industry. Neither was there a pool of experienced software developers to draw on. Many of the early implementations of software that supported the IFCs were existing commercial software that implemented an IFC interface to the existing internal data structures. This lack of knowledge has meant that, in the nearly seven years that Construction Innovation has been in existence, there have been two generations of software deliverable. The first used a commercially available object-oriented database, EDModelServer™, as the method of data storage, with the software reading and writing all data to this database. The range of software and the software architecture are shown in Figure 7.1. These computer programs (described later in the chapter) were very loosely coupled, with the database acting as the major form of communication between them.

The second generation of Construction Innovation integrated design software was built on the Eclipse open software platform, originally developed by IBM and now an open source project (Eclipse.org 2008). They built on the experience of the previous generation, and were much more tightly integrated (Figure 7.2). This is explained in more detail in the Design View section of this chapter.

Specific software is used at different times within the building project life cycle. The position of each of the computer programs within the life cycle is given in Table 7.1. The project life cycle used is based on the Generic Design and Construction Process Protocol (Kagioglou *et al.* 1998).

Frameworks for assessment

A simple framework for the factors involved in the development and use of product model-based solutions is presented in Figure 7.3. The two basic requirements for product model-based applications are the technology (the applications) and the information (basic data; constraints imposed through

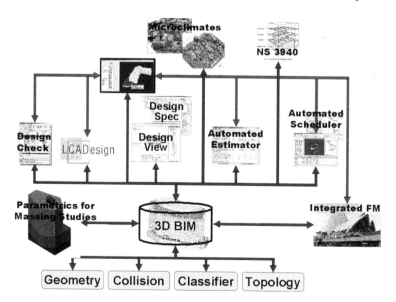

Figure 7.1 First-generation software and the software architecture.

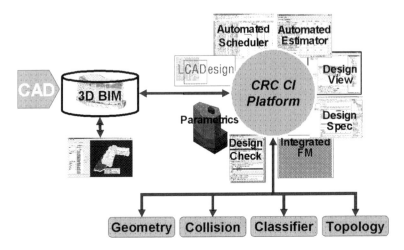

Figure 7.2 Second-generation software architecture.

Table 7.1 Construction innovation software products and the project life cycle

Stage	*Conceptual design*	*Coordinated design*	*Production information*	*Construction*	*Operation & maintenance*
Process protocol phase	Four & five	Six	Seven	Eight	Nine
		Automated Estimator			
	Design Check				
	Parametrics for Massing Studies Microclimates	LCADesign IAQEstimator Area Check to NS3940	DesignSpec	Automated Scheduler	Integrated FM

working practices, codes and standards; product libraries containing clustered information about products) required for these to be used. Project model servers are shown in the diagram as they are necessary to support full collaboration amongst the project team, but there is currently no significant penetration of these in the AECO industry. Two other factors that play an important role in the uptake of model-based work (and any other forms of innovation) are access to trained people with the skills necessary to exploit the new technologies effectively, and fitting the new technology within the workflow of the organizations in which these people work. A technology which substitutes for an existing process and hence does not

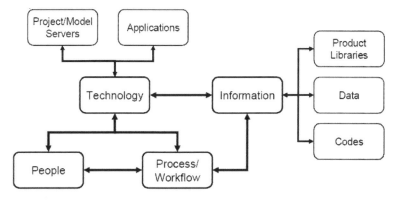

Figure 7.3 Framework for nD CAD software uptake.

require significant changes to work practices will be much easier to introduce than a revolutionary technology that requires substantial change.

One major constraint that has been identified in modelling buildings using existing CAD software is that the software itself imposes major constraints on what can be modelled. For example, creating elliptical stairs is a problem in some IFC-compliant CAD systems. Another constraint in building models for the Microclimates project (see <www.construction-innovation.info/>) was that only one CAD system could be identified that was able to handle projects with more than one building.

From an information perspective, the standard libraries supplied with the CAD systems needed to be modified to ensure that relevant data were available. For some new types of analysis software, the full information required to give a satisfactory result was not available within CAD systems. This required development of methods to supplement the CAD information. The expectation is that, as new types of software become available, a start will be made on gathering the data needed to use them. As the availability of data improves, the capability of the software will improve, leading to better data-gathering in a self-reinforcing loop. Another obvious point is that data, codes and standards are often highly location dependent. This means that software developed in Australia, for example, will need to adapt the underlying data for other countries if international markets are to be exploited.

Fox and Hietanen (2007) used a framework developed by Mooney *et al.* (1996) to define business value from the use of information technology. It includes three categories:

- *Automational.* This covers changes in efficiency deriving from the substitution of IT systems for labour. These improvements are relatively easy to obtain as no significant changes are required for them to be introduced.

- *Informational.* This is concerned with the ability of IT systems to collect, store, process and disseminate information. Changes may improve the way that a business functions, but the cost of changing processes and working methods must be balanced against the expected benefit. These can be considered as medium-term changes.
- *Transformational.* This covers the impact of process innovation and transformation. These changes can be expected to provide the most benefit, but will normally have the widest implications. Significant cooperation may be required between organizations, or between functional units within a single organization. This means that transformational change will be longer term.

The relevant areas of both of these frameworks are applied to the various Construction Innovation software deliverables below.

The first-generation software

The general software architecture and the flow of information is shown in Figure 7.4. The basic principle of operation is that the building design is developed in CAD software, exported as an IFC file and then imported into the object-oriented database. The Construction Innovation software then extracts the necessary data from the database and presents the data to the user through its own user interface. Information that must be retained is then modified in the database or added to it, and can then be exported as an IFC file for use by other software.

Automated Estimator

The Automated Estimator was one of the first IFC-based computer programs developed by Construction Innovation. The goal was to automatically extract

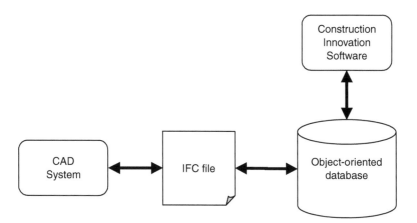

Figure 7.4 First-generation software architecture.

as much information as possible from a building information model (BIM) to present in a Bill of Quantities (BoQ) (Figure 7.5). It was not intended to build a full cost-estimating system, as there were already a number of these available. The Automated Estimator was seen as a 'bridge' between the BIM and estimating systems, automatically populating the data behind an estimating system for continued development by an estimator. The initial intention was to only build the BoQ interface, presenting the list of items broken into their standard sections, with the measured quantities. However, it was soon realized that a graphical interface was required both for debugging purposes by the software development team and for users to browse a three-dimensional view of the BIM to check results and identify issues (Figure 7.5). Many users do not want to pay for an expensive CAD system just to browse a BIM.

The trades of concrete, formwork, steelwork, masonry, reinforcement and prestressing have been implemented in the Automated Estimator.

The Automated Estimator is largely an *automational change* in that it substitutes for existing processes. It was a scoping decision for the project to develop a system that fitted cleanly within existing processes. While time savings are significant (15 minutes to take off quantities that took two weeks by manual processes during testing), very few changes in standard estimating working practices are required to use the software. The

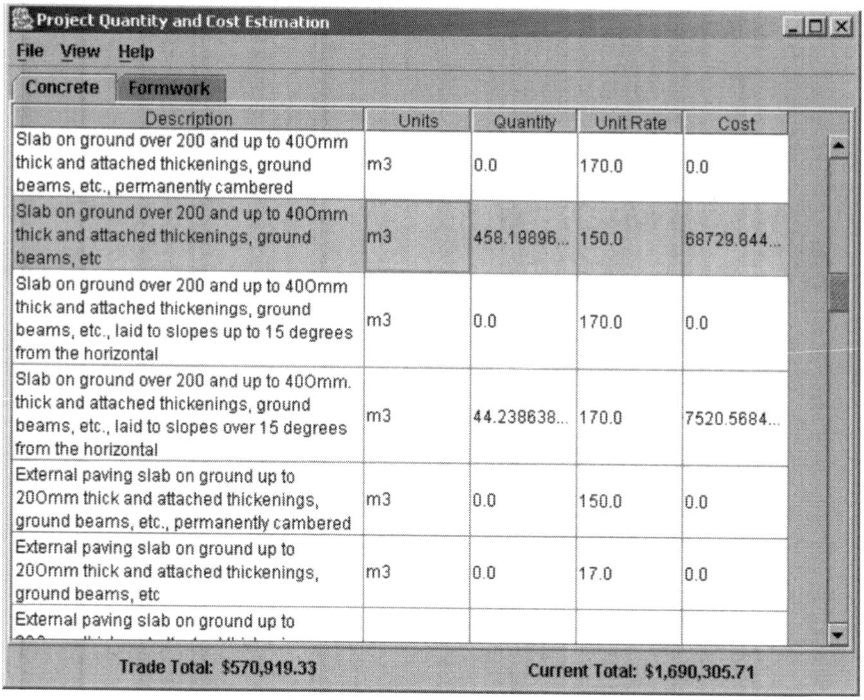

Project Quantity and Cost Estimation					_ □ ×
File View Help					
Concrete	Formwork				
Description	Units	Quantity	Unit Rate	Cost	
Slab on ground over 200 and up to 400mm thick and attached thickenings, ground beams, etc., permanently cambered	m3	0.0	170.0	0.0	
Slab on ground over 200 and up to 400mm thick and attached thickenings, ground beams, etc	m3	458.19896...	150.0	68729.844...	
Slab on ground over 200 and up to 400mm thick and attached thickenings, ground beams, etc., laid to slopes up to 15 degrees from the horizontal	m3	0.0	170.0	0.0	
Slab on ground over 200 and up to 400mm. thick and attached thickenings, ground beams, etc., laid to slopes over 15 degrees from the horizontal	m3	44.238638...	170.0	7520.5684...	
External paving slab on ground up to 200mm thick and attached thickenings, ground beams, etc., permanently cambered	m3	0.0	150.0	0.0	
External paving slab on ground up to 200mm thick and attached thickenings, ground beams, etc	m3	0.0	17.0	0.0	
External paving slab on ground up to					
Trade Total: $570,919.33			Current Total: $1,690,305.71		

Figure 7.5 Bill of Quantities and 3D Viewer interfaces for automated estimator.

method of rule encoding supports various standard methods of measurement, and is flexible enough to be adapted to a range of levels of detail as well.

There is a minor informational impact through the automatic recognition of compound building components, such as concrete beams supporting concrete slabs (Figure 7.6) and the automatic take-off of formwork, including propping heights between floors. The ability of the software to identify components that are and are not measured in a trade section also assists in tracking down errors in the BIM model.

Automated Estimator is currently being re-factored on the Design View platform (described later in this chapter) for commercialization.

LCADesign

LCADesign is possibly the most significant software deliverable to emerge from Construction Innovation. It automates the assessment of the environmental impact of a building from the extraction of the raw material to its installation in the building as a finished product. The generation and assessment of alternative designs through the selection of substitute construction systems is rapid, and there is a simple mechanism to add additional information to a CAD-generated BIM through product library data. LCADesign was started simultaneously with the Automated Estimator, with a view to delivery of a much needed automated eco-efficiency assessment tool.

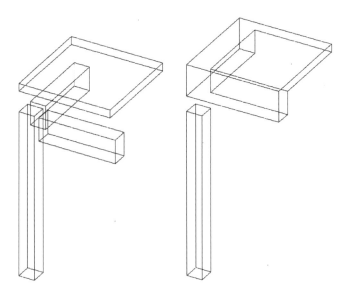

Figure 7.6 Automatic recognition of single building components modelled as separate objects.

While LCADesign is supported by one of the most comprehensive building material databases currently available, life cycle impact assessment is being held back by the lack of comprehensive, rigorous data, both nationally and internationally. When LCADesign was used to analyse a building project in the Netherlands (Figures 7.7 and 7.8), access to local data had to be negotiated with a local expert. While there are initiatives underway that will address this problem, it is an important area which urgently needs additional information to support innovation through the use of environmental impact assessment software.

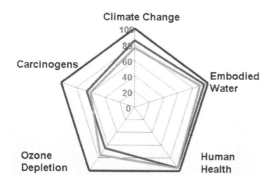

Figure 7.7 LCADesign: comparison of alternative design performance.

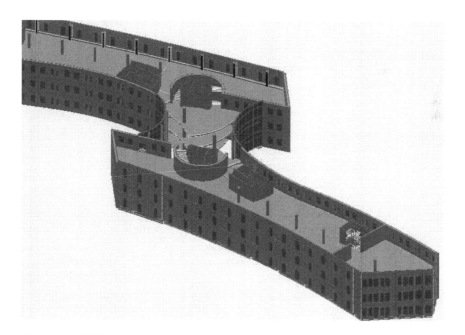

Figure 7.8 BIM for project in the Netherlands.

During testing of LCADesign, significant improvements in workflow were noted. Its use enabled the environmental assessment team to provide results to the designers within a day. Previously, using spreadsheet-based methods, the team needed three to four weeks for an assessment. This meant that the design had moved on significantly from the time that the environmental assessment was initiated, and the results were no longer very relevant. LCADesign closed this loop to provide much improved guidance to the design team and multiple environmental assessments as design and material selection progressed.

LCADesign can be considered *automational*, since it speeds up existing practices, and *informational*, since it provides more rapid access to a wider range of environmental assessment metrics. Whether it is *transformational* will depend on what type and level of impact it has on the environmental 'signature' of the building designs on which it is used.

A new version of LCADesign has been developed that runs on the Design View platform described later in this chapter.

Design Check

Design Check provides a conformance-checking environment for building designs against a set of user-defined rules stored within the software. Rules can cover any issues, not just building codes and standards. Currently, rule sets are defined that check for conformance to Australian Standard AS 1428.1–2001, Design for Access and Mobility – General Requirements for Access – New Building Work, and against the draft Part D requirements under the Building Code of Australia that covers access and mobility.

While there are a number of requirements checking programs available, Design Check is carefully built around the working practices of designers and code-compliance checkers. Users can choose to analyse a BIM against an entire code, or against selected rules within the code (Figure 7.9). They can also choose to analyse selected objects against all the rules that may apply to them. These alternative methods of analysis allow designers to focus on a particular area of the current design without being distracted by other issues.

Rather than just checking whether a particular clause passes or fails, Design Check allows a user to provide textual information as a response when an issue is identified. Someone checking a design, whether for compliance before issuing a permit or performing internal quality checks within the design office, can add comments against issues in the conformance report. This fits well with the performance-based nature of the Building Code of Australia, but can also be used as a communication mechanism among the design team. This ability to add comments and explanations also means that Design Check can be used on BIMs at various stages of design and for a range of purposes.

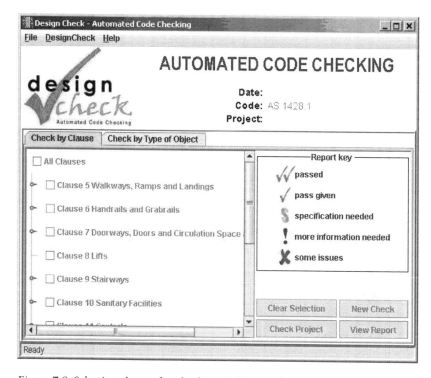

Figure 7.9 Selecting clauses for checking in Design Check.

The conformance report (Figure 7.10) lists the identified issues. This report is generated from data stored with the BIM so that it can be associated with the appropriate BIM at a later date.

An important distinguishing feature of Design Check is that all of the clauses are encoded. This means that a response is recorded against every clause, even if automatic checking is not feasible for that clause, whether there is not enough information in the BIM model or because checking is just not feasible.

When assessing Design Check against the technology framework, it fits well into existing industry work practices and supports checking against requirements, such as codes and standards. The user manual defines what attributes need to be created within the BIM for checking to be successful. Currently, some of these need to be explicitly entered by users. However, widespread use of this technology would encourage CAD vendors to add these properties to their standard libraries.

Design Check is largely automational through providing improved support for existing processes. There is a small potential informational impact through improving the exchange of information amongst designers and between designers and code-compliance checkers.

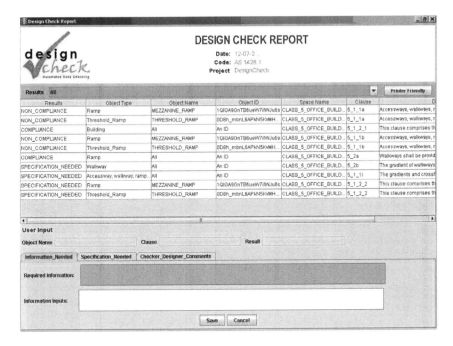

Figure 7.10 Report from Design Check.

Automated Scheduler

The Automated Scheduler scans a BIM, identifies what construction activities and resources are required, builds a construction schedule for the project and then links the activities against the relevant building components. The results can be viewed in a 4D CAD simulation package for further refinement. The major benefits are: the automation of the initial data-gathering process required to generate a construction schedule; the capturing of construction expertise explicitly in a database; and the reduction in time of the generation of the initial schedule. Consequently, Automated Scheduler is both automational (generating the schedule) and informational (retaining and exploiting construction scheduling information).

Parametric Engineering System Design at Early Design Stage

This project was one of the most ambitious undertaken within Construction Innovation from an information modelling perspective. Its aim was to support decision-making across a range of disciplines at the very early stages of building design (Figure 7.11) when the maximum range of design parameters can be explored at minimal cost (Figure 7.12).

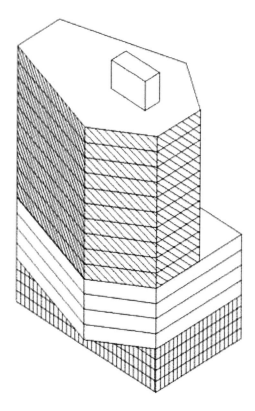

Figure 7.11 Massing model.

The strategic goal is to move decision-making forward in the design process, from curve 3 to curve 4, to allow more cost-effective decisions to be made. The discipline areas supported in the implemented 'proof of concept' system were:

- architectural spatial layout;
- structural system selection;
- hydraulics: tank capacities;
- mechanical services: sizing of plant room and major duct runs;
- electrical: substation requirements;
- cost estimates: based on unit area rates.

A method of supporting complex decision-making was needed. The Perspectors method (Haymaker *et al.* 2003) was chosen. A simple decision network is shown in Figure 7.13. The discipline-specific knowledge was gathered from a number of references (e.g. Stein 1997; Parlour 1994) and from the industry partners.

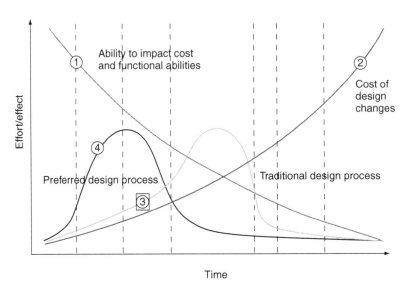

Figure 7.12 Cost-impact curve.

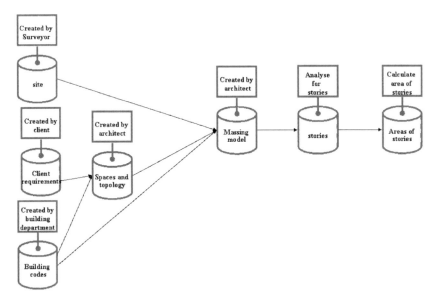

Figure 7.13 Use of Perspectors (Haymaker *et al.* 2003) to model decision points.

Simple user-interface panels that set out the relevant parameters were utilized to allow user input to the system. Figure 7.14 shows the panel for setting structural frame parameters. As each parameter is changed, all of the dependent values also change across the full range of systems. For

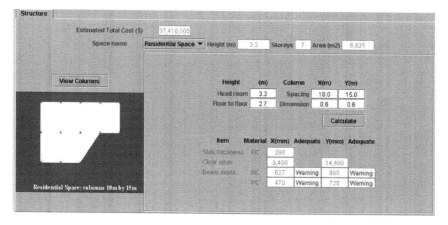

Figure 7.14 Panel for adjusting structural frame parameters.

example, an increase in the column spacing would increase the size of beams, increasing the height of the building, with possible repercussions on selection of the lift cars due to increased distances of travel for the lifts.

This system used a wide range of data from disparate sources, but did not draw on other aspects of the technology framework since it was a proof of concept. The results of the analyses lay mainly in the informational area, as they exposed decision-making across a range of disciplines.

The second-generation platform

Design View

The development of the range of software described above showed that there were many shared code modules and a shared need for a viewer interface that would allow the user to interact with the underlying data in a consistent, intuitive manner. It was decided that the best way to support these needs was to use the Eclipse Rich Client Platform application framework (eclipse.org 2008). The capabilities that Eclipse provided as standard were:

- Perspectives. These provide a named group of windows to support a particular process. Perspectives have been developed for Automated Estimator and LCADesign so that users can easily switch between these analyses for a single project.
- An underlying data representation system (Eclipse Modeling Framework). This was used to optimize the use of IFC data within the supported applications. This is also used to cache complex analysis data that are generated within the Construction Innovation software to provide faster analyses within a particular application and also to support

sharing of data between perspectives. For example, the results of the extensive processing required to analyse the IFC geometry and to convert it into usable information is shared across most perspectives.

- Problems and Tasks panels. These provide methods of communication amongst the project team. The Problems panel lists any missing data when an IFC model is imported. For example, the material of building components is often forgotten by drafters. This acts as a simple form of quality control on the input BIM. Double-clicking on a problem highlights the relevant building component in the Viewer pane. The Tasks panel allows users to list activities that they need to undertake in the future or that they expect others to complete.

The resulting computer program is called Design View. It is described as a general-purpose BIM workbench, as it does not perform analyses unless other perspectives are loaded. The use of the Eclipse platform to build Design View is a good example of the transfer of a useful piece of technology between technical fields – in this case from software engineering to building design, construction and operation.

The Design View user interface (Figure 7.15) consists of a Navigator panel to the left which allows new projects to be loaded and existing ones

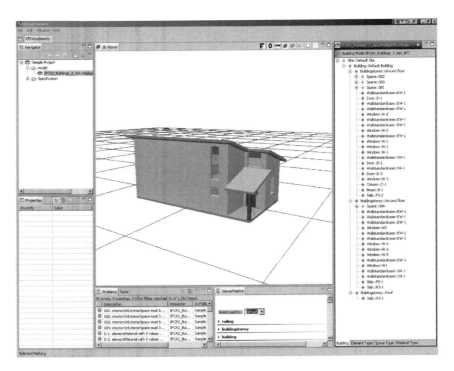

Figure 7.15 Design View user interface.

to be selected for analysis. The Problems/Tasks panel, described above, is to the left of the bottom centre. These are standard from the underlying Eclipse platform.

The central 3D Viewer panel provides visual feedback and selection of building components. This can present the geometry of the BIM as either 3D or 2D. The tree view, to the right, displays the type and name of all of the building components in the BIM. The tabs at the bottom of this panel allow the user to display the building component list as a building tree (containment hierarchy), by element type, space type or material type. The 3D Viewer and tree views are closely integrated. Selecting a building component in one highlights it in the other. Selecting a component or group of components and dragging them into the Viewer displays only those components.

The Viewer Palette allows a user to change the display properties of elements based on type. A group of elements may be displayed or hidden, have their colour set and their transparency adjusted. Particular configurations may be saved and reloaded as required. The Properties view is used to inspect and modify the properties of an individual element. Not all properties can be modified.

As mentioned previously, both Automated Estimator and LCADesign have been converted to run as perspectives (plug-ins) on the Design View platform.

The only building design operations that Design View explicitly supports are consistency checking of BIM on import into the system with subsequent resolution of some errors, and the use of the Tasks panel. At their current level of implementation, these are purely informational functions. However, further development of this concept to support full checking of incoming BIMs would be transformational, since the quality of the incoming information could be assessed and possibly checked against contractual requirements. A significant improvement in the quality of documentation would be transformational indeed through the impact on the entire building procurement process.

DesignSpec

DesignSpec is a prototype system that looks at the issues involved in directly linking a BIM and a textual building specification. The strategic goal was to resolve coordination problems when large parts of the description of a proposed building project are stored in two separate formats: BIM and structured text. This is a continuing source of errors and conflicts in building procurement.

DesignSpec was the first plug-in developed on the Design View platform. In addition to the standard Design View 'workbench' perspective, it adds 'Spec Entry' (Figure 7.16) and 'Associations' (Figure 7.17) perspectives. The Spec Entry perspective allows the user to add information to tables within a

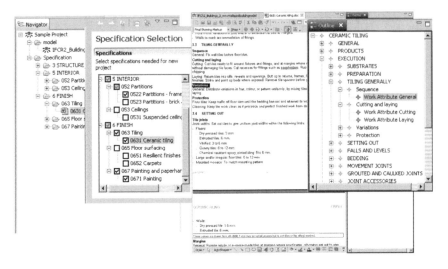

Figure 7.16 Spec Entry perspective.

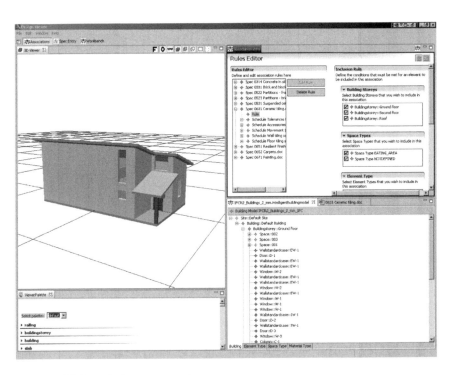

Figure 7.17 Associations perspective.

standard specification document. The current implementation supports the addition of room finishes (floor, wall and ceiling) to the appropriate schedules.

The Associations perspective is used to define filter rules to associate elements in the BIM with worksections, schedules and schedule entries in the specification. It is also used to inspect the associations that have been made with those rules.

A user can select which building storeys, space types, element types and element materials to filter by. Elements may be explicitly included or excluded from the final selection using the check boxes in the 'Filtered Elements' list.

Selecting a Spec Object (Worksection, Schedule or Schedule Entry) causes all the filtered elements to be displayed on the right. If there is no rule defined for that Spec Object, the results of the parent filter rule will be used, or the entire BIM if there are no rules defined in the hierarchy chain. The tree list on the right-hand side is linked with the 3D Viewer, so high-lighting and drag'n'drop also work. Selecting an element (or selection of elements) in the BIM Hierarchy (bottom right view in the Associations perspective) highlights all Spec Objects associated with that element in the rule editor. Double-clicking a Spec Object will open up the Spec editor and display the relevant section.

DesignSpec provides a mechanism for adding a 'builder' to the system to perform tasks based on data in the specification. Currently, only a Finishes builder has been developed (Figure 7.18). This adds finishes to elements in the building model that are associated with relevant schedules (e.g., wall and floor tiling, carpeting, paint). This process of automatically adding additional building elements to the BIM has been called 'model augmentation'. Currently there is no control over the dimensions of the finishes, and it is assumed that finishes cover the area of the element that is contained within the relevant space. Future versions of the system will allow for fine-grained control of the dimensions of the finishes.

DesignSpec uses the Australian national building specification as the base specification. The aim of the project was to demonstrate how the documentation process could be improved by linking BIM and specifications more closely. DesignSpec is both automational, linking BIM and textual

Figure 7.18 Automatic 'builder' adding finishes to the BIM.

specifications, and informational, by augmenting the BIM with new components (surface finishes) and thereby improving communication between designers and specification writers. It provides technology that could support tansformational processes, but this depends on continued development of the work and its uptake by industry.

Area calculations

A small utility was written that calculates building areas to the Norwegian Standard NS3940. This requires four areas to be determined (Figure 7.19):

- gross area: area of all floor plates;
- net area: area inside the external walls;
- usable area: area available after placement of internal partitions;
- built-on area: plan area on the site covered by the building footprint and any projections up to 5.5 m above ground.

The algorithms required to calculate this information were not particularly significant. The main effort required in this project was to handle the data in the BIM that used Norwegian rather than English terms. The BARBI server was used to identify unknown terms, such as 'betong', and the

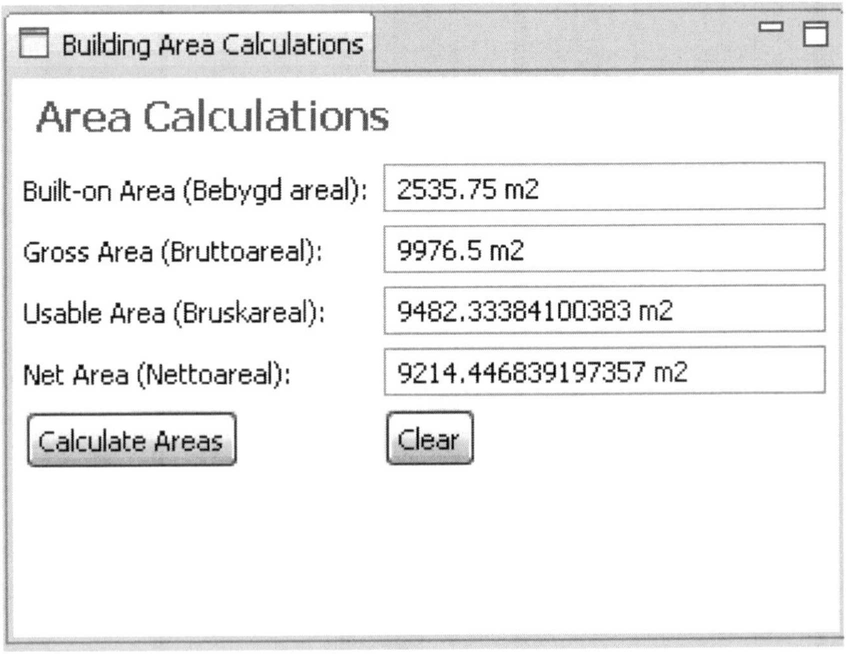

Building Area Calculations

Area Calculations

Built-on Area (Bebygd areal):   2535.75 m2

Gross Area (Bruttoareal):   9976.5 m2

Usable Area (Bruskareal):   9482.33384100383 m2

Net Area (Nettoareal):   9214.446839197357 m2

Calculate Areas   Clear

Figure 7.19 Area calculations against NS3940.

English equivalent, 'concrete', to allow further processing. The BARBI work has now been integrated into the International Framework for Dictionaries project (Bjørkhaug and Bell 2007).

This project covered a single, relatively simple process: the calculation of areas according to a standard. In order to achieve this, a special type of library – an online database matching concepts in different languages – was used. While this project used some advanced technologies, it can only be regarded as automational.

Workflow and the design office of the future

Figure 7.3 gave an overview of the position of each of the software deliverables within the overall building procurement process. There are several stages described in the Project Process Protocol before building design starts. These do not require the explicit representation of geometry and spatial relationships. They define the scope of the project: spatial requirements, financial constraints, etc. It is still a matter for conjecture how these early stages of project definition will be supported by BIM-enabled software, although some work has been done (Kiviniemi 2005; Onuma Inc. 2007).

It is now possible to give an outline of how a future BIM-enabled project procurement process may work. First, appropriate locations for a development may be identified using GIS type systems (Drogemuller 2007). A team would be formed to define the characteristics of the proposed project using integrated software that would support the identification, capture and resolution of issues. These would act as inputs to the building design process.

Once a project was scoped and funded, the design team would collaborate on defining the initial sketch designs in a similar manner to that currently employed. However, the design could be more fully resolved earlier in the process through the use of automated estimating and checking tools, such as those described in this chapter. The use of new types of analysis such as LCADesign for environmental impact assessment may allow the definition of legal requirements in new areas. This may mean that contractors, or at least those with significant construction experience, might be required to provide guidance earlier in the procurement process than is currently common. Improved integration of design and construction is necessary for design for disassembly (see Crowther, Chapter 12 in this volume).

The uptake of specialized CAD systems by subcontractors that feed directly into numerically controlled fabrication machinery (i.e., CAD Duct) may also mean that some design responsibility passes down to the subcontractors, with the design engineers defining the requirements in a more performance-based manner than the current prescriptive methods. This information would then be handed on to the constructors. The constructors would continue to add as-constructed information to the BIM for handover on completion of the construction contract. This would then be handed over to the facilities manager for use during operation of the building.

The above description covers the use of the technology to support building procurement. This would be automational and informational. There would be many contractual and management issues also involved in moving towards the above process, but it is not intended to address them here.

One of the most significant observations that has emerged from the software development work described in this chapter is that, in a BIM-enabled world, the CAD system becomes a means of entering geometrical data and spatial relationships. Thus CAD holds an important place in the building procurement process. However, it becomes another user interface into the BIM. Many other visualization tools can also contribute in parallel with current CAD software. CAD is likely to lose its primacy if distributed BIM continues to develop from the current state of technology.

Future work

Despite the large amount of effort involved in the development of the software described in this chapter, and parallel commercial and research projects internationally, there is still much more that needs to be done.

Currently, BIM is of most benefit during detailed design and construction. BIM support for project scoping and early design has not received significant attention. Its full exploitation during the tendering and construction process also presents significant technical, legal and workflow challenges. Facilities management, maintenance, refurbishment/re-lifing and demolition also need further research and development.

Within the technical framework described earlier (Figure 7.3), significant effort is required in adapting model server technology to suit workflows within the AECO industry, the definition of product libraries that support richer information and are directly accessible from design/construction software. Technologies to support checking against codes and standards also need substantially more development. All of this will need to take place within an industry and organization structure that is changing as workflows adapt to new capabilities.

Conclusion

This chapter has covered a range of software developed to support the design and construction of buildings. Indications have been given of the use of each one individually, as well as the possible cumulative impact of the use of this software and other BIM-enabled software on the AECO industry.

In ten years time, the industry is likely to have different allocations of responsibility and different processes than today. Some of this will be driven by internal forces – the need and desire to offer better services to clients and users. Some will also be external – the competitive processes of globalization and the political pressure to reduce the detrimental impacts of construction on the environment.

Acknowledgements

The work described in this chapter was funded by the Cooperative Research Centre for Construction Innovation. A wide range of researchers participated, from the CSIRO, University of Sydney, University of Newcastle, Queensland University of Technology and RMIT University. Industry partners were Rider Hunt, Project Services and the Sydney Opera House Trust.

Bibliography

Bjørkhaug, L. and Bell, H. (2007) *IFD in a Nutshell*. Online. Available at HTTP: <http://dev.ifd-library.org/index.php/Ifd:IFD_in_a_Nutshell> (accessed 30 June 2008).

Drogemuller, R. (2008) 'Virtual prototyping from need to pre-construction', in P.S Brandon and T. Kocatürk (eds) *Virtual Futures for Design, Construction & Procurement*, Malden, MA: Wiley-Blackwell.

Eclipse.org (2008) Home page. Online. Available at HTTP: <www.eclipse.org/> (accessed 21 June 2008).

Fox, I. and Hietanen, J. (2007) 'Interorganizational use of building information models: potential for automational, informational and transformational effects', *Construction Management and Economics*, 25: 289–96.

Haymaker, J., Kunz, J., Suter, B. and Fischer, M. (2003) *Perspectors: Composable, Reusable Reasoning Modules to Automatically Construct a Geometric Engineering View from Other Geometric Engineering Views*, CIFE Working Paper 082, Stanford, CA: Center for Integrated Facility Engineering, Stanford University. Online. Available at HTTP: <http://cife.stanford.edu/online.publications/WP082.pdf> (accessed 23 June 2008).

IAI (2007) Home page. Online. Available at HTTP: <www.iai-international.org/> (accessed 20 June 2007).

Kagioglou, M., Cooper, R., Aouad, G., Hinks, J., Sexton, M. and Sheath, D. (1998) *A Generic Guide to the Design and Construction Process Protocol*, Salford: University of Salford.

Kiviniemi, A. (2005) *Requirements Management Interface to Building Product Models*, CIFE Technical Report TR161, Stanford, CA: Center for Integrated Facility Engineering, Stanford University. Online. Available at HTTP: <http://cife.stanford.edu/online.publications/TR161.pdf>.

Mooney, J.G., Gurbaxani, V. and Kraemer, K.L. (1996) 'A process orientated framework for assessing the business value of information technology', *Advances in Information Systems*, 27: 68–81.

Onuma Inc. (2007) *ONUMA Planning System*, Pasadena, CA. Online. Available at HTTP: <www.onuma.com/products/OnumaPlanningSystem.php> (accessed 9 July 2008).

Parlour, R.P. (1994) *Building Services: Engineering for Architects*, Sydney: Integral Publishing.

Stein, B. (1997) *Building Technology: Mechanical and Electrical Systems*, 2nd edn, New York, NY: John Wiley.

8 Understanding collaborative design in virtual environments

Leman Figen Gül and Mary Lou Maher

The process of designing buildings has become increasingly difficult, reflecting the growing complexity of the buildings themselves and the process leading to their design, construction and management (Kalay *et al.* 1997). This complexity has required better coordination of building-related activities capable of aligning technological, economic, political and other developments (Archea 1987). Recently the developments in and extensive use of internet technologies have brought about fundamental changes in the way the building industry collaborates and designs by transforming their organizations with IT-based strategies. This initiative has been in part a response to pressure to improve efficiency, and also because of the need for communication and collaboration between AEC (Architecture, Engineering, Construction) organizations, using various computer-mediated technologies, such as video conferencing, email, the Internet, intranets and virtual worlds. Thus, computer-mediated communication technologies have become a vital medium for most of these organizations.

In general the process of designing and constructing has involved project-based alliances between several independent organizations, such as building owners, financial institutions, building users or their representatives, architectural and engineering offices, construction and subcontracting companies, and product manufacturers. The links between them are established when the need arises, and are finalized when the tasks have been completed (Kalay *et al.* 1997). Furthermore, the relationship between these temporary 'organizations' and the composition of the project team changes as the project evolves. During the collaboration process, the design teams produce their version of the design documentation in a sequential manner. Architectural firms produce drawings showing the building layout (plans, sections and facades), and these design documents then pass to other parties (for example, engineers) who need to redraw and repeat much the same process to produce their own drawings. In this sequential process, design teams need to iterate the documentation process, control the conflicts between the drawings and amend their design documents. This type of collaboration requires extra finance and time, and occurs due to the lack of:

- common conventions and format on the design documentation;
- effective communication between design teams;
- effective methods and procedures for collaboration.

The development of a standard for a building information modelling (BIM) system which provides a shared model as 'becoming almost a living organism that can be accessed asynchronously by its many contributors' (Cohen 2003, as cited in Kouider *et al.* 2007) has the potential for addressing these issues. There are some drawbacks of the BIM technology in supporting certain aspects of collaborative design, especially in relation to design management. Holtz *et al.* (2003) point out that current implementation of the BIM system does not address key questions such as who owns the data in the model, who is responsible for updating it, and how to coordinate access and ensure security in the model (as cited in Kouider *et al.* 2007). Similarly, Darst (2003) has pointed to data ownership issues in collaboration via the BIM system which do not support synchronous collaborative design.

Another parallel research direction related to project collaboration involves the study of different virtual environments that facilitate human-to-human communication. This chapter focuses on this aspect of computer-mediated collaboration – virtual environments – and their impact on remote human communication in design related to AEC processes. In the past two decades, a variety of disciplines have participated in implementing, testing and developing information technology tools that are designed specifically for human collaboration at work, commonly known as Computer Supported Collaborative Work (CSCW) systems. These can be classified into two categories:

1   computer-mediated communication technologies (facsimile, video conferencing, email, internet, wireless and satellite video, etc.), facilitating the effective communication between the parties;
2   computer-mediated collaboration technologies (application and data-sharing, intranets, wikis, virtual worlds, etc.), facilitating a shared workspace for the joint decision-making.

Computer-mediated communication is now used by virtually all construction firms in some way, for example through facsimile and email, and most now have access to online information sources to aid their decision-making processes (Dainty *et al.* 2006). These developments have led to important advances in the enabling technologies that are required to support changes in design practice.

In contrast, computer-mediated collaboration technologies are not used by many design and construction firms. This chapter reports on research funded by the CRC, Construction Innovation Program, Project 2002–024-B 'Team Collaboration in High Bandwidth Virtual Environments', to understand the impact of collaborative design tools on design practice and to provide strategies for their uptake in the construction industry. Collaboration can be achieved in two modes: asynchronous, in which the participants are not present at the same time, and synchronous, in which the participants are working at the same time. The focus in this project was on synchronous collaboration and the different ways in which high bandwidth can deliver a shared workspace. High bandwidth synchronous collaboration offers the opportunity for AEC organizations to become globally competitive because it reduces the reliance on geographical co-presence and reduces the problems of being in locations remote from company offices. The study directly addressed the industry-identified focus as one of cultural change, image, e-project management and innovative methods (see Maher *et al.* 2005b for more detail). The benefits of computer-mediated collaboration are proposed as increased opportunities for communication and interaction between people in geographically distant locations and improved quality of collaboration.

In order to analyse and document the experience amongst members of a design team using different forms of collaboration, a series of empirical studies was conducted: the three-phase study and the DesignWorld study. The three-phase study compared designing face-to-face with remote sketching and then with collaboration in a 3D modelling environment. The experiments and the results have been reported elsewhere (Maher *et al.* 2005a, 2006a). In this chapter, we report on the development of Design-World, a multi-user 3D virtual world, and the results and implications of an empirical study of designers collaborating in DesignWorld.

Studying design behaviour

With recent developments of communication and information technologies, together with their extensive use in design, understanding collaborative design as a distinct kind of design behaviour has become necessary. This includes such factors as the role that communication media play, the use of physical materials and computer tools, and the way people communicate verbally and non-verbally (Munkvold 2003). When it comes to understanding collaborative design, an additional construct becomes the key: understanding of how design models as external representations are created and shared amongst participants.

Studies of collaborative design usually assume that the participants are in geographically distant locations. Common issues in these studies involve investigating participant communication via different communication channels; analysing the components of collective thinking and team behaviour; and analysing social behaviours such as sense of community or team, level of open participation and level of participants' awareness in the computer media.

Communication is one of the key topics in collaborative design research. This behaviour involves roles and psychological modes of team members (Stempfle and Badke-Schaub 2002), shared understanding, awareness

(Schmidt 2002) and shared language. Valkenburg and Dorst (1998) said that individual designers have to tune their personal understanding about the design content to achieve a shared understanding, and proposed a conceptual theoretical framework in order to try to capture the essence of designing within a team.

In architectural design practice, Stellingwerff and Verbeke (2001) at Delft University of Technology introduced the acronym ACCOLADE (Architectural Collaborative Design) research. They defined qualities of a good collaborative design process in terms of communication behaviour and communication environment, and also proposed how those aspects should be developed.

Gabriel (2000) concluded that different communication channels produce different collaborative environments. He studied three categories of communication for architectural collaboration: face-to-face; computer-mediated collaborative design with full communication channels (audio-video); and computer-mediated collaborative design with limited communication channels (text messaging only). He found that each category has its own strengths and difficulties. He proposed that communication channels should be selected on the basis of the type of communication considered to be most effective for the stage and task of the design project.

We add to this body of research by studying not only the impact of remote communication during collaborative design, but also how different external representations, such as 2D sketches and 3D models, impact design behaviour. We develop a prototype of a collaborative 3D virtual environment that includes 2D sketching, called DesignWorld, and evaluate it using protocol analysis (Maher *et al.* 2005b, 2006b; Gül and Maher 2006a).

DesignWorld

DesignWorld is a 3D virtual world augmented with a number of web-based communication and collaborative design tools (see Maher *et al.* 2005b for more detail). It provides both sketching and 3D modelling, as shown in Figure 8.1 (the 3D virtual world client, Second Life, is on the left, and a web-based browser interface that has a link to the 2D sketching application, Groupboard, is on the right).

DesignWorld is implemented in Second Life, which is a 3D virtual world. Maher and Simoff (2000) first characterize the design activities in 3D virtual worlds as 'Designing within the Design'. Unlike the situation in computer-aided design (CAD) systems, designers in virtual worlds are represented as avatars (animated virtual characters) that are immersed within the design. This concept has also been applied to enhance remote team collaboration in design practice (Rosenman *et al.* 2005). The 3D virtual world provides an integral platform that facilitates team collaboration through direct human-to-human communication and the synchronous creation of a shared external design representation. In addition to 3D

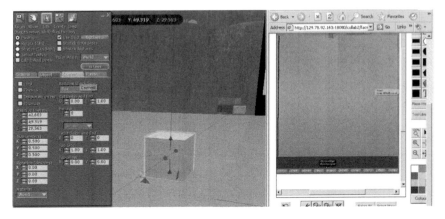

Figure 8.1 DesignWorld's interface.

modelling, DesignWorld offers a tool for collaboratively creating 2D design representations. This web-based collaborative tool (Groupboard) allows designers to communicate design ideas by sketching in addition to constructing the 3D model. The following sections highlight the basic design and communication features of DesignWorld.

Collaborative 3D and 2D design environments

Second Life supports design with a set of primitive objects whose forms are determined inside the world by selecting geometric types and manipulating their parameters. The objects are selected, designed and built in the context of a landscaped environment, as shown in Figure 8.2.

Groupboard is a set of multi-user java applets including whiteboard, chat, message board, drawing and editing tools, and file-uploading and saving on the server, as shown in Figure 8.3. The 2D drawing tools include line and shapes, colouring and hatching.

Communication and awareness

Most virtual worlds support synchronous communication by typing in a chat dialogue box. In Second Life, the text appears above the avatar's head, as illustrated in Figure 8.4. Second Life supports the presence of designers and their collaborators as avatars (awareness of self and others), uses a place metaphor (awareness of the place) and enables navigation and orientation (way-finding aids). In addition to the text-based communication features, DesignWorld augments the existing communication channels in Second Life and Groupboard with a video-audio communication channel allowing the designers to speak and to see each other in a video

Figure 8.2 The tower buildings modelled by the experiment participants.

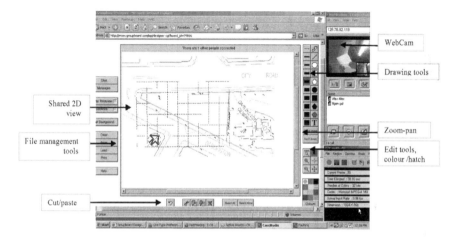

Figure 8.3 Groupboard and webcam interface.

window, as shown in Figure 8.4. While the location of the avatar provides information about what part of the 3D model the designer is looking at, the video of the person provides information about the designer's attention to the collaborative environment.

The 3D virtual world enables awareness of other designers' actions via visual feedback. For example, while a designer is modelling 3D objects in Second Life, a continuous light particle appears between the object and the designers' avatar. Thus the other designer anticipates that his or her colleague is attending to the object. In addition, DesignWorld supports collaborative creation of 3D models. In Second Life, the ownership of the objects can be flexibly arranged and shared, but one designer only can manipulate an object's properties/location at a time. In Groupboard, designers can also manipulate each others' lines and shapes.

Figure 8.4 Second Life showing avatars and text balloons on their heads, and the list of messages on the left of the screen.

Experiments in DesignWorld

In the DesignWorld study, we conducted a protocol analysis to better understand how architects collaboratively develop 3D models in response to a design brief and whether the option to work in 2D or 3D had an impact on their communication and design behaviour. We conducted three separate design sessions where three selected pairs of expert architects (from the three-phase study, see Maher *et al.* 2005a, 2006a) collaborated on a new design brief. In order to become familiar with the environment and the 3D modelling and sketching tools, the architect pairs did an initial training session in a neighbouring site. The task was to design a tower, in one hour, that includes a circulation core, a small shopping centre, a viewing area and a café/restaurant.

In order to simulate high-bandwidth audio and video, the two designers were physically in the same room and could talk to each other, but could only see each other via a webcam. The designers' activities and communication were recorded using a digital video recording (DVR) system. Typically, two cameras, two microphones and two computers were connected to the DVR, which was set to show four different views on one monitor, as shown in Figure 8.5. The design protocols comprised four continuous streams of video from different viewpoints plus audio collected for each

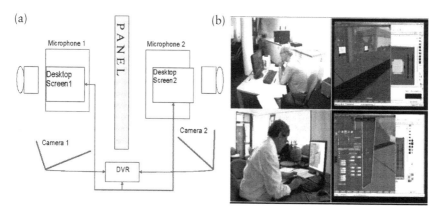

Figure 8.5 (a) Experiment set-up for the DesignWorld study; (b) DVR view of DesignWorld study.

pair's sessions, as shown in Figure 8.5. The audio data was transcribed to provide a text record of the verbal protocol. The analysis of the streams of video/audio involved segmenting the stream and then coding each segment.

We segmented the protocols based on an 'event', defined as a time interval (Dwarakanath and Blessing 1996) which begins when a new topic is mentioned or discussed, and ends when a new topic is raised. This definition is considered as the suitable one for the study, since the occurrences of designers' actions and intentions change spontaneously as they draw and communicate (see Maher *et al.* 2006c for the details on the segmentation).

A coding scheme was developed to highlight variations in the designers' verbal communication, how they manipulated the external representations, and their collective design behaviour. The codes are organized into three main categories: communication content, operations on external representations, and design problem-solving, as shown in Table 8.1.

The 'communication content' category codes the segments according to the content of the designers' conversation, focusing on the differences in the conversation topics. This category has five codes:

1 The 'software features' code captures the conversations related to how to do specific tasks with the software or problems faced during its use.
2 The 'designing' code captures the conversations on design ideas/concepts.
3 The 'awareness' code captures designer's awareness of an object/drawing or of another user.
4 The 'representations' code captures communicating a design object/ drawing to another designer, as shown in Figure 8.6.
5 The codes in the 'design problem-solving' category capture the conversations on design ideas/concepts and the analysis–synthesis and evaluation of these concepts as initially prescribed by Gero and McNeill (1998).

Table 8.1 Coding scheme

Communication content	
Software features	Software/application features or how to use such a feature
Designing	Conversations on concept development, design exploration, analysis–synthesis–evaluation
Awareness	Awareness of presence or actions of the other
Representations	Communicating a drawing/object to the other person
Context free	Conversations not related to the task
Operations on external representations	
Create	Create a design element
Modify	Change object properties or transform
Move	Orientate/rotate/move element
Erase	Erase or delete a design element
InspectBrief	Looking at, referring to the design brief
InspectReps	Looking at, attending to, referring to the representation
Design problem-solving	
Propose	Propose a new idea/concept/design solution
Clarify	Clarify meaning or a design problem, expand on a concept
AnSoln	Analyse a proposed design solution
AnReps	Analyse/understand a design representation
AnProb	Analyse the problem space
Identify	Identify or describe constraints/violations
Evaluate	Evaluate a (design) solution
SetUpGoal	Set up a goal, plan the design actions
Question	Question/mention a design issue

In order to highlight the different design behaviour in different environments, we combined some of the operation codes and the problem-solving codes into generic activity components, as shown in Table 8.2.

Create–Change

'Create' activity has different implications in different design environments; in face-to-face and remote sketching, it is drawing a line, making shapes and symbols, whereas in the 3D virtual world it is duplicating an existing object or changing an existing object to be a desired object. 'Change' activity involves carrying an object to another position or changing its properties.

Analyse–Synthesize

'Analyse–Synthesize' activity is based on Gero and McNeill's (1998) definitions of a design thinking cycle that includes analysing a problem, proposing a solution, analysing a solution and evaluating the solution. The

Table 8.2 Combined codes

Combined codes	Individual codes
Create	Create
Change	Move, Modify
Analyse	Analyse problem, Clarify, Identify
Synthesize	Propose, Analyse solution
Visual Analysis	Analyse representation, Evaluation
Manage Tasks	SetUpGoal, Question

'Analyse' code takes place in the problem space and includes the 'Analyse' problem, the 'Clarify' and the 'Identify'; the 'Synthesize' code takes place in the solution space and includes the 'Propose' and the 'Analyse Design Solution' codes.

Visual analysis

Visual analysis is purely dependent on the representation: judgements of what it should look like, how elements come together, designers' preferences on constructing it and so on. Visual analysis involves seeing or imagining what the object looks like in 3D, so the 'Analyse representation' code is included in this activity. The 'Evaluate' code is also included because we observed that evaluation was mostly based on visual analysis.

Manage tasks

The 'Manage Tasks' category refers to planning future design actions and leading the collaboration partner towards the goals to make the design. Questioning each other about design issues or knowledge is also involved in this activity. 'Manage Tasks' includes the following attributes from the coding scheme: 'setting up a goal' and 'questioning'.

Evaluation of collaborative behaviour in DesignWorld

The video/audio data from the DesignWorld study were segmented and coded. We documented how much time each participant spent on each action and category in each phase, and then compared them across the design environments using protocol analysis (see Maher *et al.* 2005a, 2006a, 2006c for more details of the results). We used the software INTERACT (Interact 2006) for our coding and analysis process (see Figure 8.6); more information on the reasons for choosing this software and how it improved our coding process can be found in Bilda *et al.* (2006).

In our evaluation of DesignWorld, we highlight four results for designers:

Figure 8.6 Video coding and analysis using Interact.

1   their focus on designing;
2   their longer intentional segments;
3   their focus on modelling;
4   their creation of new ideas in sketches and constructing in 3D.

First, the main communication topic is designing followed by software features and awareness in the DesignWorld study, as shown in Figure 8.7. Our studies have shown that designers are able to adapt to different environments, from the traditional face-to-face environment to a variety of virtual environments, and still effectively communicate and collaborate (see also Maher *et al.* 2005a, 2006a for more details of the three-phase study results). Strategically this is an important finding, because it implies that the introduction of high-bandwidth virtual environments into the design process preserves the essential aspects of designing, and allows designers to effectively communicate and collaborate while in remote locations.

Second, our studies show that the attention/intention shifts are further apart in time when designers move from face-to-face, to remote sketching, and to the 3D virtual worlds. The descriptive statistics for the segment durations for three pairs in the DesignWorld study are shown in Table 8.3. The mean duration of the segments is 10–13 seconds, and the long segment durations are observed (65–80–85 seconds). This result also reinforces our

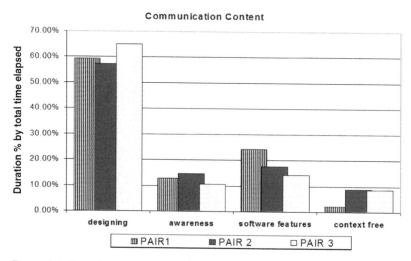

Figure 8.7 Duration percentages of Communication Content actions.

Table 8.3 Duration of segments while designing in DesignWorld

(Second)	Mean	Standard deviation	Kurtosis	Skewness	Min.	Max.	Count
Pair 1	13	9	9.28	2.35	1	65	228
Pair 2	10	8	25.08	3.78	1	80	288
Pair 3	10	9	13.89	3	2	85	259

three-phase study findings – that is, in the 3D world sessions, the designers spent more time in each segment before they became engaged in a new action or idea (average 63 seconds; see Maher *et al.* 2005a, 2006a for more details). These findings suggest that designing in virtual worlds requires a relatively long time for attending to an action or object. As a result of the precise modelling and positing of objects, the designers need to spend more time on each of the actions, due to the nature of modelling in 3D.

Third, we observed that the designers engaged with the solution rather than framing the design problem and modelling of the artefact in the Design-World study. The duration percentages of the 'synthesize' and 'manage task' actions were higher, and the duration percentages of the 'analyse' and 'visual analysis' actions were lower, as shown in Figure 8.8. This result suggests that designers had engaged with the development of the design model by proposing solutions and synthesizing them in DesignWorld as well as how/who to model. The collaborative creation of the design model requires the management of tasks, including sharing the model creation and assisting each other on the spatial adjacency.

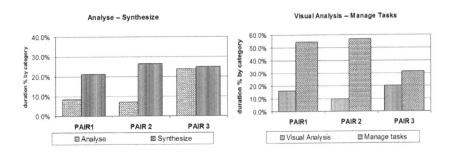

Figure 8.8 Duration percentages of Analyse–Synthesize and Visual Analysis–Manage Tasks activities.

The fourth category of impact has to do with the differences in the virtual environments studied. These differences are basically whether the designers were able to represent their design ideas/solutions in a 2D sketch representation or a 3D virtual world environment. We found that the major difference was that they focused more on creating new design depictions while using a 2D sketch (observed in the first ten minutes only); and more on constructing a design model while using 3D virtual world in the Design-World study, as shown in Figure 8.9. The combined coding category (Create–Change) and the representation category (2D–3D) are shown along the timeline of the sessions. The beginning of the session is on the left, and the length of each horizontal bar indicates how long the designer spent on each operation. Each designer's actions are coded separately, indicated by the numbers 1 and 2. For all pairs, the Create activity is lower than the Change activity in the DesignWorld study. This demonstrates that the designers were engaged more with the change activity (move, modify and transform actions) than creating new objects in DesignWorld. Our previous studies also showed similar Change behaviour in 3D virtual worlds (see also Gül and Maher 2006b; Maher *et al.* 2006a).

Conclusions

In conclusion, our evaluation of DesignWorld shows how virtual environments impact design behaviour, and highlights the advantages of collaborating in virtual environments. In this section we revisit the four main results to comment on their impact and interpretation.

Designers focus on designing

This result is significant and encouraging. While there is a risk in adopting new technologies for collaboration, our studies have shown that designers adapt to their environment and ultimately focus on the design task. We

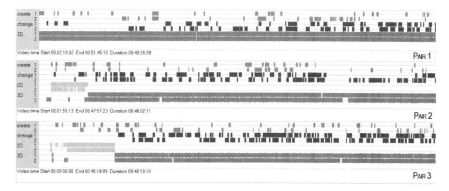

Figure 8.9 Timeline showing 2D–3D representation modes and Create–Change actions in the DesignWorld study.

noticed a decrease in the amount of time spent talking specifically about the design when compared to face-to-face designing, but this other time was spent assessing their awareness of the other designers' location in the 3D world and discussing software features. This time is not wasted, and adds to the effectiveness of the designers' activities in the virtual world. While the face-to-face design behaviour did not include this digression from the design task, the result of the face-to-face design did not include a 3D model of the result of the collaboration. Both environments encourage different types of designing: exploring design concepts and requirements (demonstrating high level of design exchanges on paper) is becoming the main focus in face-to-face, and developing a particular design idea in detail is the concern in 3D. The resulting sketches from the face-to-face collaboration were not nearly as clear or useful for future deliberations.

Designers have longer intentional segments

When the designers were using DesignWorld, it took longer for each design intention to be completed. This implies that each design intention or action was more complex, involving both the expression of an idea and the transformation of it into a new or changed 3D model. When designing in a 3D model of the design, it is possible not only to suggest a new idea or change, but also to demonstrate the change and assess it visually. The sketching environments were used very differently: to express an idea visually, it was more common for the designers to create a new line or shape. In the 3D virtual world, a new idea was typically realized by editing an object that had already been created. (Creating a model starts with copying/inserting a basic geometry from the inventory in SecondLife. This usually takes just one command. Then the user needs to modify/alter the size, shape and other properties of the object, before transferring/moving it to its place.) The result of the designers' focus on changing rather than creating in 3D is an external representation that reflects the final version of the design where a sketch may illustrate many stages of the collaborative discussion.

Designers focus on modelling

In DesignWorld, the designers spent more time modelling the geometric properties of the design than discussing the design brief or developing alternative design concepts. This is an important difference in the kind of design behaviour to be expected in 3D virtual worlds. The 3D environment is compelling as a realistic simulation of the design in context. Abstractions are hard to achieve in the modelling tools, and therefore hard to visualize. This has advantages and disadvantages in a design session. The 3D virtual world may not be a good place for brainstorming alternative ideas, but it is superior in pursuing the implications of design decisions as a simulation. This kind of simulation is not possible in a sketching environment.

Designers create new ideas in sketches and construct in 3D

This result is not surprising, and is consistent with our three-phase study. It is the main reason that we developed DesignWorld as an augmented 3D virtual world that supports sketching, rather than use an existing 3D virtual world. Having access to a sketching environment that maintains the sketch as a reference for constructing the 3D model is very different to having a sketching tool that automatically transforms to a 3D model, as in the tool SketchUp (www.sketchup.com/). The lines and shapes in a sketch are ambiguous and allow multiple interpretations, supporting the rapid generation of design ideas. The 3D model supports a visualization of the realization of the design as an object.

In summary, the development and evaluation of DesignWorld has shown that collaborative design can be effective in a remote collaborative environment. The differences in design behaviour indicate that virtual environments not only allow communication and collaboration among designers who are remotely located, but also enhance the collaboration by providing a shared modelling and simulation environment that adds value to the collaborative process.

Acknowledgements

The research was developed with the support of the Cooperative Research Centre for Construction Innovation in the University of Sydney, with the cooperation of industry partners Woods Bagot Pty Ltd, Ove Arup Pty Ltd and the CSIRO, Melbourne. The authors wish to thank Zafer Bilda for his contribution to the development of the experimental set-up, the coding scheme and the original evaluation of the results. They also thank the project partner companies and the research team: Zafer Bilda, Ning Gu, Mijeong Kim and David Marchant. This chapter was written while Mary Lou Maher was working for the National Science Foundation in the USA. Any opinions, findings, recommendations or conclusions expressed in this chapter are those of the authors and do not necessarily represent the views of the National Science Foundation.

Bibliography

Archea, J. (1987) 'Puzzle-making: what architects do when no one is looking', in Y.E. Kalay (ed.) *Computability of Design*, New York, NY: Wiley Interscience.

Bilda, Z., Gül, L.F., Gu, N. and Maher, M.L. (2006) 'Software support for collaborative data analysis in collaborative design studies', in A. Ruth (ed.) *Quality and Impact of Qualitative Research: 3rd annual QualIT Conference*, Brisbane: Institute for Integrated and Intelligent Systems, Griffith University.

Cohen, J. (2003) *The New Architect: Keeper of Knowledge and Rules*, Berkeley, CA: Jonathan Cohen and Associates. Online. Available at HTTP: <www.jcarchitects.com/New_Architect_Keeper_of_Knowledge_and_Rules.pdf>.

Dainty, A., Moore, D. and Murray, M. (2006) *Communication in Construction, Theory and Practice*, Abingdon: Taylor & Francis.

Darst, E. (2003) *The Nature of AEC Content as Applied to Building Information Modeling*. Online. Available at HTTP: <www.aecnews.com/online/Back_issues/2003/aec2003–07/05-expertsview.html>.

Dwarakanath, S. and Blessing, L. (1996) 'The design process ingredients: a comparison between group and individual work', in N. Cross, H. Christiaans and K. Doorst (eds) *Analysing Design Activity*, Chichester: John Wiley and Sons.

Gabriel, G.C. (2000) *Computer Mediated Collaborative Design in Architecture: The Effects of Communication Channels on Collaborative Design Communication*, PhD thesis, Sydney: Architectural and Design Science, Faculty of Architecture, University of Sydney.

Gero, J.S. and McNeill, T. (1998) 'An approach to the analysis of design protocols', *Design Studies*, 19(1): 21–61.

Gül, L.F. and Maher, M.L. (2006a) 'Studying design collaboration in DesignWorld: an augmented 3D virtual world', in E. Banissi, M. Sarfraz, M. Huang and Q. Wu (eds) *Proceedings of the 3rd International Conference on Computer Graphics, Imaging and Visualization Techniques and Applications (CGIV'06)*, Los Alamitos, CA: IEEE Computer Society.

—— (2006b) 'The impact of virtual environments on design collaboration', in *24th eCAADe Conference Proceedings*, Volos, Greece.

Holtz, B., Orr, J. and Yares, E. (2003) *The Building Information Model*, Bethesda, MD: Cyon Research Corporation. Online. Available at HTTP: <http://cyonresearch.com>.

Interact (2006) *Interact User Guide*, Arnstorf: Mangold Software and Consulting GmbH.

Kalay,Y., Khemlani, L. and Jinwon, C. (1997) 'An integrated model to support collaborative multi-disciplinary design of buildings', *1st International Symposium on Descriptive Model of Design*, Istanbul.

Kouider, T., Paterson, G. and Thomson, C. (2007) 'BIM as a viable collaborative working tool: a case study', in *CAADRIA 2007: Proceedings of the 12th International Conference on Computer-Aided Architectural Design Research in Asia*, Nanjing.

Maher, M.L. and Simoff, S. (2000) 'Collaboratively designing within the design', in *Proceedings of Co-Designing 2000*, London: Springer-Verlag.

Maher, M.L., Ahmed, A., Egan, S., Macindoe, O., Marchant, D., Merrick, K., Namprempree, K., Rosenman, M. and Shen, R. (2005b) *DesignWorld: A Tool for Team Collaboration in High Band Virtual Environments*, Sydney: University of Sydney.

Maher, M.L., Bilda, Z. and Gül, L.F. (2006a) 'Impact of collaborative virtual environments on design behaviour', in J. Gero (ed.) *Design Computing and Cognition '06*, Dordrecht: Springer.

Maher, M.L., Bilda, Z., Gu, N., Gül, L.F., Huang, Y., Kim, M.J., Maher, M.L., Marchant, D. and Namprempree, K. (2005a) *Collaborative Processes: Research Report on Use of Virtual Environment*, Sydney: University of Sydney.

Maher, M.L., Bilda, Z., Gül, L.F., Yinghsiu, H. and Marchant, D. (2006c) 'Comparing distance collaborative designing using digital ink sketching and 3D models in virtual environments', in K. Brown, K. Hampson and P. Brandon (eds) *Clients Driving Construction Innovation: Moving Ideas into Practice*, Brisbane: CRC for Construction Innovation.

Maher, M.L., Rosenman, M., Merrick, K. and Macindoe, O. (2006b) 'Design-World: an augmented 3D virtual world for multidisciplinary collaborative design', in *Proceedings of CAADRIA 2006*, Osaka.

Munkvold, B. (2003) *Implementing Collaboration Technologies in Industry: Case Examples and Lessons Learned*, London: Springer-Verlag.

Rosenman, M.A., Smith, G., Ding, L., Marchant, D. and Maher, M.L. (2005) 'Multidisciplinary design in virtual worlds', in *Proceedings of CAAD Futures, 2005*, Dordrecht: Springer.

Schmidt, K. (2002) 'The problem with "awareness"', *Computer Supported Cooperative Work*, 11: 285–98.

Stellingwerff, M. and Verbeke, J. (eds) (2001) *Accolade: Architecture – Collaboration – Design*, Amsterdam: Delft University Press.

Stempfle, J. and Badke-Schaub, P. (2002) 'Thinking in design teams: an analysis of team communication', *Design Studies*, 23: 473–96.

Valkenburg, R. and Dorst, K. (1998) 'The reflective practice of design teams', *Design Studies*, 19: 249–71.

9 The challenges of environmental sustainability assessment

Overcoming barriers to an eco-efficient built environment

Peter Newton

This chapter discusses issues and challenges associated with environmental sustainability assessment of the built environment. Such assessment needs to be undertaken within a commonly agreed framework from which performance objectives and criteria can be established and measured. The chapter outlines eight key challenges for environmental sustainability and eco-efficiency assessment of the built environment. It is in this context that progress towards *sustainability-oriented regulation in building and planning* must be addressed.

Although the Building Code of Australia adopted 'sustainability' as one of its goals in 2007 (BCA 2007), work on assessing the impacts of building construction on sustainability and the role of the BCA in delivering more sustainable buildings started in 2002 with the CRC for Construction Innovation report *Sustainability and the Building Code of Australia* (Pham *et al.* 2002). In that report, a variety of tools for sustainability assessment were described. More recently, the Australian Greenhouse Office has commissioned a *Scoping Study to Investigate Measures for Improving the Environmental Sustainability of Building Materials* (Centre for Design at RMIT University *et al.* 2006), and a *Scoping Study to Investigate Measures for Improving the Water Efficiency of Buildings* (GHD 2006). These two studies focused specifically on material and water usage respectively, and addressed the broader environmental issue of resource depletion. The Australian Building Codes Board has subsequently commissioned a *Study into the Suitability of Sustainability Tools* as part of a National Implementation Model and this constitutes the latest of a number of reviews of existing environmental assessment tools (e.g., Foliente *et al.* 2007; Arup Sustainability 2004; Bernstone 2003).

The objective of the National Implementation Model is to delineate and harmonize planning and building controls, to address climate change and sustainability in the built environment through a national action framework, and to make recommendations to the appropriate regulatory authorities. Of particular interest is that the *National Action Plan for Urban Australia* calls for the development of indicators of sustainability performance as a basis for reporting on specific urban development plans. The House of Representatives Standing Committee on Environment and Heritage

(2007) Inquiry into a Sustainability Charter likewise calls for the establishment of performance targets for the built environment and assessment tools for measurement and verification of performance, within a broader national sustainability charter. This chapter therefore will focus on key challenges related to environmental sustainability assessment and avenues to solutions as a necessary precursor to more extensive regulation of built environment performance.

The chapter begins by establishing the environmental imperative for interventions to ensure a more sustainable built environment. It then proceeds to discuss the eight 'bridges' that need to be crossed to deliver higher levels of eco-efficiency performance of our buildings and urban infrastructure.

The environmental imperative

There are several global and local environmental challenges that are intensifying in their impact on future urban development in Australia (discussed in detail in Newton 2008). The *global* environmental challenges include:

- *climate change*, linked to escalating concentrations of greenhouse gas (GHG) in the Earth's atmosphere and international treaties post-Kyoto that will usher in a *carbon-constrained future* (Hennessy 2008). Australia, as a world-leading generator of GHG emissions (approximately 27t/p/y) and exporter of fossil fuels, will need to confront its energy future more radically. Climate change represents a major new source of risk to property and urban functioning (Abbs 2008).
- *resource depletion*, linked primarily to escalating levels of resource consumption in advanced industrial countries (AICs) and newly developing countries (NDCs). Australia is again a leader in resource consumption (Newton 2006). With Australia's ecological footprints being of the order of 7 to 8 hectares per person, compared to 2 ha per person globally, four planet Earths would be required to sustain a global population with Australia's consumption profile. Resource depletion is expected to have its earliest and most dramatic impact on urban development in relation to oil (Dodson and Sipe 2008).
- *urbanization*, with 2010 being the year when more than 50 per cent of the world's population are expected to be living in urban as distinct from rural settings; by 2030 this will reach 60 per cent (United Nations 2007). Consequently, how cities and their built environments are planned and operated from an environmental sustainability perspective has increasing relevance. A significant proportion of the resource consumption and environmental pollution associated with cities in AICs and NDCs is actually designed into their cities and housing: 'where people live within a city and the types of dwelling they occupy will exert an impact over and above that of an individual's discretionary consumptive behaviour' (Newton 2007: 571).

Australia's environmental challenges include the following.

Water

Households account for 11 per cent of water consumption in Australia; the average household consumed 268 Kl in 2005, ranging from 209 and 219 in Victoria and New South Wales to 323 and 468 in Queensland and Western Australia (ABS 2006). Annual per capita consumption varies from 81 Kl in Victoria to 180 Kl in Western Australia. Variability in the level of water use across municipalities and socio-economic groups is significant, and Australian households consume more water than their European counterparts (OECD 2002).

A key issue for water consumption is in relation to the limits of natural supplies, namely the sustainable yield of water for Australia's urban households. Here, the statistics are challenging; water storage levels for many cities are critically low – 37 per cent (Sydney), 29 per cent (Melbourne), 18 per cent (Brisbane), 21 per cent (Perth), 31 per cent (Canberra), 10 per cent (Ballarat) and 14 per cent (Toowoomba) (Spurling *et al.* 2007).

In the face of projected growth in population and development in urban Australia, challenges exist on the supply side for transition to a portfolio of urban water systems capable of delivering a sustainable supply of water suited to end use and, on the demand side, increased efficiency of use (Inman 2008). Understanding where water is used in buildings by occupants assists in focusing opportunities for innovation in technology, policy and design. In Australia, urban water consumption by end use is in the following ranges (GHD 2006):

- Outdoor: 25% (NSW) to 55% (ACT)
- Bathroom: 15% (SA) to 26% (NSW)
- Toilet: 11% (WA) to 23% (NSW)
- Laundry: 10% (Qld) to 16% (NSW)
- Kitchen: 5% (Vic) to 10% (NSW).

Built environments and human behaviour represent the dual (linked) challenges to resource conservation. Pathways to enhanced water outcomes in Australian built environments are well summarized in GHD (2006, 2007), Spurling *et al.* (2007) and Newton (2008), and involve planning (e.g., in relation to outdoor water uses), building (indoor uses and embodied water) and lifestyle choices.

Materials and waste

Over the past decade, the volume of solid-waste generation in Australia has risen to more than 1.6 tonnes per person, placing Australia as one of the OECD 'leaders' in this area of environmental performance (Productivity

Commission 2006). The construction and demolition waste stream is the largest, at 13.7 million tonnes (42 per cent of the total). Currently, 46 per cent of waste is recycled, and there is significant variation across material streams (Newton 2006).

A recent study of materials use in Australia's built environment (Centre for Design at RMIT University *et al.* 2006) identified the criticality of resource recovery from waste streams in the light of a forecast increase in volume of material use of 40 per cent over the next 50 years, together with a 63 per cent increase in water use and a 40 per cent increase in global warming potential linked to building materials provision.

Energy

From 2004–05 to 2010–11, Australia's energy consumption is projected to grow at 2 per cent per year (ABARE 2006). Fossil fuels provide around 95 per cent and renewables 5 per cent of the energy supplied – reflected in Australia's high level of GHG emissions. The built environment accounts for almost 60 per cent of final energy consumption (comprising 39 per cent transport, 7 per cent commercial buildings and 12.5 per cent residential). The residential and commercial sectors are forecast to be the two largest contributors to electricity consumption growth through to 2030 based on current settings (ABARE 2006).

The drive for more energy-efficient buildings and appliances, underway in Australia since the first oil shock of the early 1970s, did not begin to gain traction until there were scientifically validated methods and technologies available for assessing building energy performance, and performance targets were enshrined in regulations and standards. Energy efficiency requirements for new residential buildings became operational in the BCA from 2003 for housing and 2005 for multi-residential buildings, with an equivalent system for new commercial buildings commencing in 2006. An international comparison of building energy performance standards (Horne *et al.* 2005), however, found that Australia's 5-star standard was of the order of 2 to 2.5 stars below comparable average international levels of performance for housing. Furthermore, Australia's new regulations for insulation established R-value targets at around half that of the USA's minimum standards (Ambrose 2008).

Towards an eco-efficient built environment

This environmental scorecard across water, materials and energy use would confirm the conclusions of both the 2001 and 2006 *Australian State of Environment: Human Settlement Reports* (Newton *et al.* 2001; Newton 2006), that levels of resource consumption in urban Australia are unsustainable. How buildings are designed (including design for disassembly), the selection of materials (in the context of LCA and service life performance), the assembly process, facility operation and successive refurbishment processes all contribute significantly to this current state. Transitioning to a more environmentally sustainable built environment will involve bridging from the current state of play to a future which represents significantly higher levels of *eco-efficiency performance*. This is achievable by:

- bridging the political divide;
- bridging the stakeholder divide;
- bridging the property life cycle divide;
- bridging the building and planning divide;
- bridging the divide to renewable and recyclable resources;
- bridging the as-built versus as-operated divide;
- bridging the digital divide;
- bridging the economic and environmental divide.

Bridging the political divide

It is over 20 years since the Brundtland Report (United Nations 1987) was published and became a catalyst for many governments to begin grappling with the concept of sustainability as a core operating principle for urban planning and management.

In Australia, however, the past decade has witnessed a retreat from leadership at federal government level in areas related to cities and the built environment from the perspective of sustainable development. The *Sustainable Cities* report by the House of Representatives Standing Committee on Environment and Heritage (2005), containing 32 recommendations, did not elicit a formal response from the Howard government (1996–2007), which viewed urban development primarily as a state government responsibility. This stands in marked contrast to the visionary approach taken by President John F. Kennedy, who commented in his 1962 speech to create the US Federal Department of Housing and Urban Development (which continues to operate to this day) that 'We will neglect our cities to our peril, for in neglecting them we neglect the nation' (http://home.att.net/~jrhsc/jfk.html).

Instead, a further *Inquiry into a Sustainability Charter* was initiated by the House of Representatives to

> report on key elements of a sustainability charter and identify the most important and achievable targets, particularly in relation to: the built environment, water, energy, transport, and ecological footprint.... The charter should be aspirational. It must provide targets for the Australian community to meet and, once these targets have been met they must be re-assessed so new targets can be put in place.
>
> (House of Representatives Standing Committee
> on Environment and Heritage 2006)

This inquiry (House of Representatives Standing Committee on Environment and Heritage 2007) submitted its report in November 2007. Missing, however, were any attempts at specifying the scope and requirements of *measurement systems* capable of assessing the performance of the built environment. This is in relation to both targets and a specification for performance measurement systems that need to be acceptable to state and local government, the scientific community, industry and the residents in different urban regions, all of which are central to achieving sustainable development.

In the absence of leadership in relation to sustainable urban development at federal level, there has been an explosion in the growth of urban sustainability guidelines and 'assessment systems', covering:

- local government (e.g., City of Manningham's Doncaster Hill Sustainability Guidelines, <www.doncasterhill.com/Resources_Key-Documents_Sustainability.htm>);
- development authorities (e.g., Docklands ESD Guide, <www.docklands.com/cs/Satellite?c=VPage&cid=1182927645003&pagename=Docklands%2FLayout>);
- state governments (e.g., NSW Basix Building Sustainability Index; ABGRS; NABERS; <www.deus.nsw.gov.au/Sustainability/Sustainability.asp>);
- industry (e.g., Green Building Council of Australia's Green Star environmental rating system for buildings, <www.gbca.org.au/green-star/>; Urban Development Institute of Australia's EnviroDeveloper, <www.envirodevelopment.com.au/>);
- global initiatives (e.g., ICLEI – Local Governments for Sustainability, <www.iclei.org/>).

A challenge for harmonizing urban development eco-efficiency assessment across jurisdictions relates primarily to:

- the *weighting* or level of importance that needs to be assigned to different environmental domains in different regions – for example, urban water cycles vary markedly across regions and settlements in Australia (Mitchell *et al.* 2003), as does climate. Here harmonization of *State of Environment* reporting by all three levels of government could make a fundamentally important contribution to creating databases capable of being used to set regional *targets* for environmental performance assessment.
- the political will to assign *targets* and *timeframes* commensurate with the level of environmental damage that is likely to occur in the absence of a change in behaviour from 'business as usual' (targets for CO_2 reduction and water savings are among the most important). In the absence of national leadership, it increasingly appears to be the

mayors of larger cities such as London and Melbourne who are taking leadership (City of Melbourne 2003; Ecojustice 2007). As the Mayor of Toronto put it: 'I think cities are the leaders. We're not waiting for federal and provincial governments to act – they take far too long. We're just acting. Federal policies, frankly, need to fall into line with ours' (*Toronto Star*, 23 May 2007).

Federal and state policies are all too frequently captive to the short-term nature of electoral cycles in Australian politics, which is especially problematic for challenges that are complex and intergenerational in nature, such as sustainable development, that require longer-term Horizon 3 thinking and planning and investment (Newton 2007; and Chapter 1).

Bridging the stakeholder divide

Understanding that there are multiple *stakeholders* involved in urban development, each with their own set of motivations and *value propositions* that need to be satisfied in order to embrace eco-efficiency as a key principle of business thinking, represents a key step in the sustainability transition. To ignore this fact risks inhibiting the diffusion of innovations in financing, procurement, design, construction and facility management that can drive sustainability in the property and construction sector. Key stakeholders in this sector include owners, occupiers, developers, designers, managers, investors, regulators and the public – and each will have a 'business proposition' for every development project they have a significant stake in (e.g., GBCA 2006). The Australian Sustainable Built Environment Council (ASBEC), which is a representative group of leading property development stakeholders, is developing a business case template (Table 9.1), and the Your Building portal www.yourbuilding.org/, created jointly by the CRC for Construction Innovation, the Australian Greenhouse Office (AGO) and ASBEC, is using a stakeholder group classification as a primary filter for the information base it is assembling on sustainable building knowledge and practice.

Eco-efficiency performance assessment methods and tools consequently need to have their outputs tailored in a manner that informs the different stakeholder groups on the KPIs that are of critical importance to each of them for a particular building or development project.

Bridging the property life cycle divide

For construction to be considered sustainable in the manner that manufacturing has articulated for itself (Kaebernick *et al.* 2008) requires a *cradle-to-cradle perspective*, as outlined by McDonough and Braungart (2002), where there is a closed loop from product manufacture through assembly and operation to end of life. Product stewardship is a more straightforward proposition with single manufactured products, such as automobiles,

Table 9.1 ASBEC business case value factors

	Owners	Occupiers	Developers	Designers	Managers	Public	Investors
Commissioning, operating and maintenance: reduced costs							
Energy savings	✓✓			✓	✓✓	✓✓	✓✓
Reduced capital costs of mech. system, as control systems reduce need for oversizing	✓				✓		✓
Emissions reduction	✓✓✓	✓✓✓			✓✓✓	✓✓✓	✓✓✓✓
Water savings	✓✓✓	✓✓✓					✓✓✓
Waste reduction and disposal savings	✓✓✓	✓✓✓					✓✓✓
Reduced development costs							
Accelerated planning approval process	✓✓✓	✓✓✓✓✓	✓✓✓✓✓	✓			
Lower carrying costs		✓✓✓✓✓	✓✓✓✓✓				
Compressed schedule		✓✓✓✓✓	✓✓✓✓✓				
Improved occupant productivity	✓✓			✓	✓✓		✓
Lower churn, turnover, tenant inducements	✓✓				✓✓✓✓✓		✓✓✓✓✓
Reduced occupant complaints	✓✓✓		✓✓				✓✓✓✓✓
Health and OH&S	✓✓✓			✓			✓✓✓✓✓
Sick building syndrome	✓✓✓		✓✓		✓✓✓✓✓		✓✓✓✓✓
Green premium (higher return on asset)	✓✓✓	✓✓✓✓✓		✓	✓✓✓✓✓	✓✓	✓✓✓✓✓
Improved corporate profile and community relations	✓✓✓	✓✓✓✓✓		✓	✓✓✓✓✓	✓✓✓	✓✓✓✓✓
Living corporate values through building asset	✓✓✓	✓✓✓✓✓		✓	✓✓✓✓✓		✓✓✓✓✓
Enhanced marketability	✓✓✓			✓	✓✓✓✓✓		✓✓✓✓✓
Enhanced publicity	✓✓✓			✓	✓✓✓✓✓		✓✓✓✓✓
Ability to attract and retain employees	✓✓✓				✓✓✓✓✓		✓✓✓✓✓
Tenancy benefits	✓✓✓				✓✓✓✓✓		✓✓✓✓✓
Ability to attract and retain tenants	✓✓✓				✓✓✓		✓✓✓✓✓
Reduced vacancy rates	✓✓✓				✓✓✓		✓✓✓✓✓
Higher ROI, for gross leases	✓✓✓				✓		✓✓✓✓✓
Financial incentives (accelerate depreciation, etc.)	✓✓✓		✓				✓✓✓✓✓
Insurance	✓✓✓			✓✓✓✓			✓✓✓✓✓
Reduced liability and risk	✓✓✓			✓✓✓✓		✓✓✓	✓✓✓✓✓
Future-proof buildings	✓✓✓			✓✓✓✓		✓✓✓	✓✓✓✓✓
Avoided energy infrastructure investment	✓✓✓				✓✓✓		✓✓✓✓✓
Regulatory requirement	✓✓✓						✓✓✓✓✓
Ethical investment funds and cap rates	✓✓✓						✓✓✓✓✓
Performance disclosure and building ratings	✓				✓	✓✓	✓✓✓✓✓

Source: based on ASBEC (2007).

computers and white goods. For the property and construction sector, the process is more complex but not insurmountable. The building blocks are being assembled.

The key platform is likely to be a national life cycle inventory database which will contain the environmental performance signatures (material use; emissions to land, water and air; human health impacts) of all building materials manufactured in or imported to Australia (ALCAS 2007). Additional dimensions of material performance will need to be added that relate to price; service life performance; and design for disassembly, reuse and recycling – providing the basis for a cradle-to-cradle eco-efficiency assessment of the 'building blocks' of the built environment based on individual building objects and their lifetime performance. At present, LCADesign is the only tool that can assess the eco-efficiency performance of building or infrastructure designs from an 'object' (or element) perspective as an automated process during the design process (Seo *et al.* 2008, Chapter 10 in this volume). It is the only tool capable of segmenting analysis in accordance with the principal *layers of a building* (Duffy 1989, cited in Crowther 2001) and their respective lifespans, namely the shell (foundations, structures: 50 years), the services (electrical, hydraulic, HVAC, lifts: 15 years), the scenery (internal partitioning, finishes: five to seven years) and the sets (movable items such as furniture). And each layer is of particular interest to different architectural, engineering and construction practitioner groups.

Bridging the building and planning divide

Building and planning should be mutually reinforcing in delivering maximum possible eco-efficiency outcomes for urban development and redevelopment. The principles of sustainable building and planning would suggest this, but it is not the case in practice.

BREEAM (Curwell and Spencer 1999) was the first system that sought to develop energy profiles for enterprises that captured embodied energy in building materials, operating energy of the building in use, and transport energy linked to the journey to work of employees as well as the daily travel undertaken by employees during business hours to service the needs of enterprise operations. In effect, it was a building/enterprise scale version of land-use–transport–environment modelling. The concept has come unstuck when other building ratings/assessment systems that have drawn on the BREEAM philosophy have allowed 'points' gained for a building that is well located in relation to public transport to be effectively traded off against the quality of building designed for that site. A 6-star rating for a *building* should be reflective of the attributes of that building, and not of its relative location within a city.

The challenge of developing an integrated assessment, as outlined by ISO in Figure 9.1, relates to establishing key linkages. For buildings, it is with the material products that are embodied in the design of the structures, and

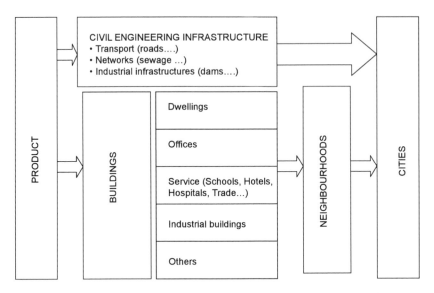

Figure 9.1 The urban sustainability framework: products, buildings, infrastructure, neighbourhoods and cities (source: ISO/TC 59/SC 17/N 172, 2005–11–02, ISO/CD2 15392, ISO/TC 59/IC 17/WG 1 (doc N041) Sustainability in Building Construction – General Principles).

the linkages of a building to key neighbourhood infrastructures (Newton 2002). For example:

- *Energy*. The relatively benign nature of energy performance requirements in current building regulations (insulation can almost single-handedly deliver the required ratings) has meant that builders and designers have not needed to exert pressure on urban planners to produce more energy-efficient *subdivisions* (Ambrose 2008). Also, there is an absence of guidelines that can be used by those who want to evaluate the possibilities of integrating *energy generation* into their building proposal, and to sell excess electricity onto the grid.
- *Water*. The issue of water capture, storage and reuse *on site* versus neighbourhood/subdivision scale remains to be fully scoped, modelled and costed as a basis for guiding future urban development and redevelopment in directions other than the current linear system of divert, dam, distribute, use and dispose to receiving waters (see Diaper *et al.* 2008 and Maheepala and Blackmore 2008 for most recent progress).

Bridging the divide to renewable and recyclable resources

Urban development in the mid- to late twentieth century was based on a paradigm of resource availability linked to land, water, energy, minerals and forest products. In the twenty-first century, the challenge will be achieving sustainable development in a *carbon-constrained and resource-constrained world*, where much greater focus will be on *renewable resources* as well as achieving maximum recovery, reuse and recycling of those resources that are already part of the existing urban fabric (estimated to be of the order of one-third of total resources, with a further third in landfill).

Driving energy efficiencies across all sectors of industry can make significant contributions to GHG reductions, as can a shift to more compact cities and more integrated mixed land use–transport development. But without a concerted shift to incorporating *renewable energy* as well as *energy efficiency* objectives into urban development planning and design, targets of 60 per cent reduction in CO_2 by 2050 are unlikely to be achieved (much less the higher targets foreshadowed by the Garnaut Climate Change Review 2008).

Avenues for achieving higher levels of renewable energy generation and use can be via large-scale developments, e.g., biofuel production, wind farms, large-scale solar plants (Newton 2008), that can substitute for the capacity of centralized coal-based plants as they are progressively removed from the grid, as well as smaller-scale *distributed energy generation* (DEG) that supplies energy for individual properties or neighbourhoods, with capacity for grid connection (DEG capacity in Australia is currently 2 per cent versus 7 per cent in Europe (Jones 2008)). The London Borough of Merton (2004) has adopted a policy ('The Merton Rule'), which is to be extended to all UK local governments, requiring all new non-residential developments above $1,000\,m^2$ to incorporate renewable energy systems capable of generating at least 10 per cent of predicted energy use. It is an area requiring coordination between building and planning, but in the UK context, approval via the building route is viewed as being too inflexible and slow, with most councils favouring the planning route.

In the UK, where the shift to renewables is most advanced in the context of integration with the built environment – as a result of government renewables targets – there are good examples of advanced thinking in relation to DEG and master planning, including DEG toolkits for planners that discuss design and planning issues involving PV, biomass, wind, and how to navigate the application, design, construction and verification processes. For large-scale renewable energy projects, key challenges are primarily land-use planning-related, and involve securing and (re)zoning sufficient areas of land to support viable projects (Newton and Mo 2006).

For Australia's building sector to respond appropriately to the greenhouse and peak-oil challenges will require a new approach which *combines* design for energy efficiency (building shell plus appliances) with distributed energy generation via a range of micro-generation options (which include solar photovoltaics, wind turbines, solar thermal hot water, ground source heating/cooling, bio-energy, micro-combined heat and power, fuel cells), creating what we term 'hybrid buildings'.

This bridge to renewables must also accommodate other key resource transitions related to water (i.e., from linear 'divert–distribute–use–dispose to receiving waters' to closed loop systems which integrate treated stormwater and wastewater) and materials (i.e., from linear 'extract–process–manufacture–use–dispose to landfill' to cradle-to-cradle systems utilizing industrial ecology principles). These waste-to-resources transitions are discussed in detail in Newton (2008), and all have major challenges for building and planning.

Bridging the as-built versus as-operated divide

In an attempt to delay the introduction of a 5-star energy rating system for housing across all states and territories in Australia, there was a measure of obfuscation by some industry associations who attempted to confuse issues of 'as built' versus 'as operated' dwellings (see Figure 9.2) in lobbying the federal minister (Campbell *et al.* 2005).

Dwellings rated as 5-star in Melbourne can be expected under 'average' household operating conditions to consume around $150\,\text{MJ/m}^2$ each year in heating and cooling energy. Clearly, there will be a range of energy consumption statistics around this average which reflects a mix of lifestyle and thermal comfort 'settings' among occupants. To address above-average levels of consumption requires a range of attitude and behaviour change initiatives, such as the black balloons advertisements, the *Carbon Cops* TV

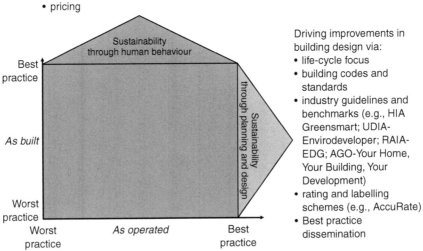

Figure 9.2 Twin drivers of environmental performance of buildings over the life cycle.

series, EPA Victoria's Greenhouse Calculator and CSIRO's *Handbook on Low Carbon Living*. The new version of the Greenhouse Calculator (www.epa.vic.gov.au/GreenhouseCalculator/calculator/default.asp) provides an opportunity to assess both as-built and as-operated elements of household energy use (including transport). There is no good purpose served in attempting to inhibit attempts to improve the thermal efficiency of the building 'shell' while seeking ways to change attitudes and behaviour in relation to energy use. Indeed, a shift to 7-star dwellings in Melbourne would result in annual average operating energy consumption declining to approximately $80\,MJ/m^2$ – almost half that of 5-star dwellings.

Bridging the digital divide

The next decade should witness the rapid convergence of IT with design science, engineering and sustainability science to a point where all of the key stakeholders in the built environment will be able to obtain assessments of urban development projects in real time for those performance parameters that interest them. *City of Bits* (Mitchell 1996) represents a powerful metaphor and a goal for digital design initiatives which for the first time can begin to model complex systems – like buildings, neighbourhoods, infrastructure – as collections of objects (i.e., all those thousands of different material products that are assembled into built forms) together with all their attributes and behaviours. This provides for automation, visualization and simulation of design options – one of the key platforms for delivering more sustainable built environments: *virtual building*, and an ability to examine life cycle performance *before* construction.

It means moving from current Horizon 1 thinking (see Newton 2007) in sustainability performance assessments that involve checklists and points systems, to Horizon 2 innovations represented by building information models and automated eco-efficiency assessment tools such as LCADesign (Seo *et al.* 2007), to Horizon 3 innovations which will include systems for multi-criteria sustainability assessments, virtual construction for urban development projects positioned within their wider spatial (neighbourhood and infrastructure) context (4D CAD plus GIS), and embedded sensors for real-time monitoring and feedback to building occupants of the environmental impacts of their use of buildings and appliances while in use.

Bridging the economic and environmental divide

Up until now it has been a requirement of federal and state governments that before a new regulation relating to the built environment can be introduced (e.g., as a BCA provision) it must undergo a mandatory Regulation Impact Statement (RIS) process in which the costs and benefits of particular options are assessed, followed by a recommendation supporting the most effective and efficient option (ORR 1998; ABCB 1999). The process at present is focused strongly on current or 'first cost' economic costs and benefits; and it does not provide any guidance on how to assess future life cycle-related eco-efficiency costs or benefits.

There are strong parallels with the World Bank and how it, as a bastion of economists, wrestled with the concepts of sustainable growth and sustainable development as key factors to be incorporated in shaping its lending policies and practices. Herman Daly's (1996) account of his attempt at the World Bank to have the economy seen as a sub-system of the Earth's total ecosystem – recognizing that growth is ultimately physically limited, and that focus needs to shift from growth to development that is *within the carrying capacity of the environment* – is a clear articulation of the paradigm shift required to incorporate environmentally sustainable development principles within Australian building and planning codes of practice. The paradigm shift will have succeeded when achieving *ecological efficiency* (i.e., minimizing ecological costs of resource depletion and pollution impact) gains equivalence with *economic efficiency* (i.e., minimizing market costs and maximizing profit) as key goals for development.

Following the receipt and national workshopping of the Pham *et al.* (2002) report, the Australian Building Codes Board (ABCB) adopted sustainability as one of its goals for the future Building Code of Australia (BCA) in 2007 (BCA 2007). In its Annual Business Plan 2006–07, the deliverables included undertaking 'scoping work and consultation with industry and the community, noting the Board's previously agreed focus on the areas of energy, water, materials and indoor environment quality' (ABCB 2006).

The principal environmental factors selected as a focus for the future BCA relate primarily to resource consumption and waste generation, encompassing the broader issue of *resource depletion*. The assessment of resource use in buildings can be addressed now in eco-efficiency assessment tools such as LCADesign. The addition of indoor environment quality extends a connection to occupant health and productivity (see Figure 9.3), but can be addressed initially by focusing on those building materials that can be demonstrated as having negative human health impacts during manufacturing (as reflected in LCI signatures), assembly or building occupation. An *IAQ Estimator* prototype has been developed by the CRC for Construction Innovation (Tucker *et al.* 2007; Brown *et al.* 2008, Chapter 11 in this volume) that can operate in a stand-alone assessment mode, or be linked into LCADesign to provide broader social and eco-efficiency performance assessment.

Conclusions

The pathways for transitioning to more sustainable urban development are being articulated in response to the looming vulnerabilities that built environments face in the twenty-first century if their development

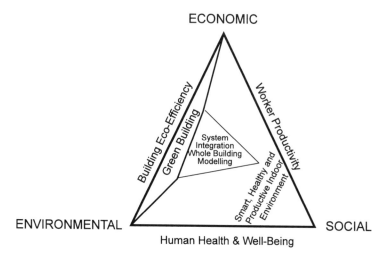

Figure 9.3 Sustainability triangle for commercial building performance assessment.

continues to be based on a paradigm of the Earth providing both *unlimited resources* ('Earth as quarry' or 'the Magic Pudding') and a *limitless sink* (atmosphere, rivers, oceans, land as receivers of waste).

The barriers that must be overcome to enable society to bridge to a more sustainable form of urban development are numerous. Eight have been outlined in this chapter. In order to be in a position, however, to adopt sustainability targets in urban development at the 'project' level (and beyond energy as the sole environmental dimension) requires at least three practical initiatives:

1   Development of a replacement RIS for ABCB's current (1999) *Economic Evaluation Model for Building Regulatory Change*, that enables a *valuing of environmental issues and impacts* (e.g., as indicated in Kats 2003; OECD 2003; SDC 2006; Your Building 2007) to be incorporated in cost–benefit assessments as well as an approach for engaging with *future values* via life cycle assessment and costings (Kishk *et al.* 2003; Kats 2003; Stern 2006; Centre for International Economics 2007). Currently RIS constitutes a major inhibitor to progress on sustainable development. It is the antithesis of the emerging Californian model that operates on the principle of using all available knowledge to formulate policy that can provide a regulatory framework that unleashes creativity and innovation among the scientific and business community to address sustainability challenges.

2   Development of *integrated eco-efficiency assessment tools* that are capable of meeting the requirements of government, industry, science and

community. The proposed national sustainability charter would represent an important stimulus towards a national initiative for achieving more sustainable urban development. Without assessment tools, however, that incorporate indicators, weightings, benchmarks, targets – and a facility for undertaking tradeoffs (e.g., via multi-criteria assessment models) – all in a *transparent* manner, we continue to live with greenwash. And this is irrespective of whether performance-based assessments will be applied in the context of *regulatory mechanisms, market-based policies* or the more experimental *targeted transparency systems* (Fung *et al.* 2007).

3   Developing a *protocol* capable of enabling comparative evaluation of the performance of competing building and planning assessment and rating tools. The protocol would need to cover the spectrum of eco-efficiency criteria relevant to all stakeholder groups, all classes of building (and infrastructure), existing as well as new building, and the range of performance issues outlined in the body of this chapter. It is a process which has served the software industry well. It is also a means by which the property, planning and development industry, and government can be openly informed about the relative strengths and weaknesses of each tool, obviating the call for a 'single rating tool' which merely serves to stifle innovation. It will also help eliminate 'camps of uncooperative stake-holders' that tend to presently surround particular 'green building' tools.

Bibliography

ABARE (2006) *Australian Energy: National and State Projections to 2029–30*, Research Report 06.26, Canberra: Australian Bureau of Agricultural and Resource Economics.

Abbs, D. (2008) 'Flood', in P.W. Newton (ed.) *Transitions: Pathways Towards Sustainable Urban Development in Australia*, Melbourne: CSIRO and Dordrecht: Springer.

ABCB (1999) *Economic Evaluation Model: Building Regulatory Change*, Canberra: Australian Building Codes Board.

—— (2006) *Annual Business Plan 2006–07*, Canberra: Australian Building Codes Board.

ABS (2006) *Water Account 2004–5*, Cat. no. 4610.0, Canberra: Australian Bureau of Statistics.

ALCAS (2007) Australian Life Cycle Assessment Society, Melbourne. Online. Available at HTTP: <www.alcas.asn.au/alcas2/index.php?option=com_content&task=section&id=3&Itemid=10>.

Ambrose, M. (2008) 'Energy efficient housing and subdivision design', in P.W. Newton (ed.) *Transitions: Pathways Towards Sustainable Urban Development in Australia*, Melbourne: CSIRO and Dordrecht: Springer.

Arup Sustainability (2004) *Overview of Sustainability Rating Tools*, report prepared for Brisbane City Council.

ASBEC (2007) *Template for a Business Case for Sustainable Buildings*, Australian Sustainable Built Environment Council, Melbourne (unpublished, to appear in Your Building portal).

BCA (2007) *Building Code of Australia*, Canberra.

Bernstone, R. (2003) 'Rating the rating systems', *Building Australia*, 24–25 Oct.

Campbell, I., Macdonald, I. and Macfarlane, I. (2005) *Proposed 'Five-Star' Energy Ratings Seriously Flawed*, media release, Canberra, 2 Dec.

Centre for Design at RMIT University *et al.* (2006) *Scoping Study to Investigate Measures for Improving the Environmental Sustainability of Building Materials*, report prepared for Department of the Environment and Heritage, Canberra.

Centre for International Economics (2007) *Capitalising on the Building Sector's Potential to Lessen the Costs of a Broad-Based GHG Emissions Cut*, report prepared for ASBEC Climate Change Task Force, Canberra.

City of Melbourne (2003) *Zero Net Emissions by 2020: A Roadmap to a Climate Neutral City*, Melbourne. Online. Available at HTTP: <www.melbourne.vic.gov.au/rsrc/PDFs/EnvironmentalPrograms/ZeroNetEmissionsFull.pdf>.

Crowther, P. (2001) 'Developing an inclusive model for design for deconstruction', in A.R. Chini (ed.) *Deconstruction and Materials Reuse: Technology, Economic and Policy*, Proceedings of the CIB Task Group 39 – *Deconstruction* meeting, Wellington, New Zealand, 6 April.

Curwell, S. and Spencer, L. (1999) *Environmental Assessment of Buildings: Survey of W100 Members*, Salford: Research Centre for the Built and Human Environment, University of Salford (unpublished).

Daly, H.E. (1996) *Beyond Growth: The Economics of Sustainable Development*, Boston, MA: Beacon.

Diaper, C., Sharma, A. and Tjandraatmadja, G. (2008) 'Decentralised water and wastewater systems', in P.W. Newton (ed.) *Transitions: Pathways Towards Sustainable Urban Development in Australia*, Melbourne: CSIRO and Dordrecht: Springer.

Dodson, J. and Sipe, N. (2008) 'Energy security, oil vulnerability and cities', in P.W. Newton (ed.) *Transitions: Pathways Towards Sustainable Urban Development in Australia*, Melbourne: CSIRO and Dordrecht: Springer.

Ecojustice (2007) *The Municipal Powers Report*, Vancouver. Online. Available at HTTP: <www.ecojustice.ca/publications/reports/the-municipal-powers-report/>.

Foliente, G., Tucker, S., Seo, S., Hall, M., Boxhall, P., Clark, M., Mellon, R. and Larsson, N. (2007) *Performance Setting and Measurement for Sustainable Commercial Buildings*. Online. Available at HTTP: <www.yourbuilding.org/display/yb/Performance+setting+and+measurement+for+sustainable+commercial+buildings>.

Fung, A., Graham, M. and Weil, D. (2007) *Full Disclosure: The Perils and Promise of Transparency*, Cambridge: Cambridge University Press.

Garnaut Climate Change Review (2008) *Interim Report to the Commonwealth, State and Territory Governments of Australia*, Melbourne. Online. Available at HTTP: <www.garnautreview.org.au/CA25734E0016A131/pages/reports-and-papers>.

GBCA (2006) *The Dollars and Sense of Green Buildings: Building the Business Case for Green Commercial Buildings in Australia*, Sydney: Green Building Council of Australia.

GHD (2006) *Scoping Study to Investigate Measures for Improving the Water Efficiency of Buildings*, report prepared for Department of the Environment and Heritage, Canberra.

—— (2007) *Water Efficiency*. Online. Available at HTTP: <www.yourbuilding.org>.

Hennessy, K. (2008) 'Climate change', in P.W. Newton (ed.) *Transitions: Pathways Towards Sustainable Urban Development in Australia*, Melbourne: CSIRO and Dordrecht: Springer.

Horne, R.E., Hayles, C., Hes, D., Jensen, C., Opray, L., Wakefield, R. and Wasiluk, K. (2005) *International Comparison of Building Energy Performance Standards*, report prepared for Department of the Environment and Heritage, Canberra.

House of Representatives Standing Committee on Environment and Heritage (2005) *Sustainable Cities*, Canberra: Parliament of the Commonwealth of Australia. Online. Available at HTTP: <www.aph.gov.au/house/committee/environ/cities/report.htm#chapters>.

—— (2006) *Inquiry into a Sustainability Charter*, Discussion Paper, Canberra: Parliament of the Commonwealth of Australia. Online. Available at HTTP: <www.aph.gov.au/house/committee/environ/charter/discussionpaper.pdf>.

—— (2007) *Sustainability for Survival: Creating a Climate for Change: Inquiry into a Sustainability Charter*, Canberra: Parliament of the Commonwealth of Australia. Online. Available at HTTP: <www.aph.gov.au/house/committee/environ/charter/report.htm>.

Inman, M. (2008) 'The water efficient city: technological and institutional drivers', in P.W. Newton (ed.) *Transitions: Pathways Towards Sustainable Urban Development in Australia*, Melbourne: CSIRO and Dordrecht: Springer.

ISO (2005) *Sustainability in Building Construction – General Principles*, ISO TC59, Geneva (doc. N041).

Jones, T. (2008) 'Distributed energy systems', in P.W. Newton (ed.) *Transitions: Pathways Towards Sustainable Urban Development in Australia*, Melbourne: CSIRO and Dordrecht: Springer.

Kaebernick, H., Ibbotson, S. and Kara, S. (2008) 'Cradle to cradle manufacturing', in P.W. Newton (ed.) *Transitions: Pathways Towards Sustainable Urban Development in Australia*, Melbourne: CSIRO and Dordrecht: Springer.

Kats, G. (2003) *The Costs and Financial Benefits of Green Buildings: Report to California's Sustainable Building Taskforce*, Washington, DC: Capital E.

Kishk, M., Al-Hajj, A., Pollock, R., Aouad, G., Bakis, N. and Sun, M. (2003) 'Whole life costing in construction: a state of the art review', *RICS Research Papers*, 4 (18): 1–38.

London Borough of Merton (2004) *The 10% Renewable Energy Policy (The Merton Rule)*. Online. Available at HTTP: <www.merton.gov.uk/living/planning/planningpolicy/mertonrule.htm>.

Maheepala, S. and Blackmore, J. (2008) 'Integrated urban water management', in P.W. Newton (ed.) *Transitions: Pathways Towards Sustainable Urban Development in Australia*, Melbourne: CSIRO and Dordrecht: Springer.

McDonough, W. and Braungart, M. (2002) *Cradle to Cradle: Remaking the Way We Make Things*, New York, NY: North Point.

Mitchell, V.G., McMahon, T.A. and Mein, R.G. (2003) 'Components of the total water balance of an urban catchment', *Environmental Management*, 32 (6): 735–46.

Mitchell, W. (1996) *City of Bits: Space, Place, and the Infobahn*, Cambridge, Mass: MIT Press.

Newton, P.W. (2002) 'Urban Australia: review and prospect', *Australian Planner*, 39 (1): 37–45.

—— (2006) *Australia State of the Environment 2006: Human Settlements: Theme Commentary*, Canberra: Department of the Environment and Heritage. Online. Available at HTTP: <www.environment.gov.au/soe/2006/publications/commentaries/settlements/index.html>.

—— (2007) 'Horizon 3 planning: meshing liveability with sustainability', *Environment and Planning B: Planning and Design*, 34: 571–5.

—— (ed.) (2008) *Transitions: Pathways Towards Sustainable Urban Development in Australia*, Melbourne: CSIRO and Dordrecht: Springer.

Newton, P.W. and Mo, J. (2006) 'Urban energyscapes: planning for renewable-based cities', *Australian Planner*, 43 (4): 8–9.

Newton, P.W., Baum, S., Bhatia, K., Brown, S.K., Cameron, A.S., Foran, B., Grant, T., Mak, S.L., Memmott, P.C., Mitchell, V.G., Neate, K.L., Pears, A., Smith, N., Stimson, R.J., Tucker, S.N. and Yencken, D. (2001) *Australia State of the Environment 2001: Human Settlements*, Melbourne: CSIRO Publishing.

OECD (2002) *Towards Sustainable Household Consumption? Trends and Policies in OECD Countries*, Paris: OECD.

—— (2003) *Environmentally Sustainable Buildings: Challenges and Policies*, Paris: OECD.

ORR (1998) *A Guide to Regulation*, Melbourne: Office of Regulation Review, Productivity Commission.

Pham, L., Hargreaves, R., Ashe, B., Newton, P.W., Enker, R., Bell, J., Apelt, R., Hough, R., Thomas, P.C., McWhinney, S., Loveridge, R., Davis, M. and Patteson, M. (2002) *Sustainability and the Building Code of Australia: Final Report*, Brisbane: CRC for Construction Innovation.

Productivity Commission (2006) *Waste Management*, Report no. 38, Melbourne: Productivity Commission.

SDC (2006) *Stock Take: Delivering Improvements in Existing Housing*, London: Sustainable Development Commission.

Seo, S., Tucker, S.N. and Newton, P.W. (2007) 'Automated material selection and environmental assessment in the context of 3D building modelling', *Journal of Green Building*, 2 (2): 51–61.

Spurling, T., Srinivasan, M., Coombes, P., Cox, S., Dillon, P., Langford, J., Leslie, G., Marsden, J., Priestley, T., Roseth, N., Slatyer, T., Wong, T. and Young, R. (2007) *et al.* (2007) *Water for Our Cities: Building Resilience in a Climate of Uncertainty*, report prepared for Prime Minister's Science, Engineering and Innovation Council, Canberra.

Stern, N. (2006) *Stern Review on the Economics of Climate Change*, London: HM Treasury.

Tucker, S.N., Brown, S.K., Egan, S., Morawska, L., He, C., Boulaire, F. and Williams, A. (2007) *The Indoor Air Quality Estimator*, Report 2004–033-B-01, Brisbane: CRC for Construction Innovation.

United Nations (1987) *Our Common Future: Report of the World Commission on Environment and Development*, New York. NY: United Nations.

—— (2007) *World Urbanization Prospects: The 2007 Revision Population Database*, New York, NY: Population Division, United Nations. Online. Available at HTTP: <http://esa.un.org/unup/>.

Your Building (2007) *Water Use and Sustainable Commercial Buildings*. Online. Available at HTTP: <www.yourbuilding.org/display/yb/Water+use+and+sustainable+commercial+buildings>.

10 Automated environmental assessment of buildings

Seongwon Seo, Selwyn Tucker and Peter Newton

A key factor in the transition to more sustainable built environments will be the availability of software tools that enable design professionals to assess the eco-efficiency performance of buildings before they are constructed – as virtual buildings at the design stage. Research reported here relates to LCADesign, a software tool developed at the CRC for Construction Innovation for automated environmental impact assessment of materials selected for assembly into a new or regenerated building. Initially developed for application to commercial buildings, it has the functionality that would enable its extension to housing and urban infrastructure, thereby providing the platform for eco-efficiency assessment of the entire built environment (Newton 2002).

The process of sustainable building requires the integration of a number of complex strategies during the design (as well as construction and operation) stage of building projects. Foremost among these should be careful selection of building materials – together with design of key space and layout configurations – and an assessment of their combined impact on the physical environment and on the health, comfort and productivity of the building occupants. Integrated building design and materials selection offers considerable potential for substantially reducing the environmental impacts of urban development projects (AboulNaga and Amin 1996; Kim and Rigdon 1998; Seo *et al.* 2006; Department of the Environment and Heritage 2006). There are significant socio-technical challenges to be overcome, however. Foremost among these is a scientifically validated assessment tool that is acceptable across all stakeholder groups in the sector.

In order to assist the building and property industry progress towards more sustainable urban development options, a number of tools have been developed over the past decade to assess the impact that choice of materials has on energy consumption and other specific environmental impacts of buildings. Most have limitations and weaknesses and, in a review of such tools, many common problem areas have been identified (Seo 2002; Todd *et al.* 2001). The weaknesses include having a narrow focus, lacking in-depth assessment, needing professional assessors, requiring time-consuming data input, considering minimal economic criteria, and lacking transparency in weighting environmental indicators.

Successful implementation of a tool capable of performing the required tasks involves not only the development of computer software and related databases but also paying considerable attention to the needs of the potential users. The technological advances made in producing a unique and versatile tool – such as a real-time automated eco-efficiency assessment tool based on a building information model (BIM) – constitute a paradigm shift in the ability to effectively assess the environmental impacts of buildings, but will be successfully implemented only if the tool addresses the problems faced by those who currently assess the environmental impacts of building and their materials contributions (Watson *et al.* 2004).

The above problem has begun to be addressed by developing integrated building design and evaluation tools. This integration can be grouped into two classes. The first class deals with the development of software dedicated to an integration of BIMs and building material inventories, while the second class focuses on the transfer of data between design applications by means of a central database (Ellis and Mathews 2002). Currently, however, the complexity of existing tools and their integration into the design process seem to constitute the biggest barriers. To be attractive to users, the tools should be able to provide answers quickly, calculations should require the minimum amount of input data so as to be useful at any stage during the design process, and they should be able to provide quantitative feedback regarding the influence of particular design and material selection decisions on environmental performance of the building and its components.

A new integrated eco-assessment tool, LCADesign, was developed to fulfil the above requirements by addressing the needs identified by the stakeholder groups involved in building performance evaluation. This chapter gives a brief overview of the tool, which enables building design professionals to make more informed decisions in real-time about a building and its material products during the design stage. Also featured is its application to a case-study building to demonstrate how a tool such as LCADesign can be applied in the sustainable building design and material selection process, to satisfy the requirements of building design professionals and commercial clients – and future building regulators.

LCADesign: eco-efficiency assessment tool

The objective of LCADesign is to integrate building environmental assessment into a 3D CAD model to avoid the need for manual transcription of data from one step to another in the evaluation process. The essential steps in the process are shown in Figure 10.1, and involve:

- creating a 3D CAD model of a building;
- tagging each object in the 3D CAD model by selecting a rule identifying both the materials in that object and its method of calculation;

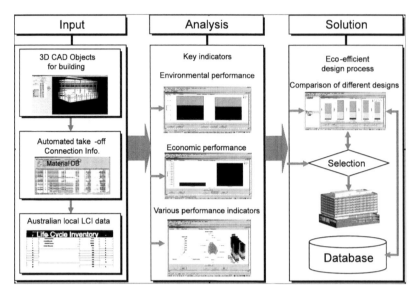

Figure 10.1 LCADesign essential steps.

- using the dimensional information in the 3D CAD model to automatically estimate quantities of all materials in the building;
- estimating all material and gross building environmental burdens by factoring each material quantity with results of their contribution to emissions generation and resource depletion – from a database;
- calculating a series of environmental indicators based on life cycle analysis;
- providing a facility to undertake detailed analysis of alternative designs and material selections, including benchmarking over time to enable design professionals to create buildings with least environmental impact in the context of their service delivery requirements.

To achieve this integration, information has to flow seamlessly from the 3D CAD model to the evaluation stage without interruption or intervention from the designer or environmental assessor. This enables the designer to receive almost instant feedback on whether a particular building design iteration provides a better environmental outcome compared to any of its predecessors. Unlike almost all other environmental assessment tools in existence, evaluation can occur while a design evolves and not, as is typically done, as a post-design evaluation to check whether a required benchmark has been achieved.

The CAD information associated with a building design together with the quantities of all related building materials are stored in a database along with formulae to calculate their environmental impacts accrued in materials product manufacturing. The database also contains the environmental burdens associated with all building materials per unit material from what is termed the Life Cycle Inventory (LCI).

3D CAD objects for building and building information models

Obtaining and entering building data into an assessment tool is a time-consuming process, and is a significant disadvantage of contemporary procedures in building environmental assessment. One obvious source of information is a CAD drawing of a building, which traditionally has consisted of simple line representations with no associated information as to what the lines represent (walls, windows, roofs, etc.). A number of current object-orientated CAD systems now offer an effective solution for rapid data transfer, as they contain detailed attribute information that provides an opportunity to develop automated analysis software. To enable a complete exchange of information about what objects represent, as well as their dimensions, an approach based on Industry Foundation Classes (IFCs) is currently being developed by an international consortium. These IFCs are being implemented worldwide for information exchange from proprietary CAD systems. They are sets of electronic specifications (Wix and Liebich 1997) that represent objects in built facilities, such as doors, walls, fans, etc., and abstract concepts such as space, organization, process, etc. Such objects are defined in such a way as to allow analytical software calculating performance measures to obtain most, if not all, the required characteristics directly from a BIM. These specifications represent a data structure supporting an electronic project model that enables sharing of data across a wide variety of applications. A significant advantage of IFC technology is that it facilitates analysis of building models that have been produced from several software vendors.

Data related to quantities of the building object are extracted directly from the 3D CAD model (Figure 10.2). Quantities of all building components whose specific materials are identified to calculate a complete list of the quantities of all materials such as concrete, steel, timber, plastic, etc. are automatically taken off. This information is linked with the LCI database to estimate key environmental indicators via Reasoning Rules (RR; see below). Thus 3D CAD Object reasoning rules need to be specified in terms of 'known' component product manufacturing processes in order to estimate a comprehensive inventory of building environmental burdens.

Life cycle inventory data

The LCI is used to calculate environmental burdens associated with resource depletion and emissions to air, land and water from cradle to grave (ISO 1998) for a product. While currently conducted in a 'cradle to construction gate' system boundary, the scope of assessments will ultimately become

Figure 10.2 3D CAD model of a building.

'cradle to grave'. This means that over a given building and component design life, the scope of work is to include all known environmental flows of resources from, and emissions to, air, land and water in acquisition, manufacture, construction, operation and final disposal. Capital equipment, employee facilities and activities are not included in an LCI product database as standard practice. The integrated assessment tool LCADesign includes in its LCI database the following processes involved in producing each and every product:

- mining, crushing and chemical use in extraction and processing of raw materials;
- acquisition of cultivated, collected or harvested agricultural product;
- fuel production to supply power and process energy and transport of materials;
- process energy and transport for raw, intermediate and ancillary materials;
- resources consumed in processing such as lubricants, tyres, energy;
- packaging, maintenance, renewal, recycling and disposition operations.

An LCI inventory for a product contains all the resources used and emissions generated by each process required to create that product per functional unit (usually kg, but can also be m, m^2, m^3, etc., if these units are commonly used in practice). This inventory is a result of modelling all the direct and indirect inputs and outputs of all the processes involved using specific-purpose software such as Boustead (1995) or SimaPro (PRé 2006). All resource usage and emissions data from each process are aggregated for the entire manufacturing process to derive gross totals for any product per functional unit, and it is these values which appear in an LCI database. Figure 10.3 shows part of a typical process flow model for dry process bagged cement used for mortar, mapped from raw material acquisition to manufacturing gate.

   For the process map, LCI data quantify the environmental emissions, resource consumption and waste flows as shown in Table 10.1.

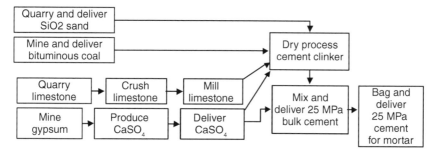

Figure 10.3 Process map for cement mortar.

Table 10.1 Materials database: selected contents

Raw material 30 items, e.g.	Emission to air 43 items, e.g.	Emission to water 44 items, e.g.	Solid waste 16 items, e.g.
Bauxite	CO_2	Ammonia	Ash
Coal	Hydrocarbons	Cadmium (Cd)	Industrial waste
Limestone	Methane	Fluoride	Solid waste
Gypsum	N_2O	Phosphate (PO_4^{3-})	Slags/ash

Linking to 3D CAD

Since 3D CAD objects do not contain all the required data for LCA analysis and users are inconsistent in entering the available attributes, a system is required to provide the link between the components in the building model database and the resource usage and emissions associated with the materials. One form of link considered was to evaluate the proportions of materials in a range of sub-systems of a building, e.g., walls, windows, staircases, ceilings, etc. This approach, however, requires many approximations, and is not consistent with a full modelling approach where relationships are identified and results are fully scalable as the object varies in size. Thus the idea of reasoning rules (RRs) was created. These use the dimensional parameters from 3D CAD to calculate quantities of every material in a 3D CAD object in terms of the functional unit of that material in the LCI database. Each and every resource usage and emission of that material is then multiplied by the quantities to obtain the totals of the environmental impact items of each material in that 3D CAD object. RRs thus exploit minimally defined building elements by relating them to product definitions in an LCI database of construction materials.

There are two types of RRs: Product Reasoning Rules (PRR) and Object Reasoning Rules (ORR) (Figure 10.4):

- A PRR connects the CAD Object dimensional information to a single LCI material/product using the formula defined within it to calculate

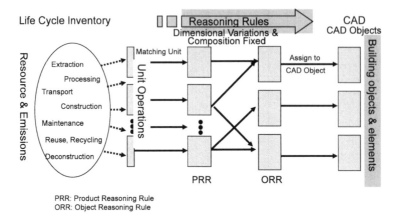

PRR: Product Reasoning Rule
ORR: Object Reasoning Rule

Figure 10.4 Reasoning Rules link from life cycle inventory to 3D CAD objects.

the amount of that material in the 3D CAD object (e.g., volume of concrete or mass or reinforcement steel).

- An ORR is an assembly of PRRs which includes all the materials in a particular CAD object and which is attached/tagged to the CAD object (e.g., concrete, reinforcement steel, formwork, membranes, fixings, finishes such as render or paint, etc.). Any one PRR can be used by multiple ORRs where the same formula and material is required (e.g., volume of concrete) and may be multiplied by a factor to represent repeated use in a 3D CAD object (e.g., coats of paint) or part usage.

This association of 3D CAD elements with quantity data and RRs specified in terms of categorized components in the ORRs enables different perspectives on building performance to be evaluated, ranging from product to component, to sub-assembly to entire building. To cope with the varying levels of detail required in the design process, the RRs must be defined down to the finest level specified by the product/service specifications and the LCI.

RRs combine all relevant materials from the LCI to create a real building component or product. For example, a 3D CAD representation of a window has little other than the name to identify it as aluminium framed, so the RRs must contain relationships to calculate the quantities of all the materials in the window, such as aluminium, glass, sealant, fixings, etc., from the one set of dimensions. The amount of material in the frame is calculated from the perimeter of the window and the typical cross-section of the extrusion, while the amount of glass uses the approximate net area and the typical glass thickness for the particular window type. Thus, while the rules are scalable for size, needing only one rule, different window types (e.g., fixed, awning, sliding) require additional rules. In most buildings, however, there is a fairly limited set of standard window types.

Environmental analysis

This step estimates all material and gross building environmental burdens by factoring each material quantity with metrics of their emissions generation and resource depletion. LCADesign uses the life cycle assessment (LCA) methodology to quantify the environmental impacts for building products. The system boundary applied in LCADesign comprises cradle to construction, and Life Cycle Impact Assessment (LCIA) methods defined by the ISO 14042.3 Standards (1998, 2000) are used to assess product impacts on the environment.

Many environmental indicators can be estimated, including key internationally recognized indicators such as Eco-indicator 99, which is one of the popularly used endpoint approaches in LCIA methods. The application of the Eco-indicator 99 method can involve a single score, but it is also capable of generating separate impact indicators, such as the three 'damage' indicators (human health, ecosystem quality and resource depletion): the human health category consists of carcinogens, respiratory organics, respiratory inorganics, climate change, radiation and ozone layer impacts; the ecosystem quality category consists of ecotoxicity, acidification, eutrophication and land use; while the resource depletion category consists of minerals and fossil fuel use (Goedkoop and Spriensma 2001). Each impact category is calculated and individually viewable in LCADesign. LCADesign allows the user to investigate one or more of the characterized damage impacts, or just to use a single indicator.

LCADesign also includes additional indicators which are not included in other models of environmental impact assessment. The set of environmental impact indicators available are:

- all indicators used in Eco-indicator 99;
- embodied energy;
- embodied water;
- embodied carbon emissions;
- total greenhouse gas emissions.

Application of LCADesign to commercial building assessment

To illustrate how the decision support tool LCADesign can be used, an example of its application in a recent project on the environmental assessment of a building proposed for regeneration is described. The case-study building was Council House One (CH1), a multi-storey building constructed in the early 1970s and owned by the Melbourne City Council.

198   *S. Seo* et al.

The site runs on a north–south axis and has a surface area of 1,960 m². There are three floors of car park for 230 cars, a retail area of 400 m², offices on seven floors each of 1,070 m² per floor totalling 7,490 m², and a roof-level plant room.

The structure is a concrete sway frame with horizontal bracing at gable ends and between columns in one southern bay. The core walls are not structural. Floor to floor height is 3,150 mm with a slab thickness of 235 mm, and a down-stand beam in the core area of 465 mm which gives a clearance of 2,450 mm. There is an up-stand beam around the perimeter of 560 mm in height. Windows are formed within pre-cast facade units of size 3,150 mm H×1,565 mm W and have a glass area of 2,110 mm H×1,380 mm W. This gives a glass area of 69 per cent of the total facade area. The west-side glass is covered with a reflective film.

3D CAD modelling

The structure of the existing building has been drawn as a 3D CAD model in IFC-compliant ArchiCAD, as illustrated in Figure 10.5. Each pre-cast facade unit had to be drawn individually, making the model very extensive in terms of number of objects.

The 3D CAD model of the case-study building is an early schematic design stage showing the two basements, main ground level and mezzanine and seven typical office floors, and the plant room on the roof. All struc-

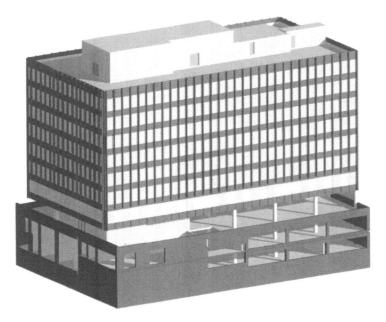

Figure 10.5 3D CAD view of case-study building (Council House 1).

tural elements, such as columns, slabs, roof and walls, are drawn in the model. Staircases are included, as are all interior walls and doors (such as illustrated in Figure 10.6). Sanitary facilities are included, but the associated hydraulic services are not drawn in the 3D CAD model and neither are HVAC items. The model is considered an elaborate envelope model only, suitable for an early schematic design stage primarily to support early architectural design decisions related to regeneration and refurbishment.

Measures of environmental impact

To identify the environmental impact of the existing case-study building, Eco-indicator 99 was chosen as the basis for deriving the single environmental indicator (ecopoints/m^2), which could be broken down further, if required, into three damage categories. For the case-study building, its environmental signature in its original (pre-regeneration) condition, derived from LCADesign, is shown in Figure 10.7.

For LCADesign analyses, the CH1 building has been classified into several layers related to longevity of built components as a means of identifying where the greatest environmental benefits may lie in the regeneration process (after Brand 1994). These layers comprise the shell (structure), services (cabling, plumbing, air conditioning, lifts), scenery (layout of partitions, dropped ceilings, etc.), set (shifting of furniture by the occupants) and site. For this analysis, only two layers are subject to LCADesign assessment, namely, the shell and scenery. Of these two, more than half the environmental impact (as measured by Eco-indicator 99) is due to the

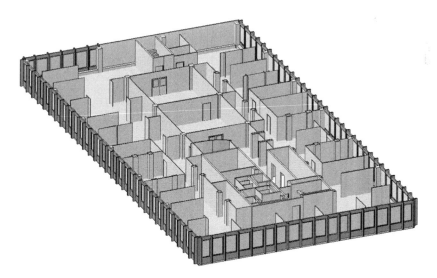

Figure 10.6 3D CAD view of a single floor of case-study building (Council House 1).

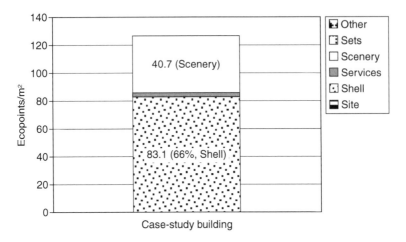

Figure 10.7 Environmental impact by layers for case-study building.

shell (66 per cent of total impact), which further divides into two parts: superstructure and substructure. Table 10.2 shows the key environmental indicators for the building and its further breakdown into material level. The superstructure is the most dominant part of the building from the perspective of environmental impact.

Figure 10.8 shows the environmental impact for the shell and scenery parts of the case-study building, and a further breakdown for the impacts according to building element levels. As seen in Figure 10.8(a), the super-structure contributed more than 95 per cent of environmental impact for the shell part. The superstructure can be further broken down into more detailed elemental groups, such as windows, roof, external doors, internal/external walls, columns and staircases. The largest contributions to the superstructure are shown as upper floors (41 per cent), internal walls (23 per cent) and external walls (17 per cent). The scenery element contributed 32 per cent to the total environmental impact of the build-ing, and can be classified into two parts: fittings and finishes. The fin-ishes element can be further broken down into wall, ceiling and floor finishes.

This set of baseline results suggests that the most likely route for redu-cing the total environmental impact as a result of the building regeneration is by reducing the environmental impacts of key elements like upper floor and internal/external walls.

Design alternatives

Possible alternative design options for a regenerated CH1 building comprised the following elements (among many others not dealt with in this chapter):

Table 10.2 Key environmental indicators for case-study building and sub-elements

Indicator	Unit	Building layer			
		Building	Shell	Superstructure	Upper floors
Eco-indicator 99	Ecopoints/m²	127	83	80	33
Embodied energy	J/m²	31,968	19,300	18,603	6,316
Embodied water	Litre/m²	17,557	11,641	11,109	4,272
GHG emissions	Carbon/m²	1,150	994	937	533

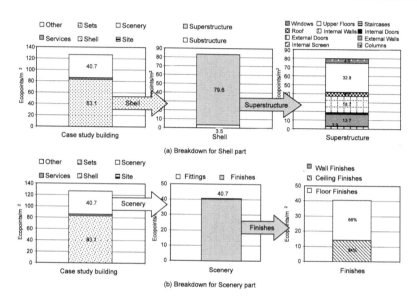

Figure 10.8 Environmental impact by layers and sub-layers for case-study building.

- upper floors for superstructure part;
- internal walls;
- external walls;
- floor finishes.

Environmental impacts were evaluated using LCADesign in relation to alternatives which focused on use of recycled content in some of the building elements (see Table 10.3). Use of recycled materials was considered as one of the possible options to reduce environmental impact. However, recycling may not always be the most appropriate option and, as such, building materials containing recycled contents should be evaluated in a manner consistent with a quantitative assessment of the overall environmental impacts.

There are many options available for applying building products or elements that contain recycled materials. By becoming more aware of

Table 10.3 Alternative regeneration options for case-study building

Alternative	Design option
Alternative (Alt) 1	Replacing material in the upper floors for superstructure part with recycled contents*
Alternative (Alt) 2	Replacing material in the upper floors, internal wall and external wall for superstructure part with recycled contents*
Alternative (Alt) 3	Alternative 2 + replacing floor finishes material with alternative product (wool blended carpet)

Note
* Reinforcement bars (up to 99 per cent recycled) and 7 per cent fly-ash concrete are considered in the superstructure part of the building.

which building materials and elements have the lowest environmental impact, building designers can encourage the development of more sustainable buildings by specifying the more environmentally friendly products and redesigning buildings to reduce the environmental impact of those elements which contribute most to a building's negative environmental signature. For example, with the upper floor structure contributing 41 per cent of the impact of the superstructure, it is worth attempting to reduce the impact of this element first before considering any reduction in much smaller contributors such as the roof, which is only 5 per cent of the total superstructure contribution. A 10 per cent reduction in the impact of the upper floor structure would almost equal the whole contribution of the roof.

The LCADesign analyses enable a greater focus on achieving good design *and* material specification, rather than being influenced by the market to apply specific proprietary *products* as 'building solutions'.

Comparisons

Preliminary comparisons of the alternatives with the baseline building show that choosing alternative systems and/or materials offers considerable potential for environmental improvement (Figure 10.9). By replacing the non-recycled material components with materials incorporating recycled content (99 per cent recycled reinforcement bars and 7 per cent fly-ash concrete) for the shell part (particularly the superstructure part), the total environmental impact could be reduced by 6 per cent (120 ecopoints/m^2 for alternative 1) and 9 per cent (116 ecopoints/m^2 for alternative 2), respectively. Furthermore, when the eco-preferred material (i.e., wool blended carpet) replaced the polypropylene carpet for floor finishing under the scenery part, the comparison shows alternative 3 has a much lower environmental impact compared to the baseline building (that is, a reduction of 19 per cent of total environmental impact to 102.2 ecopoints/m^2).

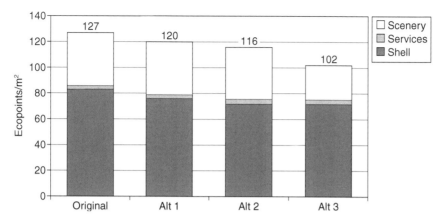

Figure 10.9 Comparison of the baseline building with alternative regeneration options.

A further set of environmental indicators relating to embodied energy, embodied water and greenhouse gas emissions are shown in Figure 10.10 for the baseline building and some regeneration alternatives.

While the overall macro indicator (Eco-indicator 99) decreased as a result of alternative regeneration options, as seen in Figure 10.9, the embodied energy component has risen, driven by the big increase in the scenery (finishes) component (Figure 10.10). This rise is due to the contribution that biomass energy sources such as wood and animal products make to the production of the wool carpet material alternative.

For water consumption, alternative 3 shows more consumed in the scenery by applying wool blended carpet compared to other alternatives. However, considering total embodied water consumption, it shows still less consumed water compared to the case-study building.

Overall, the results demonstrate that total environmental impact will decrease by applying recycled building materials (6 per cent fly-ash concrete and recycled reinforcement steel) in the shell part of the case-study building. The outputs also illustrate how useful the finer breakdowns are in revealing which components have smaller environmental impact signatures than others. Clearly, *materials matter* in the environmental impact of buildings.

Conclusions

There is an increasing drive on the part of building designers and developers to reduce the environmental impacts of their projects. This will intensify further as a result of the pressures on development in a carbon-constrained and resource-constrained world (Newton 2008).

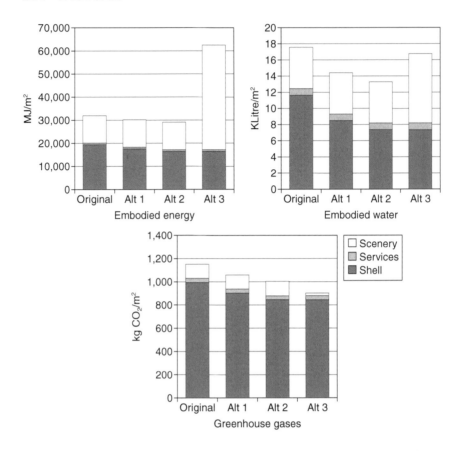

Figure 10.10 Comparison of environmental indicators for the baseline building and the alternatives (energy, water, greenhouse gases).

Real-time assessment tools for use in determining the environmental performance of buildings – as designed, as built, as commissioned, as operated and as regenerated – will be critical in driving a transition to more sustainable built environments.

Building material selection is not straightforward. To deal with this challenge effectively, LCADesign was developed for the design professions to support their decision-making by providing integrated assessment and comparison of environmental impacts of material and design alternatives for a new or regenerated building. This chapter has outlined the key characteristics of LCADesign and has provided a case study to show how it can be applied in procuring a greener building with lower environment material signatures.

LCADesign was created to meet a growing need from designers and regulators for real-time appraisal of design performance of constructed assets by providing:

- automated and integrated environmental assessment direct from 3D CAD drawings/BIM;
- a range of environmental impact and performance assessment indicators;
- sketch and detailed design evaluation;
- comparative ratings of environmental impacts of alternative designs and materials;
- assessment of buildings at all levels/layers of design analysis;
- comprehensive graphical and tabular outputs;
- a capacity for extension to eco-efficiency assessment by automated linkage to product and assembly cost databases.

As such, LCADesign has the functionality to become the basis for a perform-ance assessment tool for future built environment regulations that establish targets for energy, water and material use in buildings – key elements in a National Sustainability Charter (House of Representatives Standing Committee on Environment and Heritage 2007).

Bibliography

AboulNaga, M. and Amin, M. (1996) 'Towards a healthy urban environment in hot-humid zones: information systems as an effective evaluation tool for urban conservation techniques', *Proceedings of the 24th International Association of Housing Science World Congress*, Ankara.

Boustead, I. (1995) *The Boustead Model for LCI Calculations*, Vol. 1, Horsham: Boustead Consulting.

Brand, S. (1994) *How Buildings Learn: What Happens After They Are Built*, New York, NY: Viking.

Department of the Environment and Heritage (2006) *ESD Design Guide for Australian Government Buildings*, Canberra: Department of the Environment and Heritage.

Ellis, M.W. and Mathews, E.H. (2002) 'Needs and trends in building and HVAC system design tools', *Building and Environment*, 37 (5): 461–70.

Goedkoop, M. and Spriensma, R. (2001) *The Eco-indicator 99: A Damage Oriented Method for Life Cycle Impact Assessment*, 3rd edn, Amersfoort: PRé Consultants.

House of Representatives Standing Committee on Environment and Heritage (2007) *Sustainability for Survival: Creating a Climate for Change: Inquiry into a Sustainability Charter*, Canberra: Parliament of the Commonwealth of Australia. Online. Available at HTTP: <www.aph.gov.au/house/committee/environ/charter/report.htm>.

ISO (1998) *ISO/CD 14042.3: Life cycle Assessment – Impact Assessment*, Geneva: International Standards Organization.

—— (2000) *ISO 14042: Environmental Management – Life Cycle Assessment – Life Cycle Impact Assessment*, Geneva: International Standards Organization.

Kim, J.-J. and Rigdon, B. (1998) *Sustainable Architecture Module: Qualities, Use, and Examples of Sustainable Building Materials*, Ann Arbor, MI: National Pollution Prevention Center for Higher Education, University of Michigan.

Newton, P.W. (2002) 'Environmental assessment systems for commercial buildings', *Annual Report 2001–02*, Brisbane: CRC for Construction Innovation.

—— (ed.) (2008) *Transitions: Pathways Towards More Sustainable Urban Development in Australia*, Melbourne: CSIRO and Dordrecht: Springer.

PRé (2006) *SimaPro 7 Life Cycle Assessment Software Package*, Amersfoort: PRé Consultants. Online. Available at HTTP: <www.pre.nl/simapro/simapro_lca_software.htm#SimaProFamily>.

Seo, S. (2002) *International Review of Environmental Assessment Tools and Databases*, Report 2001–006-B-02, Brisbane: CRC for Construction Innovation.

Seo, S., Tucker, S. and Newton, P. (2006) *Sustainable Decision Support Tool for Building Materials*, 5th Australian Life Cycle Assessment Conference: Achieving Business Benefits from Managing Life Cycle Assessments, Melbourne, 22–24 November.

Todd, J.A., Crawley, D., Geissler, S. and Lindsey, G. (2001) 'Comparative assessment of environmental performance tools and the role of the Green Building Challenge', *Building Research and Information*, 29 (5): 324–35.

Watson, P., Mitchell, P. and Jones, D. (2004) *Environmental Assessment for Commercial Buildings: Stakeholder Requirements and Tool Characteristics*, Report 2001–006-B-01, Melbourne: CSIRO.

Wix, J. and Liebich, T. (1997) 'Industry Foundation Classes: architecture and development guidelines', *IT Support for Construction Process Re-Engineering: Proceedings of CIB Workshop W078 and TG10*, Cairns, Australia: James Cook University.

11 Estimating indoor air quality at design

Stephen Brown, Selwyn Tucker,
Lidia Morawska and Stephen Egan

Current assessment of indoor air quality (IAQ) focuses on the measurement of pollutants in the constructed building to assess compliance with mandatory or advisory guidelines. Pollutant emissions from indoor materials and products are known to exert a significant influence on IAQ (Brown 1999a; Brown *et al.* 1994):

- while the building is new (up to about six months old), from paints, adhesives, floor coverings and furniture;
- over the life of the building, from office equipment and some reconstituted wood-based panels.

At the design stage, experience is used to select low-emission materials, if available, but generally the prediction of IAQ is poor. No model or tool exists which is specifically aimed at predicting the IAQ of a building at this stage, yet such a method/tool would assist designers in creating optimum indoor air environments. A method for estimating IAQ will allow key decisions on the selection of materials during design such that environmental and occupant health consequences can be minimized. Such a method should also support the implementation of IAQ air quality guidelines in building codes (Brown 1997).

This chapter describes the development of an innovative design tool, the IAQ Estimator (Tucker *et al.* 2007), with which the impacts of major pollutant sources in office buildings are predicted and minimized. The IAQ Estimator considers volatile organic compounds, formaldehyde and airborne particles from indoor sources, and the contribution of outdoor urban air pollutants according to urban location and ventilation system filtration. The estimated pollutant loads are for a single, fully mixed and ventilated zone in an office building, with acceptance criteria derived from Australian and international health-based pollutant exposure guidelines. The prediction is made using a software tool for building designers so that they can select materials and appliances that, in combination, are sufficiently low emitting to prevent guidelines for indoor air pollution from being exceeded. The tool acquires its dimensional data for the indoor spaces either manually or from a 3D

CAD model via Industry Foundation Classes (IFC) files, and its emission data from an in-house building products/contents emissions database. This chapter describes the underlying development of IAQ Estimator and discusses its application by the designers of buildings.

IAQ Estimator was developed by:

- creating a database of air pollutant emission rates for typical large area building materials and contents, focusing on representative examples of paints, adhesives, floor coverings, plasterboard, reconstituted wood-based panels, office furniture and copiers/printers;
- utilizing this product database to estimate the effects of product selection, product loading, building age and ventilation scenarios on IAQ for a single level of an office building (simplified to one zone) in a 3D CAD model;
- estimating urban particle and air toxic levels in mechanically ventilated office buildings for different conditions of urban air pollution, emissions from copiers/printers and ventilation system filter efficiency;
- integrating the above three factors to estimate indoor air pollutant levels within a building directly from the products information available in a 3D CAD model or from information introduced manually;
- promoting design decisions on product selection according to the relative impacts of products on IAQ estimates in comparison to health-based IAQ exposure guidelines.

Indoor air pollution

The primary sources of indoor air pollution in office buildings (see Figure 11.1) are considered to be:

- emissions from large-area building products;
- emissions from office furniture and equipment;
- pollutants from urban air introduced by ventilation.

Emissions from large-area building products are volatile organic compounds (VOCs) and formaldehyde from paints, floor-covering systems, painted plasterboard and wood-based panels (Brown 1999a, 2002), and from office furniture such as workstations (Brown 1999b). These emissions are generally proportional to the area of product exposed indoors, and are expressed as an Emission Factor in units of pollutant mass/area/time. Emission Factors of building products are highly variable from product to product and generally decrease rapidly to background levels in the first weeks to months after construction, as in Figure 11.2. Formaldehyde emission from wood-based panels is an exception to this behaviour, with emissions reducing only to elevated steady-state levels within a few months of manufacture, and then persisting for some years

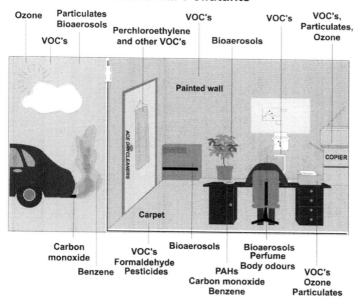

Indoor Air Pollutants

Figure 11.1 Primary sources of indoor air pollution.

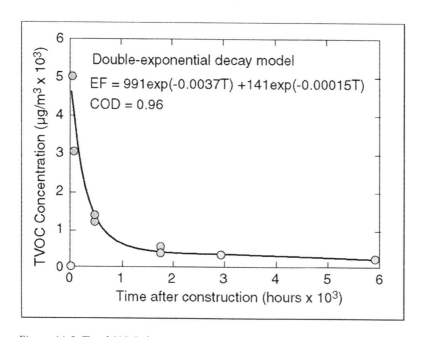

Figure 11.2 Total VOC decay in a new building after construction (source: Brown (2001), reproduced with permission).

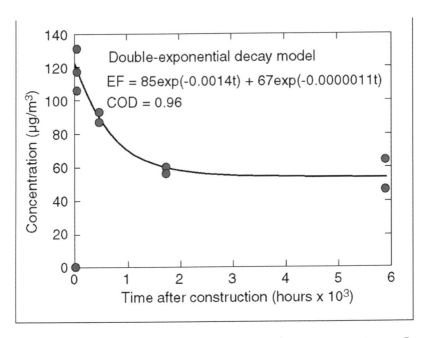

Figure 11.3 Formaldehyde decay in a new building after construction (source: Brown (2001), reproduced with permission).

(Figure 11.3). For IAQ Estimator, it was considered that estimates needed to be considered for the first 1–28 days for product emissions that decay rapidly, and additionally at six months for persistent product emissions.

Emissions from office equipment are VOCs and respirable and sub-micrometre particles (Brown 1999c, 1999d; He *et al.* 2007). Such emissions occur predominantly while the equipment is operating, linked to the number of copies produced. Emission Factors for these products are in units of pollutant mass/copy. The impact on IAQ will depend on the frequency of operation of office equipment, but will be independent of equipment age. For IAQ Estimator, the designer will need to know what type of office equipment will be used and its frequency of operation.

Air pollutants in mechanically ventilated office buildings are from outdoor urban respirable and sub-micrometre particles and air toxics pollution. In Australia, there are health-based National Environmental Protection Measures for these pollutants (NEPC 2007). Generally, these urban air pollutants will occur at higher levels in city centres or close to busy roads. Their ingress into buildings will depend on the type of ventilation used (e.g., natural or mechanical), the ventilation rate and the efficiency of filtration. For IAQ Estimator, the designer will need to input the building's location, the ventilation rate and the filter efficiency.

IAQ Estimator modelling

IAQ Estimator utilizes a methodology that considers the basic IAQ factors described above but also simplifies the building scenario by:

- considering only key, large area materials and contents;
- identifying the dominant VOCs and airborne particles present in indoor air of office buildings;
- developing a database of Emission Factors for the selected product types and pollutants;
- loading the building space, considered as a single fully-mixed zone, at quantified ratios with materials and contents;
- interfacing with pollutants introduced from outdoor air and the effect of ventilation filtration;
- aggregating each pollutant contributed from the indoor sources and outdoor air to estimate a profile of pollutant levels over time after construction;
- comparing the estimated pollutant levels with guidelines derived from health-based criteria, such that products causing guidelines to be exceeded can be easily identified and substituted.

Emissions for selected building materials and pollutants

A list of 20 key VOCs (including formaldehyde) was derived from existing knowledge of the VOC species found in Australian buildings and emitted from materials and equipment (Brown 1999a, 1999b, 1999c, 1999d, 2002). It was essential that a health-based environmental guideline existed for each (NHMRC 1996; WHO 2000; NEPC 2007; CEPA 2005; Calabrese and Kenyon 1991; Nielsen *et al.* 1998; ISIAQ/CIB Task Group TG 42 2004); where more than one guideline existed, values were averaged to provide a criterion for IAQ Estimator. No pollutant was included if such a guideline was not established. The compounds and maximum concentration goals within IAQ Estimator are presented in Table 11.1. Available Australian air emission data for building and furniture products for these 20 VOCs were collated into a database of representative products, covering emissions considered to be low, typical and high:

- paints (zero emission, low odour and acrylic, solvent-based) on plasterboard or other substrates;
- floor-covering systems (carpet/underlay/low and high-emitting adhesives, tile, wood panel floorboards, timber pre-coated with lacquer);
- wall boards (plasterboard and reconstituted wood-based panels, including medium density fibreboard (MDF));
- fixed furniture materials (shelf units, workstations).

Table 11.1 Pollutants and goal values for IAQ Estimator tool

Pollutants	IAQ Estimator goal ($\mu g/m^3$ unless stated)
Acetaldehyde	300
Benzene	60
2,6-Di-tert-butyl-4-methylphenol (BHT)	500
1,4-Dichlorobenzene	800
1,2-Dichloroethane	700
Dichloromethane	1,100
Diethylene glycol ethyl ether	6,000
Ethylbenzene	800
Ethylene glycol ethyl ether	200
Formaldehyde	40
Isobutyl methyl ketone	500
Naphthalene	30
Phenol	300
Styrene	500
Tetrachloroethylene	100
Toluene	300
Trichloroethylene	150
Total VOC (TVOC)	500
m-/p-Xylene	300
o-xylene and o-/m-/p-xylene	300
PM2.5 particles	25
PN1 particles	5,000 particles/cm^3

Emissions from operating equipment

Previous research (Brown 1999c, 1999d) developed a room chamber methodology for assessing emissions from office equipment, and showed that VOCs (ethylbenzene, xylene isomers, styrene, toluene) and respirable particles were the dominant emissions from dry-process copiers and printers, these being proportional to the number of copy operations. Hence, Emission Factors can be expressed as pollutant mass/copy. However, office equipment changed shortly afterwards to digital copier technology, in which documents were scanned one time (instead of one scan per copy) and then reproduced in multiples as needed. Also, a recent study of laser printer emissions (He *et al.* 2007) has reported the emission from some printers of high levels of sub-micrometre particle numbers (PN), 0.02 to 1 micrometre diameter, and referred to here as PN1.

Thus, IAQ Estimator emission data for office equipment include VOCs, respirable fine particles PM2.5 (mass concentration of particles smaller than 2.5 micrometre (μm) cut-point) and sub-micrometre particle numbers as PN1 from the studies described above (goals for the latter two are also presented in Table 11.1). Since emission data were lacking for current digital copiers, further assessment of these was undertaken with the same chamber method used previously, but including PN1 emission. Pollutant emissions for the 1998 copier and a 2007 digital multifunctional copier are summarized in Table 11.2. While both copiers were below detection for formaldehyde emission and exhibited very low ozone emissions (old printer emissions were previously found to emit ~70 μg/copy), they exhibited similar emission levels for VOCs. However, the 2007 copier exhibited lower respirable particle emissions by an order of magnitude. This copier was also a high emitter of PN1, emitting at a similar level to the high-emission printers reported by He *et al.* (2007).

All data discussed above were included in the products emission database, expressed as pollutant mass per copy. Generally, estimates should be based on actual emission data for specific office equipment, but default values based on the higher emission equipment were recommended for equipment absent from the database.

IAQ Estimator requires an estimate of copy rate per hour for copier or printer operation. Ideally, actual copy rates would be imported into the estimate, but in practice (especially at the design phase) IAQ Estimator provides the following guidance on specifying copy rates:

- a high usage value (2,000 copies per hour) should be applied to multifunction copiers unless actual usage data are available;
- medium-use (18 to 35 copies/minute) and low-use (15 to 20 copies/minute) copiers and personal printers should apply a usage value based on the average copy rate for office workers of 50 copies per day for each person sharing the equipment.

Emissions from office furniture

Typical air emissions of the 20 key VOCs from office furniture were included in the database. Office furniture emissions can be expressed as pollutant mass/workstation/time, but the approach used here was to estimate workstation areas (a typical workstation included desk, desk return, drawers, shelf unit and chair) and express the emissions as pollutant mass/area/time. This proved a useful approach where additional furniture items (fixed shelving, workbenches, etc.) were included.

Table 11.2 Pollutant emissions from copiers

Copier	Emission factor (μg/copy for all but PN1 as particles/copy)					
	Ozone	Formaldehyde	PM2.5	PN1	TVOC	VOCs
1998 copier	0.4	<1	2.5	n.m.	92	Ethbz (33), Xyl (27), Sty (20)
2007 copier	3	<1	0.3	6×10^8	74	Tol (13)

Pollutants in urban air used for building ventilation

Three real-world categories of outdoor pollution were incorporated into IAQ Estimator according to building location: busy road, urban and rural (Table 11.3).

The particle filtration system of mechanically ventilated office buildings was linked to the above PN1 and PM2.5 levels to estimate the impact of different filtration performances on indoor particle levels (Table 11.4). Particle deposition was ignored in estimates for simplification purposes.

Note that urban air levels for VOCs of health concern, commonly referred to as BTEX (benzene, toluene, ethylbenzene, xylenes), and total VOC (TVOC) levels were also included in IAQ Estimator, but with no removal process considered.

Product emission database

A product emission database was constructed for individual products within the classifications:

- paints;
- floor coverings;
- furniture/wood-based panels;
- copiers/printers.

For each product, an Emission Factor (EF, mass of pollutant/area/time, or mass of pollutant/copy) was documented for each pollutant at specific

Table 11.3 Pollutant levels outdoors ($\mu g/m^3$ unless specified)

Location	PN1 (p/cm³)	PM2.5	Benzene	Toluene	Ethylbenzene	Xylenes	TVOC
Busy road	100,000	25	6	17	<2	10	150
Urban area	10,000	15	4	4	<1	3	60
Rural area	1,000	10	1	1	<0.5	1	20

Table 11.4 Filter efficiencies used in IAQ Estimator

Filter type	Filter efficiency (E)	
	PN1	PM2.5
Bag (95%)	0.65	0.65
Bag (85%)	0.55	0.58
Bag (65%)	0.23	0.28
Pleated	0.035	0.092

times considered relevant to occupancy of new buildings – one day, three days, seven days, 14 days, 28 days and six months – with data at the latter two periods often being unavailable and having to be extrapolated from measurements at earlier times. IAQ Estimator derived an indoor air concentration for each pollutant at each of these times by aggregating the pollutants contributed by the indoor products and outdoor air, with consideration of the product loading, the filter efficiency, and the ventilation rate and air recirculation strategy. Generally, for VOCs and formaldehyde this required a simple summation for products x_1, x_2,..., as follows:

Estimated Concentration = $\Sigma(EF_x.Area_x)/$(Indoor Volume. Ventilation Rate)

but a more complex treatment was required for particles due to filtration in the mechanical ventilation system.

Particle filtering of air

Most mechanical ventilation systems are considered to operate in an air recirculation mode whereby:

- outdoor air, at a flow rate (Q_{OA} m³/h) specified in building codes, is drawn into the building through one filter (Filter 1, particle efficiency E_1);
- building air is recirculated through the system at a return air rate (Q_{RA} m³/h) with the combined return air plus the outdoor air being passed through a second filter (Filter 2, particle efficiency E_2).

IAQ Estimator is based on a model that incorporates this general ventilation system design, as in Figure 11.4. Note that an estimate can be made for a natural ventilation scenario by setting particle efficiencies to zero.

The indoor particle concentration (C) over time (t) for this ventilation system can be simplified to (Jamriska *et al.* 2003):

$dC/dt + \alpha C = \beta$

$\alpha = [Q_{RA} + Q_{OA} - Q_{RA} (1 - E_2)]/V$

$\beta = [C_{OA} Q_{OA} (1 - E_1)(1 - E_2) + \Sigma_i G_i]/V$

and at steady state conditions:

$C_\infty = \beta/\alpha = [C_{OA} Q_{OA} (1 - E_1)(1 - E_2) + \Sigma_i G_i]/[Q_{RA} + Q_{OA} - Q_{RA} (1 - E_2)]$

where:

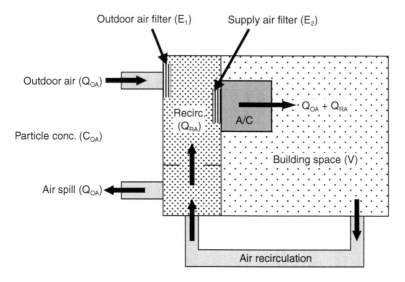

Figure 11.4 Schematic diagram of air circulation and filtering (source: Tucker *et al.*
(2007), reproduced with permission).

C_{OA} = concentration of particles in outdoor air (mass/volume),
Q_{OA} = outdoor air flow rate into building (volume/time),
Q_{RA} = recirculation air flow rate into building (volume/time),
$\Sigma_i G_i$ = sum of indoor particle source emissions (mass/time),
E_1, E_2 = particle removal efficiencies (0 = no removal, 1 = 100% removal) of
outdoor air and supply air (OA + RA) filters, respectively.

Note that this model ignores deposition losses of particles to interior surfaces since this was considered to be too variable a factor, and such simplification prevents the underestimation of particle levels. Similarly, as
discussed earlier, it was assumed that there would be no removal of VOCs
and formaldehyde by the ventilation system or by surface losses ('sink'
effects) within the building. As for particles, this prevents underestimation
of pollutant levels.

Computer software

The computer model was assembled as an integration of:

- acquiring dimensional data for the indoor spaces from a 3D CAD building information model via IFC files, an application of DesignView™
  software (Egan *et al.* 2007);
- pollutant emissions from products within office buildings;
- impacts of outdoor air quality, ventilation rate and filtration efficiency;

- comparison of IAQ estimates with goals such that the designer can identify unacceptable products.

The prototype software acquires its dimensional data for the indoor spaces either from a 3D CAD model or alternatively from manual entry of building components and their sizes. Specific building products in the model and the types of office equipment are identified by the user. The user also inputs outdoor air and return air flow rates, which then link to the building volume as ventilation rates. Ventilation system filtration was limited to three scenarios of typical particle filtration. Outputs are indoor air pollutant concentrations that are estimated for six periods after construction. The tool then designates a pass/fail scenario for each period by comparison with IAQ goal concentrations (Table 11.1).

The use of the DesignView™ platform for IAQ Estimator provided powerful functions besides the ability to view IFC files, such as:

- a plug-in architecture based on the Eclipse Rich Client Platform (Eclipse 2007a) and Eclipse Modeling Framework (Eclipse 2007b) which allowed multiple analysis applications to sit alongside and interact with DesignView™;
- a navigator panel which allowed selection of a particular model for viewing;
- a 'tree view' which allowed the user to rapidly select sections of the building model to visualize;
- a properties panel which displayed tabular information attached to a selected building component in the viewer panel or the tree view;
- a problems list which displayed a list of missing information that had been identified by DesignView™ on loading a building model;
- a tasks list that allowed users to enter 'to do' lists to ensure that items were not forgotten during design development.

Design View™ was developed as a 'next-generation' viewer which embodied lessons learned in implementing and using previous versions of IFC viewers. The open and flexible plug-in architecture provides a foundation for other tools to be linked to it, such as LCADesign (Seo *et al.*, Chapter 10 of this volume), allowing developers to concentrate on design-specific problems.

The key feature of DesignView™ which was attractive to IAQ Estimator was the ability to add finishes to any building object in the 3D CAD model and to see it visually. The system was modified to add paints and panels from the material emissions database. It was also expanded in order to be able to add office furniture and equipment into the space. These finishes could be visually inspected in the 3D Viewer (Figure 11.5).

The key features of DesignView™ most useful in IAQ Estimator are the 3D Viewer, the Model Outline and the Navigator views.

Figure 11.5 3D Viewer of DesignView™ (source: Tucker *et al.* (2007), reproduced with permission).

- The 3D Viewer displays a real-time, fully rendered 3D view of the active building model. The user may use the mouse and keyboard to explore and interact with the model. The camera can be panned, zoomed and orientated in any direction, and the user can select individual building elements.
- The Model Outline displays the hierarchy of the building model in a tree structure. The model may be traversed according to four hierarchies: Building, Element Type, Space Type and Material Type. The Model Outline view is linked with the other views in a Workbench perspective. Selecting elements in the Outline will highlight their visual representations in the 3D Viewer (if they are visible). Conversely, selections made in the 3D Viewer will be reflected in the Outline.
- The Navigator view is used to display and navigate through the workspace, and functions just like Explorer in MS Windows.

Proof of concept

Validation of the IAQ Estimator tool was limited to comparing estimates with published building measurements (Brown 2002) and specific data collected for one test building (see below). As currently available, it is considered a proven concept tool and a significant step towards a commercial product. For example, for hypothetical buildings:

- it showed the impacts of high-, medium- and low-polluting sources;
- it was possible to refine the building design to meet goals not only by selecting low-emission products but also by reducing surface areas, increasing ventilation rates or delaying time to occupancy.

Proof of the tool in a test building is a complex task, since it requires documentation of *all* of the following:

- building materials used, their quantities and date(s) installed;
- building furniture used, their quantities and date(s) installed;

- office equipment and its frequency of use;
- pollutant emission properties of all the above;
- ventilation rate (fresh air exchange rate) of the building;
- air filtration used in the ventilation system, its efficiency for the IAQ pollutants;
- levels of pollutants in urban air;
- time schedule for the building construction in relation to its occupancy;
- IAQ measurements for the building during several periods after construction.

Proof-of-concept for IAQ Estimator was sought with application to one building (Brown 2007), and was based on a restricted number of materials and pollutants for which data were available or were specifically measured for this project. The building was a nine-storey office block in a central city location, designed to be a landmark green building. Most wall and ceiling surfaces were bare concrete, and it used a low-VOC paint on a small proportion of walls. Low-emission carpet tiles, mechanically fixed (i.e., no adhesive), were used throughout. Shelf units and workbenches consisted of a low-formaldehyde plywood sealed with a water-based lacquer. Workstations were constructed from powder-coated, low-formaldehyde MDF panels. Pollutant concentrations estimated for one level of the building, and based on emissions from only the above products, are presented in Table 11.5.

Formaldehyde and VOC measurements in the building are presented in Table 11.6. Note the limitation that the time from occupancy will differ from the time after installation due to construction schedules; it is considered likely that delays of one to six weeks occurred between installation of some products and building occupancy.

Generally, IAQ Estimator predicted that the concentrations of formaldehyde, acetaldehyde and TVOC would be well below the IAQ goals. The formaldehyde measurements agreed with this, the levels being significantly below those typically found in office buildings ($20–100\,\mu g/m^3$) (Brown 1997), due to the low-emission building products used. Overall,

Table 11.5 Pollutant concentrations estimated for one level of a green office building

Pollutant	Pollutant concentration (µg/m³) at time after product installation					
	1 day	3 days	7 days	14 days	28 days	6 months
Formaldehyde	0.7	0.8	0.8	0.6	0.8	0.8
Octyl acetate	2	3	1	0.4	0.4	0.4
Ethyl hexanol	>0.2	0.2	0.4	0.4	0.4	0.4
Acetaldehyde	n.a.	n.a.	n.a.	1	0.5	0.3
4PC	n.a.	n.a.	n.a.	0.2	0.2	0.2
TVOC	26	>6	>4	>3	>2	>2

Table 11.6 Pollutant measurements for one level of a green office building

Pollutant	IAQ goal ($\mu g/m^3$)	Concentrations ($\mu g/m^3$) at ~time from occupancy			
		15 days		4.5 months	
		Range	Average	Range	Average
Formaldehyde	40	–	9	<5–16	11
Octyl acetate	n.a.	–	<6*	<2	<2
Ethyl hexanol	n.a.	–	n.a.	3–4	3
Acetaldehyde	300	–	n.a.	n.a.	n.a.
4PC	n.a.	–	n.a.	<2	<2
TVOC	500	–	18*	110–160	130

Note
* It is considered likely that the 15-day VOC measurements are underestimated due to use of a different analytical procedure.

however, IAQ Estimator has underestimated the formaldehyde and VOC levels, because there are other sources of these pollutants in the building that were not considered. Tables 11.5 and 11.6 also show the following:

- ethyl hexanol was found in the low-emission paint used in the building; its six-month estimate was $0.4 \mu g/m^3$ in comparison to a measurement of $3 \mu g/m^3$, both considered to be very low indoor air concentrations;
- octyl acetate was also emitted from the low-emission paint (in fact at higher amounts than ethyl hexanol), but was not detected in measurements;
- 4-phenyl cyclohexene (4PC), the odorant commonly found with new carpets, was below detection ($<2 \mu g/m^3$) in both estimates and measurements;
- despite the differences described above, in no case did estimates and measurements differ in showing that IAQ goals were met;
- overall, it is clear that proof-of-concept for IAQ Estimator requires:
  - a building where medium to high pollutant levels will be measured so that the comparison to estimates will have a broader range of pollutants, more of which will be in a measurable range;
  - a scenario where the materials used in the building are fully characterized for emissions over a six-month period.

Implications for industry

IAQ Estimator will enable building designers to estimate the impacts of different materials, finishes, office equipment and ventilation practices on

IAQ. By selecting different scenarios, the possibility of IAQ goals being exceeded can be understood, different strategies can be adopted (short-term increase in ventilation, delayed occupancy) and pollutant exposures can be reduced. The present tool has many simplifying assumptions, such as:

- a building level is treated as one fully-mixed zone;
- only large-area materials are included;
- the emissions database is not extensive for the materials considered (though it can grow with applications);
- only the 'dominant' VOCs found in building air or product emissions are included;
- pollutants without health-based goals are not included;
- losses of pollutants to surfaces are not considered;
- filtration efficiencies are for new filters.

In general these assumptions will lead to overestimates of indoor air pollution, and so this should not be considered a fully predictive tool. The key benefit of IAQ Estimator is that it will allow avoidance of polluting products at an early stage in the design process. It is not expected to replace the need for IAQ assessment of new buildings after construction, but it should reduce the likelihood of IAQ being found to be unacceptable.

In general, IAQ Estimator is:

- an office design tool for selection of materials, office equipment and ventilation;
- useful in optimizing IAQ early in the design process;
- a tool that allows control of indoor air pollutants:
  - from new materials (aimed at first six months of construction);
  - from long-term factors such as office equipment, filtration and urban air.

IAQ Estimator is currently not:

- a means of distinguishing a priori whether indoor air presents health risks;
- a means of predicting IAQ with precision;
- a means of dealing with all aspects of IAQ (e.g., provision/distribution of ventilation air, maintenance factors, indoor activities, other pollutants such as micro-organisms, combustion gases, etc.);
- a tool for use in regulations.

However, it is a pointer to the future in relation to the emergence of a suite of tools for automated performance assessment of buildings at the design stage.

222   *S. Brown* et al.

Acknowledgement

The work reported in this chapter was part of a research project conducted by the Cooperative Research Centre for Construction Innovation which receives funds from the Australian government's Cooperative Research Centre Program. Also, the authors acknowledge the technical contributions received from Ms Fanny Boulaire, CSIRO, and Dr Congrong He, Queensland University of Technology.

Bibliography

Brown, S.K. (1997) *National State of the Environment Report: Indoor Air Quality*, SoE Technical Report Series, Canberra: Department of the Environment, Sport and Territories.

—— (1999a) 'Occurrence of VOCs in indoor air', in T. Salthammer (ed.) *Organic Indoor Air Pollutants: Occurrence, Measurement, Evaluation*, Weinheim: Wiley-VCH.

—— (1999b) 'Chamber assessment of formaldehyde and VOC emissions from wood-based panels', *Indoor Air*, 9: 209–15.

—— (1999c) 'Assessment of pollutant emissions from dry-process photocopiers', *Indoor Air*, 9: 259–67.

—— (1999d) 'Pollutant emissions from a dry-process photocopier and laser printers', in *Proceedings of the 8th International Conference on Indoor Air Quality and Climate*, Vol. 5, Edinburgh.

—— (2001) 'Air toxics in a new Australian dwelling over an 8-month period', *Indoor and Built Environment*, 10: 160–66.

—— (2002) 'Volatile organic pollutant concentrations in new and established buildings from Melbourne, Australia', *Indoor Air*, 12: 55–63.

—— (2007) 'Specification and assessment of high quality indoor environments for sustainable office buildings', in *25th Annual Conference of Australian Institute of Occupational Hygiene*, Melbourne.

Brown, S.K., Sim, M.R., Abramson, M.J. and Gray, C.N. (1994) 'Concentrations of volatile organic compounds in indoor air: a review', *Indoor Air*, 4: 123–34.

Calabrese, E.J. and Kenyon, E.M. (1991) *Air Toxics and Risk Assessment*, Chelsea, MI: Lewis.

CEPA (2005) *All Chronic Reference Exposure Levels Adopted by OEHHA as of February 2005*, Sacramento: Office of Environmental Health Hazard Assessment, California Environmental Protection Agency. Online. Available at HTTP: <www.oehha.org/air/chronic_rels/AllChrels.html>.

Eclipse (2007a) *Eclipse Modeling Framework*. Online. Available at HTTP: <www.eclipse.org/modeling/emf/>.

—— (2007b) *Eclipse Rich Client Platform*. Online. Available at HTTP: http://wiki.eclipse.org/index.php/Rich_Client_Platform.

Egan, S., Boulaire, F., Drogemuller, R., James, M., McDonald, K. and McNamara, C. (2007) *SpecNotes and Viewer Extension*, Report 2004–014-B, Brisbane: CRC for Construction Innovation.

He, C., Morawska, L. and Taplin, L. (2007) 'Particle emission characteristics of office printers', *Environmental Science & Technology*, 41: 6039–45.

ISIAQ/CIB Task Group TG 42 (2004) *Performance Criteria of Buildings for Health and Comfort*, Helsinki: International Society of Indoor Air Quality and Climate, and International Council for Research and Innovation in Building and Construction.

Jamriska, M., Morawska, L. and Ensor, D.S. (2003) 'Control strategies for sub-micrometer particles indoors: model study of air filtration and ventilation', *Indoor Air*, 13: 96–105.

NEPC (2007) *Review of the National Environment Protection (Ambient Air Quality) Measure*, Canberra: National Environment Protection Council.

NHMRC (1996) *Ambient Air Quality Goals and Interim National Indoor Air Quality Goals*, Canberra: National Health and Medical Research Council.

Nielsen, G.D., Hansen, L.F., Nexo, B.A. and Ponken, O.M. (1998) 'Indoor air guideline levels for 2-ethoxyethanol, 2-(2-ethoxyethoxy)ethanol, 2-(2-butoxyethoxy) ethanol and 1-methoxy-2-propanol', *Indoor Air*, Suppl. 5: 37–54.

Tucker, S.N., Brown, S.K., Egan, S., Morawska, L., He, C., Boulaire, F. and Williams, A. (2007) *Indoor Air Quality Estimator*, Report 2004–033-B-01, Brisbane: CRC for Construction Innovation.

WHO (2000) *Guidelines for Air Quality*, 2nd edn, Geneva: World Health Organization.

12 Designing for disassembly

Philip Crowther

With increasing levels of awareness about ecosystems constraints on development, and a growing desire to create green buildings, there is a general acceptance of the need to reduce the quantity of building material consumption and to reduce construction and demolition waste. Much of the construction industry activity, in seeking to achieve these goals, is focused on the efficiency of material use and waste minimization on the construction site. At the other end of the building's service life, in the demolition industry, we see some attempts to recycle materials, though generally this is in the form of down-cycling (such as crushing concrete to create road base). Higher levels of recycling and reuse are less common. The basic problem is that buildings are not generally designed to be taken apart.

Current practice in the deconstruction of existing buildings has shown that there are numerous technical barriers to the successful recovery and reuse of materials and components. These barriers stem mainly from current construction practices that view the assembly of materials and components as a unidirectional practice with an end goal of producing a final building. Such a linear and truncated view of the built environment severely limits the end-of-life options when a building has reached the end of its service life. A more cyclic or closed-loop view of the built environment and the materials within it will recognize the need to consider, at the design stage of a project, the deconstruction process as well as the construction process. Such consideration can be expressed as the need to design for disassembly.

While current industrialized building practice pays little attention to the issues of reuse and recycling, there are numerous historic examples of buildings that have been successfully deconstructed for reuse and that have been specifically designed for such disassembly. Analysis of these examples highlights common strategies that can offer useful information to designers seeking to improve the rates of future material and component recovery. Such information can also be sourced from related fields such as industrial design.

Broad design themes for disassembly

A review of architectural history and of related industries, such as industrial design, shows that there are two types of knowledge of relevance to design for disassembly. First are the broad themes that address the issues of why, what, where and when to disassemble. Second are the specific design principles of how to design for disassembly.

There are three broad themes that significantly impact the decision-making process of designing a building for future disassembly. These are:

1   a holistic model of environmentally sustainable construction;
2   the understanding of a building as a series of layers with different service lives;
3   a recycling hierarchy that recognizes the cost-benefits of different end-of-life scenarios.

A model of environmentally sustainable construction

Before attempts are made to design for disassembly, the consequences must be understood within a wider picture of the built environment, and indeed within the global environment. While designing for future reuse will have obvious environmental benefits there are also potential environmental costs and, while these are almost certainly of a smaller impact, they must be recognized and considered. To manage this process, a model is required that allows the place and role of design for disassembly to be seen within the overall scheme of environmentally sustainable construction.

The notion of life cycle assessment (LCA) is a well recognized way of understanding, assessing and planning a reduction in the environmental consequences of our actions in materials and product manufacturing. A life cycle assessment of a system or product identifies all of the inputs and outputs, both beneficial and otherwise, during the life of that system or product. It is usual to visualize this analysis as a two-dimensional graph or matrix that plots environmental resources against the stages of the system or product life. In this way, all of the cumulative environmental impacts can be seen and analysed. This model, with its two axes of environmental resources and life cycle stages, does not, however, offer solutions or strategies for dealing with the unwanted impacts.

To propose solutions to these problems, a third axis of principles for environmental responsibility can be added. In this way, a three-dimensional matrix can be created. Charles Kibert (1994), of the University of Florida, proposes such a model (see Figure 12.1). This model, with the three axes of environmental resources, life cycle stages and principles of sustainability, can be used to graphically illustrate the large number of issues that pertain to a sustainable construction industry and the interrelationships between them. It can also be used as a tool to manage the decision-making process

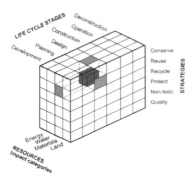

Figure 12.1 A conceptual model for sustainable construction (source: based on Kibert (1994: 11)).

during a construction project. At any point of intersection along the three axes there will be a range of decisions to be made, each with further impacts to be considered along these axes.

It can be seen in Kibert's model that there is a time and place for the design of a building to maximize resource reuse for materials in the future – that is, it has identified an explicit place for design for disassembly. This model highlights the fact that such an activity should exist within the general field of sustainable construction, and also shows the potential relationships with other environmental issues, strategies and stages.

This model locates design for disassembly within a larger picture of sustainable construction. In particular, it shows where the concepts and principles of design for disassembly belong. Such an understanding of where design for disassembly sits, in comparison with other environmental sustainability strategies, forms an important part of the conceptual knowledge base for design for disassembly. This model assists in an understanding of why to design for disassembly. It can therefore be used as a design tool to manage the conflicts that will occur between alternative principles and between design for disassembly guidelines (discussed later in this chapter).

Time-related building layers

When we discuss a building, we tend to think of it as just that – a single building: conceived, designed, constructed, used and disposed of as a complete entity. However, this notion is flawed, in part resulting from our conceiving of the building over a limited timeframe. Most buildings have long lives in some form or another, and usually change in various ways over their lifetime. This results in a series of different buildings over time that may or may not share certain physical parts. Typically, the structure of a building may be retained while the internal-space making components are removed and replaced, or while services are upgraded.

Analysis of vernacular architecture, especially by the noted writer John Habraken (1998), typically identifies two layers of building: first, the structural frame, which has a long service life; and second, the space-making elements of partitioning walls that may be removed and reused or replaced over time as spatial needs change. In these buildings, steps are taken to allow for such changes over time through the design for disassembly of those components with a shorter service-life expectancy.

Recognition of the different service lives of different parts of a building was a popular topic with architects in the 1960s, when groups such as Archigram in Britain and the Metabolists in Japan were experimenting with building systems where such disassembly was not only possible but also highlighted in the aesthetic character of the projects. These architects often proposed specific service lives for different parts of buildings, depending on their life expectancy.

For a similar but much expanded analysis of the different service lives of the layers of buildings, the work of Stewart Brand (1994) is noteworthy. He dissects the layer of a building into Structure, Skin, Services, Space Plan and Stuff, and also adds the layer of the Site on which the building stands.

- The Site is defined as geographical setting, the ground on which the building sits. 'Site is eternal.'
- The Structure is the foundations and load-bearing components of the building – those parts that make the building stand up. Structure is expected to last from 30 to 300 years.
- The Skin of the building is the cladding and roofing system that excludes (or controls) the natural elements from the interior. This will last an expected 20 years due to wholesale maintenance, changing technology and fashion.
- The Services – electrical, hydraulic, data, etc. – have an expected life of seven to 15 years.
- The Space Plan, the internal partitioning walls and systems, will change every three years in commercial buildings, and up to every 30 years in domestic buildings.
- The Stuff – the furniture and other non-attached space-defining elements – will change daily to monthly. Brand points out that furniture is called *mobilia* in Italian for good reason.

Brand goes to great lengths to explain the technical and social benefits of designing and constructing buildings in a layered manner. Like Habraken (1998), he recognizes the lessons already learned by vernacular builders, and further suggests specific lessons for building designers based on the historic study of buildings and their adaptation, addition and relocation over time.

As we have seen, Brand suggests typical service-life expectancies for each of these layers, but a range of other writers have also suggested times for the service-life expectancies of different parts or layers of buildings. A synthesis of these proposals suggests the following typical lifespans (Crowther 2001) as:

- Space Plan: three to ten years;
- Services: five to 30 years;
- Skin: 15 to 40 years;
- Structure: 20 to 65 years;
- Site: eternal.

There is significant relevance in these time-related building layers to the concerns of design for disassembly. It is at the junctions of layers that disassembly will need to occur. As such, these junctions need to be designed to facilitate appropriate disassembly at the places where it will be required – that is, between components with different service-life expectancies. Facilitating such disassembly will allow buildings to develop over time in an environmentally and socially responsible way. An understanding of time-related building layers will assist in understanding where and when to design for disassembly. To achieve a better level of environmental sustainability, buildings need to be designed for disassembly between layers.

Recycling hierarchy

The typical mode of operation in our industrialized society is one of use once and dispose. Materials are extracted from the natural environment, processed, manufactured, used once and then disposed of, usually back into the natural environment. In the building industry, this mode of operation is certainly the dominant one. This so-called life cycle is in fact not at all cyclic but rather linear, starting with material extraction and ending in the dumping of unwanted waste. Such a model for how materials pass through the built environment identifies a number of life cycle stages: extraction, processing, manufacture, assembly, use, demolition and disposal (see Figure 12.2).

This model is not the only available option for construction and operation of the built environment, and it is not difficult to reconfigure these stages into a true cycle of material life whereby building materials and components, or indeed whole buildings, can be recycled or reused. With the appropriate disassembly strategy such recycling can occur in many different ways, as shown in Figure 12.3. It can be seen that there is a range of possible 'recycling scenarios' with a range of outcomes. If the technical outcomes of the disassembly process are considered, four differently scaled outcomes are possible:

1   the reuse of a whole building;
2   the production of a new building;

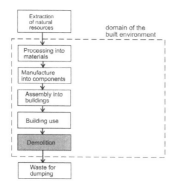

Figure 12.2 Dominant life cycle of the built environment.

3   the production of new components;
4   the production of new materials.

These would relate to four possible recycling strategies or, perhaps more appropriately named, end-of-life scenarios:

1   building reuse or relocation;
2   component reuse or relocation in a new building;
3   material reuse in the manufacture of new components;
4   material recycling into new materials.

If the strategy of deconstruction were applied to the built environment, the life cycle stage of demolition could be replaced with a stage of disassembly. The typical once-through life cycle of materials in the built environment could then be altered to accommodate the range of possible end-of-life scenarios and produce a range of alternative life cycles (see Figure 12.3).

The most significant aspect of these scenarios is that some of them are more environmentally desirable than others; for example, the reuse of a building component has the added advantage of requiring less energy or new resource input than the recycling of base materials (Seo *et al.* 2008). In a society where all energy has some environmental cost, and indeed where most is produced through major environmentally damaging processes such as the burning of fossil fuels, any strategy that reduces energy and resource use has potential advantages.

If we plan at the design stage to disassemble a building for environmental benefit, we can design for greater levels of environmental benefit as well as regeneration efficiencies. Buildings might, for example, be better designed for the reuse of components rather than simply the recycling of materials. In reality it will be advantageous for buildings to be designed for all of these levels of recycling, since the future reuse possibilities of a

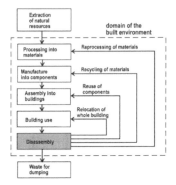

Figure 12.3 Possible end-of-life scenarios for the built environment.

building cannot be accurately predicted decades before eventual disassembly. An understanding of a hierarchy of recycling offers guidance of what to disassemble for any given end-of-life scenario. It must be noted that it may not always be preferable to design for disassembly at building or component level. It is quite possible that for a particular project there are other environmental concerns that outweigh the benefits from a design for disassembly strategy. It may, for example, be highly beneficial to construct a house using straw bales and mud, or rammed earth, both of which have significant environmental benefits due to low-impact technologies, but which do not allow for complete component disassembly. This is why the holistic picture of a sustainable construction industry is needed to guide this decision-making process.

These three broad themes of a model for environmentally sustainable construction, time-related building layers and a recycling hierarchy are important in managing the process of design for disassembly. They do not, however, answer the question of how to design for disassembly. For that, a number of design principles or design guidelines are required.

Design principles for design for disassembly

While the design for disassembly of buildings is not common practice at the present time, there are a number of important historic examples of buildings that have been disassembled, either by design or otherwise, that offer significant information about the technical aspects of such disassembly. Review of these buildings, some realized projects and some conceptual investigations reveals a pattern of common solutions or approaches to the difficulties of design for disassembly. These offer recurring principles that may be seen as design guidelines or design techniques.

1 Use recycled and recyclable materials – to allow for all levels of the recycling hierarchy. Increased use of recycled materials will also encourage industry and government to develop new technologies for recycling, and to create larger support networks and markets for future recycling. 'One of the most important facets of design for disassembly is the initial selection of materials that are easy to recycle' (Rosenberg 1992: 17). The use of recyclable materials will increase the end-of-life value of a product by allowing for all levels of the hierarchy of recycling (material recycling), even if it is expected that the component in question will be reused or relocated (Dowie and Simon 1994).

2 Minimize the number of different types of materials – this will simplify the process of sorting during disassembly and reduce transport to different recycling locations, and result in greater quantities of each material. Research into industrial design for disassembly has identified minimizing the number of types of materials as an important criterion for simplifying the recycling process due to an increase in the ease of sorting for different and separate recycling processes (Dowie and Simon 1994).

3 Avoid toxic and hazardous materials – this will reduce the potential for contaminating materials that are being sorted for recycling, and will reduce the potential for health risks that might otherwise discourage disassembly. Graedel and Allenby (1995: 263) note that 'the presence of toxics is a deterrent to detailed disassembly'. Research into design for disassembly and design for recycling in buildings acknowledges the potential future problems associated with hazardous and toxic materials (Sassi and Thompson 1998: 2).

4 Avoid composite materials and make inseparable subassemblies from the same material – in this way, large amounts of one material will not be contaminated by a small amount of a foreign material that cannot be easily separated. Research into the obstructive factors to the reuse of demolition waste in Japan has identified the use of composite materials and assemblies as a major deterrent to disassembly for recycling (Yashiro and Yamahata 1994: 589). This may diminish as a barrier in the future as recyclability becomes a performance requirement for composite and hybrid materials.

5 Avoid secondary finishes to materials – such coatings may contaminate the base material and make recycling difficult. Where possible, use materials that provide their own suitable finish or use mechanically separable finishes. Applied coatings and painted finishes are best avoided to allow better levels of disassembly (Pawley 1998: 99). An analysis of the recycling and reuse potential of building materials does note, however, that in some applications applied coatings will not greatly interfere with the recycling process (Guequierre and Kristinsson 1999: 2030). Moreover, future developments in composites, as outlined in point (4) above, may remove or diminish this as a barrier to recycling and reuse.

6 Provide standard and permanent identification of material types – many materials such as plastics are not easily identifiable, and should be provided with a non-removable and non-contaminating identification mark to allow for future sorting. This could indicate material type, place and time of origin, structural capacity, toxic content, etc. The particular problems associated with identifying plastics have resulted in the development of the International Standard Organization (ISO) recommendations for their marking with ISO terminology and symbols to allow identification for recycling (Graedel and Allenby 1995: 359–62). Progress with product declarations and eco-labelling will assist with this current obstacle.

7 Minimize the number of different types of components – this will simplify the process of sorting and reduce the number of disassembly procedures to be undertaken. It will also make recycling and reuse more attractive due to greater numbers or volumes of fewer components. Research into design for disassembly has identified the problems with sorting many different parts during the process (Andreu 1995: 10–16). The same research calls for a reduction in the number of parts and the number of types of parts.

8 Use mechanical connections rather than chemical ones – this will allow the easy separation of materials and components without force, reduce contamination of materials and reduce damage to components. In research into the obstructive factors to reusing building materials and components, Yashiro and Yamahata (1994: 589) list 'applying non-reversible jointing detail of building components' as one of six major problems.

9 Use an open building system where parts of the building are more freely interchangeable and less unique to one application – this will allow alterations in the building layout through relocation of components without significant modification. Open systems, as developed in the mid-twentieth century (Cowan 1977: 161), use components that are universally standard and are typically available off-the-shelf or in catalogues.

10 Use modular design – use materials and components that are compatible with other systems both dimensionally and functionally. The modern-day dimensional coordination of building materials and components is largely due to the efforts of Albert Farwell Bemis, who first proposed modular design in his 1936 publication *Rationalized Design*. Such coordination would 'improve the inefficient methods of assembly of unrelated materials and reduce the cost of building by applying industrial production techniques' (Cowan 1977: 320).

11 Use construction technologies that are compatible with standard building practice and common tools – specialist technologies will make disassembly difficult to perform and a less attractive option, particularly for the user. Examples of temporary and portable buildings have also used

the principle of designing for use of standard and low-tech tools and equipment. Portable colonial cottages and other temporary buildings of the nineteenth century were typically 'tailored to the limited resources of skill and tools available to the emigrant' (Herbert 1978: 11).

12 Separate the structure from the cladding, internal walls and services – to allow for parallel disassembly such that some layers of the building may be removed without affecting other layers. The system of separate frame and in-fill walls is by far the more compatible construction system for a range of disassembly requirements (Mark 1995: 182–92).

13 Provide access to all parts of the building and to all components – ease of access will allow ease of disassembly. Allow access for disassembly from within the building if possible. Many designers in the 1960s proposed highly adaptable buildings, such as those of Archigram and the Japanese Metabolists. These projects were often designed as 'inside-out' buildings consisting of large-scale superstructures onto which exposed modules were attached (Kurokawa 1977).

14 Make materials and components of a size that suits the intended means of handling – allow for various handling operations during assembly, disassembly, transport, reprocessing and re-assembly. Buildings designed to be fully transportable, such as tents and yurts, utilize a system in which all components are easily handled by one person (Kronenburg 1995: 16–21).

15 Provide a means of handling and locating components during the assembly and disassembly procedure – handling may require points of attachment for lifting equipment, as well as temporary supporting and locating devices. Examples of the handling of large components can be seen in construction that utilizes pre-cast concrete elements that have some form of connection or lifting points, as is the case in some tilt-up construction. Though the form of construction does not usually allow for it, these lifting points could also facilitate easy disassembly of the panels (Hassanain and Harkness 1998: 100).

16 Provide realistic tolerances to allow for manoeuvring during disassembly – the repeated assembly and disassembly process may require greater tolerance than for the manufacture process or for a one-off assembly process. Research into buildability recognizes the need to 'detail for achievable tolerances' as a general principle of good buildability (CIRIA 1983: 9). In a building designed for disassembly, the system of construction may rely more on the assembly of products rather than on the working of materials, as in traditional building construction.

17 Use a minimum number of fasteners or connectors – to allow for easy and quick disassembly and so that the disassembly procedure is not complex or difficult to understand. Research in industrial design has identified that minimizing the number of connectors or joining elements will have a positive effect (Magrab 1997: 153).

18 Use a minimum number of different types of fasteners or connectors – to allow for a more standardized process of assembly and disassembly without the need for numerous different tools and operations. Research into partially automated disassembly in industrial design has identified a number of basic guidelines for design for disassembly (Kiesgen *et al.* 1996: 13). These included minimizing the number of different types of fasteners used within a product.

19 Design joints and connectors to withstand repeated use – to minimize damage and deformation of materials and components during repeated assembly and disassembly procedures. Temporary deployable structures, that operate much like machines, are designed to withstand the stresses created during the assembly and disassembly processes and to minimize the problems of material fatigue (Kronenburg 1995).

20 Allow for parallel rather than sequential disassembly – so that materials or components can be removed without disrupting other materials or components. Where this is not possible, make the most reusable or valuable parts of the building most accessible to allow for maximum recovery of those materials and components that are most likely to be reused. Where 'certain operations are dependent on others being completed before they can be undertaken', it is necessary to design details to suit the sequence (Miller 1990: 35). If such serial sequences and their associated limitations are to be avoided, it will be necessary to detail components to allow parallel disassembly.

21 Provide permanent identification of component type – in a coordinated way with material information and total building system information, ideally electronically readable to international standards. Finch *et al.* (1994: 1) discuss the importance of information retention in the construction industry in detail, stating that 'a major obstacle to any recycling initiative in construction is the inability to carry out the dismantling and segregation process in a routine manner akin to a manufacturing process'. They note that the lack of information results in a lack of homogeneity in the recovered components, which in turn results in a lack of value.

22 Use a structural grid – the grid dimension and orientation should be related to the materials used such that structural spans are designed to make the most efficient use of material type and allow coordinated relocating of components such as cladding. Even ancient building types have used a grid to their advantage in making temporary buildings or buildings that can be disassembled easily, such as the traditional timber houses of Japan based on the dimensions of the Tatami mat (Itoh 1972: 42–55).

23 Use prefabricated subassemblies and a system of mass production – to reduce site work and allow greater control over component quality and conformity. The large-scale use of prefabrication and mass production became possible only after the Industrial Revolution. The

notion of prefabrication 'inherent in the manufacture of cast-iron objects, arose simultaneously with the construction of the first cast-iron buildings' (Strike 1991: 39).

24 Use lightweight materials and components – this will make handling easier and quicker, thereby making disassembly and reuse a more attractive option. Kronenburg's (1996) review of portable architecture presents a wide range of buildings designed for various levels of disassembly, including exhibition pavilions, factories, theatres and space-ships. All exhibit concern for the weight of materials and components used in order to facilitate ease of disassembly.

25 Permanently identify points of disassembly – so as not to be confused with other design features and to sustain knowledge on the component systems of the building. As well as indicating points of disassembly, it may be necessary to indicate opening or disassembly procedures as instructions (*Guidelines* 1998: 5).

26 Provide spare parts and on-site storage for them – particularly for custom-designed parts, both to replace broken or damaged components and to facilitate minor alterations to the building design. Many of the projects from the 1960s groups Archigram (Cook *et al.* 1972) and the Japanese Metabolists (Kurokawa 1977) proposed highly adaptable buildings that incorporated strategies for dealing with spare components.

27 Sustain all information on the building construction systems and assembly and disassembly procedures – efforts should be made to retain and update information such as 'as built' drawings including all reuse and recycling potentials as an assets register. In defining build-ability, Adams (1989: 2–4) identifies three principal criteria: simplic-ity, standardization and communication. Lack of information can easily result in poor assembly and disassembly. The Construction Industry Research and Information Association also notes that 'com-plete project information should be planned and coordinated ... to facilitate the best possible communication and understanding on site' (CIRIA 1983: 10).

Conclusions

It is apparent from this list of design for disassembly principles that there will be many occasions when there will be a conflict between some of them. The principles in themselves offer guidance on how to design for future disassembly, but, as already noted, there are broader themes that must be engaged with in order to answer the more challenging questions of what, where, when and, indeed, if to disassemble.

Any comprehensive strategy to design an individual building for future disassembly must operate within the (changing) framework of the con-struction industry and the quickly developing recycling and reuse industry. The individual nuances of any architectural project, and the expected long-term environmental outcomes, make it difficult to propose generic principles that will always be appropriate. The guidance offered here must be taken as a starting point for the development of individual strategies for particular buildings.

It can be seen that the technological steps that might be taken, through design, to improve the rates of material and component recovery in the future are neither complex nor alien to current industry practice. Furthermore, they are compatible with general good design practice, and with attempts to improve the environmental sustainability of the construction industry. Design for disassembly should be an important consideration in any built environment project. It offers significant opportunity to reduce our impact on the built and natural environments, now and into the future.

Bibliography

Adams, S. (1989) *Practical Buildability*, London: Butterworths (and CIRIA).

Andreu, J.J. (1995) *The Remanufacturing Process*, Manchester Metropolitan University. Online. Available at HTTP: <http://sun1.mpce.stu.mmu.ac.uk/pages/projects/dfe/pubs/dfe24/report24.htm> (accessed 13 August 1998).

Brand, S. (1994) *How Buildings Learn: What Happens After They're Built*, New York, NY: Viking.

CIRIA (1983) *Buildability: An Assessment*, London: Construction Industry Research and Information Association.

Cook, P., Chalk, W., Crompton, D., Greene, D., Herron, R. and Webb, M. (eds) (1972) *Archigram*, London: Studio Vista.

Cowan, H.J. (1977) *An Historical Outline of Architectural Science*, London: Applied Science Publishers.

Crowther, P. (2001) 'Developing an inclusive model for design for deconstruction', in A.R. Chini (ed.) *Deconstruction and Materials Reuse: Technology, Economic and Policy*, Proceedings of CIB Task Group 39 – *Deconstruction* meeting, Wellington, New Zealand, 6 April.

Dowie, T. and Simon, M. (1994) *Guidelines for Designing for Disassembly and Recycling*, Manchester Metropolitan University. Online. Available at HTTP: <http://sun1.mpce.stu.mmu.ac.uk/pages/projects/dfe/pubs/dfe18/report18.htm> (accessed 13 August 1998).

Finch, E., Flanagan, R. and March, L. (1994) 'Using auto-ID to enable efficient recycling of building materials', in CIB Task Group 8, *Buildings and the Environment*, First International Conference, Watford, UK, May.

Graedel, T.E. and Allenby, B.R. (1995) *Industrial Ecology*, Englewood Cliffs, NJ: Prentice Hall.

Guequierre, N.M.J. and Kristinsson, J. (1999) 'Product features that influence the end of a building', in M.A. Lacasse and D.J. Vanier (eds) *Durability of Materials and Components 8*, Ottawa: Institute for Research in Construction.

Guidelines (1998) Manchester Metropolitan University. Online. Available at HTTP: <http://sun1.mpce.stu.mmu.ac.uk/pages/projects/dfe/guide/guideline3.html#5> (accessed 20 August 1998).

Habraken, N.J. (1998) *The Structure of the Ordinary*, Cambridge, MA: MIT Press.

Hassanain, M.A. and Harkness, E.L. (1998) *Building Investment Sustainability: Design for Systems Replaceability*, London: Minerva.

Herbert, G. (1978) *Pioneers of Prefabrication*, Baltimore, MD: Johns Hopkins University Press.

Itoh, T. (1972) *Traditional Domestic Architecture of Japan*, New York, NY: Weatherhill/Heibonsha.

Kibert, C.J. (1994) 'Establishing principles and a model for sustainable construction', in *Sustainable Construction*, Proceedings of CIB TG 16 conference, Tampa, Florida, 6–9 November.

Kiesgen, G., Emenako, M.E. and Slawik, F. (1996) 'Development of partially-automated disassembly plants for regaining reusable parts', *Industrial Robot*, 23 (3): 11–15.

Kronenburg, R. (1995) *Houses in Motion*, London: Academy Editions.

—— (1996) *Portable Architecture*, Oxford: Architectural Press.

Kurokawa, K. (1977) *Metabolism in Architecture*, Boulder, CO: Westview.

Magrab, E.B. (1997) *Integrated Product and Process Design and Development: The Product Realization Process*, New York, NY: CRC Press.

Mark, R. (ed.) (1995) *Architectural Technology up to the Scientific Revolution*, Cambridge, MA: MIT Press.

Miller, G. (1990) 'Buildability: a design problem', *Exedra*, 2 (2): 34–8.

Pawley, M. (1998) 'XX architecten', *World Architecture*, 69: 96–9.

Rosenberg, D. (1992) 'Designing for disassembly', *Technology Review*, 95 (Nov/Dec): 17–18.

Sassi, P. and Thompson, M.W. (1998) *Summary of a Study on the Potential of Recycling in the Building Industry and the Development of an Index System to Assess the Sustainability of Materials for Recycling and the Benefits from Recycling*, Eurosolar Conference, Bonn, 1998.

Seo, S., Tucker, S. and Newton, P. (2008) 'Automated environmental assessment of buildings', in P. Newton, K. Hampson and R. Drogemuller (eds) *Technology, Design and Process Innovation in the Built Environment*, London: Spon.

Strike, J. (1991) *Construction into Design: The Influence of New Methods of Construction on Architectural Design*, Oxford: Butterworth Architecture.

Yashiro, T. and Yamahata, N. (1994) 'Obstructive factors to reuse waste from demolished residential buildings in Japan', in *Sustainable Construction*, Proceedings of CIB TG 16 conference, Tampa, Florida, 6–9 November.

13 Energy-efficient planning and design

Michael Ambrose

Energy consumption is the dominant source of greenhouse gas emissions in Australia, contributing 70 per cent of the total (Australian Greenhouse Office 2007). The buildings we live and work in are major users of this energy, and addressing how they can improve their energy efficiency through better planning, design and operation is a key element to reducing greenhouse emissions. Energy end-use in both residential and commercial buildings has grown by 11 and 9 per cent respectively since 1998–99, and residential buildings now account for 12 per cent of Australia's total energy use, while commercial buildings are responsible for a further 7 per cent (Trewin 2006). Electricity use represents the dominant source of the energy used, with natural gas the other major source.

Most of Australia's electricity generation is sourced from large-scale fossil fuel fired power stations, and delivered to consumers through an extensive national power grid. Only 7.4 per cent comes from renewable sources, with the majority of this from large-scale hydro power stations located predominantly in Tasmania. Hydro power accounts for 86 per cent of renewable electricity generation, with the remaining renewable energy sources accounting for only 1 per cent (ABARE 2007).

Within buildings, energy use is dominated by heating and cooling systems used to regulate the indoor environment. Both residential and commercial buildings have these systems as their primary energy user, representing 39 per cent of residential energy use (Reardon 2001) and 66 per cent in commercial buildings (Australian Greenhouse Office 1999). Water heating, appliances and refrigeration are the other major contributors in residential dwellings, while lighting is the other main contributor in commercial buildings. The dominance of heating and cooling systems in the energy use make-up of buildings means that these systems should be targeted.

Passive design

The amount of energy a heating and cooling system utilizes is dictated by the efficiency of the system and of the building envelope. Passively

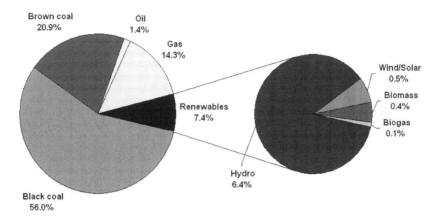

Figure 13.1 Australian electricity generation source.

designed buildings are designed to maximize their use of natural energy flows to maintain thermal comfort and minimize, if not eliminate, the need for mechanical heating and cooling systems. The principles of passive design can be applied to both residential and commercial buildings, and basically encompass the maximizing of cool air flows and shading of sun during summer months, and the trapping and storing of heat from the sun and minimizing heat loss during the winter months. The exact mechanisms for achieving this vary depending on the climate zone, but the general principles are the same.

Employing passive design principles has mostly been seen in residential buildings, but recently several landmark commercial buildings have been constructed in Australia that demonstrate how passive design can work in an office building environment. The CH2 building in Melbourne (Figure 13.2) is regarded as Australia's 'greenest' office building, and uses both passive and active systems to control the indoor environment. Natural ventilation plays a large part, both during occupied hours and to cool the thermal mass during night-time and at weekends. The concrete ceiling slab absorbs excess heat during the daytime and releases it at night, while in the cooler parts of the year an underfloor convective heating system allows warm air to percolate into the working environment without the use of fans. Much of this heat has been recovered from waste heat produced by a co-generation plant. The building has been in operation for two years, and so far has been adequately cooled by the passive thermal mass of the concrete ceilings and fresh air supply into the office, with the active cooling system yet to be needed (Jones 2008).

In residential buildings, passive design can virtually eliminate the need for active systems. Many homes across Australia have demonstrated this in a range of climate zones. Using a combination of thermal mass, insulation, good orientation and good design will result in a home that will dramatically

Figure 13.2 Council House 2 (CH2) building, Melbourne.

reduce heating/cooling costs and generally be a more pleasant environment to live in.

One such example is Research House, which was built by Queensland's Department of Public Works in Rockhampton as a live working example of sustainability principles. Since its completion in 2001 it has had tenants living in it, and its performance has been extensively monitored. The house has no active heating or cooling system, and instead relies on its passive design features to control the internal temperatures. The roof space is highly insulated, and this has resulted in significantly reducing the heat gain to the rooms below. The contribution of heat through the roof to total heat gain is around 45 per cent, compared to 90 per cent for an uninsulated roof and 62 per cent for a house meeting current building standards (Built Environment Research Unit 2004).

The high insulation levels, combined with good cross-ventilation, shaded windows and careful use of thermal mass, have resulted in a home that performs well above the norm for the area. The tenants' anecdotal feedback on thermal comfort suggests that Research House has been comfortable for about 95 per cent of the time. Considering Rockhampton's tropical location, this is an excellent result for a non-conditioned building.

Figure 13.3 Research House's passive cooling techniques, with good cross-ventilation and well-ventilated roof space.

Regulating energy efficiency

Building regulators have been slowly recognizing the benefits of passive designed homes, and steadily increasing the energy efficiency requirements of the building envelope. Within Australia, a star-based rating system is used to assess the relative energy efficiency of residential construction. The number of stars achieved is determined through a computer-based thermal modelling program that determines the average annual energy usage required to maintain a house within a particular thermal comfort range, over a set time period and in a particular climate zone. The resulting energy total, in MJ/m^2 per annum, determines which star band a house falls into (Ambrose and Miller 2005). The Building Code of Australia now requires that new homes achieve a 4-star rating, while some states require a 5-star rating on the 10-star scale. Before the introduction, most homes were only achieving the equivalent of two stars. The regulations have resulted in an improvement in the thermal performance of these new homes of around 65 per cent. This not only benefits the home owner in reduced energy costs; it also reduces the overall demand for energy, thus reducing the need to expand generating capacity. An analysis in Victoria estimated that the 5-star requirements in that state will reduce the growth rate in residential energy by 46 per cent (Energy Efficient Strategies 2002).

Regulations covering commercial buildings' energy efficiency have been slower to appear than for residential buildings, but are now part of the current regulations. Reduction in greenhouse gas emissions is the main driving force behind the BCA 2006 energy-efficiency provisions. The Australian Building Codes Board (2006) anticipates that, within the first year of operation, the environmental benefits of more energy-efficient buildings will include annual savings of around 300,000 tonnes of greenhouse gas emissions. This is equivalent to removing around 70,000 cars from our roads, or planting 450,000 trees every year. Within ten years, energy-efficient commercial and public buildings will save Australians over 18 million tonnes of greenhouse gas emissions.

Subdivision design

Although good design of a building is paramount to achieve energy efficiency, the ease with which this is achieved is greatly enhanced if the site on which the building is constructed is orientated correctly. This is especially true with residential land, where lots are generally smaller and allow less flexibility in design options than larger commercial building sites. Solar access is a key element in achieving good energy-efficient subdivision design, and involves manipulating the key variables of aspect, shape and density, in combination with site characteristics such as topography and slope, to achieve an optimum mix of lot sizes that are appropriately orientated. The characteristics of a subdivision correlate with good solar access for new housing. An effective energy-efficient subdivision will passively direct that an overall development is significantly more energy-efficient than a conventional development. When lots are correctly aligned and proportioned, individual energy-efficient housing can be provided with comparatively less effort due to the suitability of the lot to site a dwelling with good solar access.

A subdivision design needs to maximize and protect solar access for each dwelling. Thus, consideration needs to be given to the fundamental basics – orientation, shape, size and width of the lot, solar setbacks and building heights. A study in Brisbane found that an optimal lot orientation can improve the energy performance of a dwelling by up to 32 per cent (Miller and Ambrose 2005).

Good orientation of lots within a development will enable well-orientated homes to be more easily achieved. Ideally, the aim should be for north–south streets to be within 20° west and 30° east of true north, and for east–west streets to be within 30° south and 20° north of true east (see Figure 13.4). In the southern hemisphere, north-facing slopes improve opportunities for solar access; small lots are best suited to north-facing slopes with gradients of less than 15 per cent (or 1:9). South-facing slopes

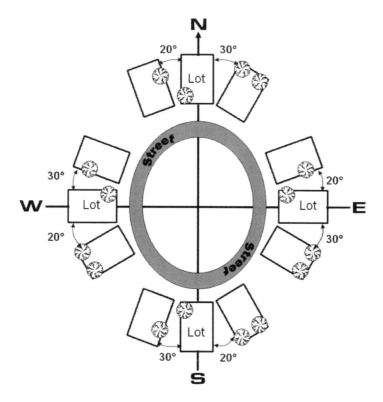

Figure 13.4 Ideal lot orientations for solar access.

impose a penalty on solar access; large lots/lowest densities are therefore best suited to south-facing slopes. The situation is reversed for the northern hemisphere.

Well-orientated lots also enable the future home to have potentially greater roof space correctly orientated for solar hot water systems and photovoltaic arrays. A large expanse of north-sloping roof space means that installing such systems is greatly simplified and reduces the need for expensive modifications. Concerns about the aesthetic impact of such devices is often reduced, as ideally the north-facing aspect should be the rear of the property and consequently not visible from the streetscape.

Economic benefits of energy efficiency

The need to reduce our greenhouse gas emissions is the main driver for energy-efficient design. If Australia sourced all its energy from clean renewable resources, then it could be argued that energy efficiency was not an issue and that consumers could continue to use energy at ever-increasing

Figure 13.5 Well-orientated lots allow easier installation of solar systems.

rates. Of course, in this seemingly ideal world, this approach would still have a significant economic cost, and in our real world this approach is having not only an economic cost but also a significant environmental cost. Despite the rhetoric that Australia will achieve its Kyoto target of an 8 per cent increase in CO_2 emissions from 1990 levels by 2012, the reality is that our emissions from stationary energy production have already increased by 43 per cent based on 2005 data and will have increased by 65 per cent by the end of the Kyoto reporting period in 2012, if the current trend continues (see Figure 13.6) (Australian Greenhouse Office 2008).

In Australia, energy users currently spend around $50 billion annually on energy. Government program experience, advice from energy auditors and independent analysis suggest that many businesses and households can save 10 to 30 per cent on their energy costs without reducing productivity or comfort levels. In many cases, these savings have very short paybacks under current energy prices. Achieving these reductions could deliver from $5 billion to $15 billion in potential savings from energy, but would require significant investment in new equipment and changes to existing practices. Experience and analysis indicate that these investments would have a positive net present value over the life of the investment, and that many have paybacks in as little as six months (Commonwealth of Australia 2004).

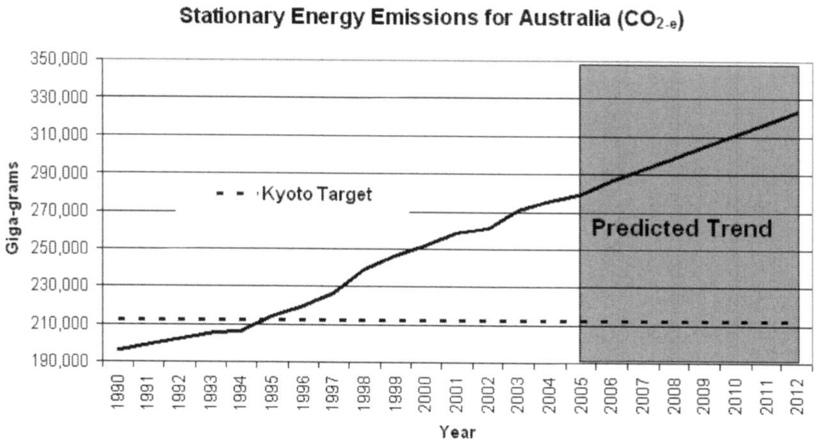

Stationary Energy Emissions for Australia (CO$_{2-e}$)

Figure 13.6 Stationary energy emissions for Australia.

Recent analysis done as part of the National Framework for Energy Efficiency (NFEE) identified substantial areas where commercial energy-efficiency opportunities are not being taken up, and found that significant opportunities with paybacks of four years or less exist across the commercial, residential and industrial sectors. The analysis estimated that if half of these gains were commercially attractive, implementing them would increase GDP by around $975 million a year once fully implemented (Commonwealth of Australia 2004).

Energy-efficiency opportunities

The building envelope is the most critical part of ensuring a successful energy-efficient building. How it controls thermal transfer, daylight and natural ventilation is of primary importance. Optimizing these natural passive systems will reduce, and sometimes even eliminate, the need for active and artificial systems. Nevertheless, most buildings will require some degree of alternative systems to supplement the natural systems. Artificial lighting is a typical example, and of course at night no real alternative exists. In residential buildings, lighting makes up only a small percentage of the overall energy profile (about 5 per cent); however, in commercial buildings (especially retail and office buildings) it can represent up to 37 per cent of energy consumption. Selecting an efficient, well-controlled system can reduce lighting energy consumption by at least 50 per cent. In addition, the reduced heat loading from these more efficient lighting systems can help cut heating and cooling energy requirements. In one case, a retailer was able to reduce its heating/cooling loads by 75 per cent due to the installation of an energy-efficient lighting system, resulting in an overall energy use reduction of 50 per cent (Lighting Council Australia 2006).

In commercial buildings, the heating, ventilation and air-conditioning system (HVAC) is the major energy consumer. The key to saving energy lies in sizing, installing, commissioning, maintaining and operating the HVAC system correctly. It is important to choose the correct technology and type of system for a particular building. Since heating and cooling are major contributors to peak demand for energy, they contribute disproportionately to energy bills, which increasingly include extra charges for high energy demand. This means great savings can be made by increasing HVAC efficiency. Other benefits include better working conditions and health for occupants.

Incorporating natural ventilation with an HVAC system is usually called a 'mixed-mode' (sometimes referred to as hybrid) system, and refers to an approach to space conditioning that uses a combination of natural ventilation from operable windows (either manually or automatically controlled) and mechanical systems that include air distribution equipment and refrigeration equipment for cooling (Center for the Built Environment 2005).

In residential buildings, hot water heating represents the highest energy consumer after heating and cooling systems. Fortunately, a simple solution that has been available for decades is able to slash this part of the energy equation. Solar hot water units are probably the best known use of solar power for dwellings. Traditional hot water services that use either electricity or gas can account for 30 per cent of a household's total greenhouse gas emissions (even more in warmer climate zones), whereas solar hot water can provide up to 90 per cent of a household's hot water requirements, depending on the type of system and the climate it is located in (Reardon 2001).

Energy-efficiency costs

The greatest single barrier to the adoption of energy-efficient design principles by builders and designers is the perception that it is a costly exercise that will not be accepted by the client. This is particularly true within the residential building industry. Figure 13.7 shows results from a series of surveys of residential designers and builders which found that 93 per cent believed an energy-efficient house would be at least 5 per cent more expensive than a standard new house, with 37 per cent believing the cost would exceed 10 per cent (Ambrose *et al.* 2005).

An independent report commissioned by Victoria's Building Commission found that the additional cost of improving the average 250-m^2 Victorian house to achieve the 5-star house energy rating was $1,500 – equivalent to 0.6 per cent of the cost of the average new house, or 0.4 per cent of the cost of the average house-and-land package (Jettaree Pty Ltd 2005).

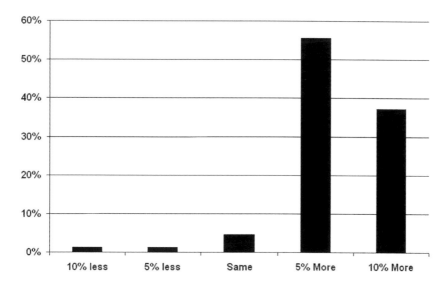

Figure 13.7 Perceived cost difference between standard and energy-efficient residential house design.

The reason that costs are much lower than builders expect is that many improvements are possible with no additional cost, through better orientation and internal layouts. Some alterations actually save money, such as reducing the amount of glazing, while the remaining improvements are generally related to improving the insulation levels throughout the building – a relatively inexpensive exercise. Another study that looked at the most effective solutions for improving energy efficiency found that improving insulation levels in ceilings and walls was among the top solutions, along with improved external shading and tinting windows (Ambrose *et al.* 2005). Like the previous study, this also found that most improvements cost less than $2,000 to implement.

The future of energy-efficient planning and design

Australia is one of the highest, if not *the* highest, per capita emitter of greenhouse gases in the world. The most recent data show that the per capita emission rate was 27.6 tonnes $CO_{2\text{-e}}$ (Australian Greenhouse Office 2007) compared to the USA's emission rate of 24.1 tonnes $CO_{2\text{-e}}$. Australia has been slow to respond to the need for cutting our emissions, as reflected in our growing energy consumption, but public opinion is shifting as awareness increases, and governments are responding by improving mandatory energy requirements through building regulations and providing financial incentives to improve building performance. New residential and commercial buildings are showing improvements in their performance over traditional designs. This has been achieved through a combination of mandatory and voluntary schemes, often driven by a client's desire to have a high-performing building. This is particularly true with commercial buildings, such as CH2. Voluntary schemes, such as the Green Building Council of Australia's Green Star program, are now being utilized by many commercial building clients (both private and government) to define the performance criteria for new developments. At the time of writing, there were 47 Green Star certified commercial buildings in Australia (Green Building Council of Australia 2008).

Sustainable land development is now also being addressed. Land developers are starting to incorporate a range of sustainability principles into their developments, including water-sensitive urban design, third-pipe systems (for reticulated greywater) and solar lot access. Assessment strategies are being developed, often by industry bodies rather than regulators, including the Urban Developers Institute of Australia's EnviroDevelopment scheme. Originally developed for the booming south-east Queensland area, it is now being assessed for nationwide usability. Like commercial buildings, many of these sustainable developments are being driven by the industry's desire for creating more efficient developments rather than by regulatory requirements. Nevertheless, just as regulators have moved to incorporate energy-efficiency standards into building regulations, it would seem likely that governments will investigate the opportunities to incorporate sustainability principles into guidelines and ultimately into regulations.

Energy-efficient and sustainable buildings and developments play a critical role in lowering Australia's overall environmental impact. The inevitable changes brought on through climate change will see many buildings and developments being forced to adapt to a new environment. However, our new developments have the opportunity to be designed and built with climate change adaptation, thus creating communities that are more sustainable as key design parameters. The result will be communities that are well prepared to face the inevitable sustainability challenges of the future.

Bibliography

ABARE (2007) *Energy in Australia 2006*, Canberra: Australian Bureau of Agricultural and Resource Economics.

Ambrose, M.D. and Miller, A. (2005) 'How to achieve sustainability: regulatory challenges', in F. Shafii and M.Z. Othman (eds) *Proceedings of the Conference on Sustainable Building South East Asia*, Kuala Lumpur: Universiti Teknologi Malaysia.

Ambrose, M.D., Tucker, S.N., Delsante, A.E. and Johnston, D.R. (2005) 'Energy efficiency uptake within the project house building industry', in J. Yang, P.S. Brandon and A.C. Sidwell (eds) *Smart and Sustainable Built Environments*, Oxford: Blackwell.

Australian Building Codes Board (2006) *Regulation Impact Statement: RIS 2006–02*, Canberra: Australian Building Codes Board.

Australian Greenhouse Office (1999) *Australian Commercial Building Sector Greenhouse Gas Emissions 1990–2010*, Canberra: Australian Greenhouse Office.

—— (2007) *National Greenhouse Gas Inventory 2005*, Canberra: Australian Greenhouse Office.

—— (2008) *Australian Greenhouse Emissions Information System*, Canberra: Australian Greenhouse Office. Online. Available at HTTP: <www.ageis.greenhouse.gov.au/GGIDMUserFunc/QueryModel/Ext_QueryModelResults.asp#resultStartMarker> (accessed 18 February 2008).

Built Environment Research Unit (2004) *Research House Annual Temperature Study Report: December 2002 to November 2003*, Brisbane: Department of Public Works.

Center for the Built Environment (2005) *About Mixed Mode*, Berkeley, CA: University of California. Online. Available at HTTP: <www.cbe.berkeley.edu/mixedmode/aboutmm.html> (accessed 18 February 2008).

Commonwealth of Australia (2004) *Securing Australia's Energy Future*, Canberra: Department of Prime Minister and Cabinet.

Energy Efficient Strategies (2002) *Comparative Cost Benefit Study of Energy Efficiency Measures for Class 1 Buildings and High Rise Apartments*, Melbourne: Sustainable Energy Authority of Victoria.

Green Building Council of Australia (2008) *Green Star Certified Projects*, Sydney: Green Building Council of Australia. Online. Available at HTTP: <www.gbca.org.au/green-star/certified-projects/> (accessed 6 March 2008).

Jettaree Pty Ltd (2005) *Research Report Summary on the Direct Cost of Compliance with the 5 Star Standard for New Housing*, Melbourne: Building Commission.

Jones, W. (2008) 'Australia's greenest office building', *Building*. Online. Available at HTTP: <www.building.co.uk/sustain_story.asp?sectioncode=747&storycode=3104100&c=2> (accessed 14 February 2008).

Lighting Council Australia (2006) *Australia's First Retail Lighting Experiment Dramatically Reduces Lighting Energy Costs and Greenhouse Emissions*, Canberra: Lighting Council Australia.

Miller, A. and Ambrose, M.D. (2005) *Sustainable Subdivisions: Energy-Efficient Design*, Brisbane: CRC for Construction Innovation.

Reardon, C. (2001) *Your Home Technical Manual*, Canberra: Australian Greenhouse Office.

Trewin, D. (ed.) (2006) *2006 Year Book Australia*, Canberra: Australian Bureau of Statistics.

14 Design for urban microclimates

Robin Drogemuller, Medha Gokhale and Fanny Boulaire

Climate change and global warming represent new factors for consideration in building design (Hennessy 2008). Research undertaken previously (Thomas and Moller 2007; Moller and Thomas, Chapter 22 in this volume) has indicated that there are existing issues in optimizing heating, ventilating and air conditioning (HVAC) system design in buildings. One factor in designing efficient HVAC systems is to ensure that the design conditions are as accurate as possible. This includes consideration of the urban heat island effect (Arnfield 2003), the effect extending well into the suburban morphology (Gokhale-Madan 1994; Gokhale 1997, 1999), which is expected to be accentuated under future climate change scenarios (Howden and Crimp 2008).

Current concerns about sustainability and building service performance have led to the expansion of this concern across the range of building services, particularly as the currently available heat flow calculation software packages do not have the capacity to estimate localized design conditions (Szokolay and Gokhale 1998). Work that takes a more detailed analytical approach only became available after completion of the project (Erell and Williamson 2006).

A suggestion to improve the information available for building services design was to develop a software tool that would assist in identifying the climatic parameters around proposed buildings in urban environments. This was aimed at the early stages of design where a massing model, consisting of the bounding surfaces of the proposed building, would be used as the basis for exploring the local microclimatic effects (buildings on microclimate and microclimate on buildings). This would build on the extensive experience developed in the analysis of single buildings as described by Drogemuller *et al.* (Chapter 7 in this volume).

The overall aim of the project was to quantify and model the potential microclimatic influences and impacts of a proposed building, within a precinct of buildings, through assessment of a 3D CAD model. This assessment would include the influences of key microclimate criteria, such as wind, rainfall, air temperature, radiation, humidity, solar access and day lighting, and their combined impact on the operating energy performance of buildings, adopting the 'pocket heat flow' calculation methodology and algorithms (Gokhale 2002) based on the admittance procedure, adequate for calculations at a massing level. The duration of this project was one year.

An area of the central business district (CBD) of Brisbane, Australia, was chosen as a test site for the work. Plan and perspective views of the selected area are shown in Figures 14.1 and 14.2. The site is located at the latitude 27°28′ 11.52′S, longitude 153°1′ 21.97′E, with the north-west boundary being Anne Street, the north-east boundary Albert Street, the south-east boundary Elizabeth Street and the south-west boundary George Street. The precinct measures approximately 90 by 200 metres.

The precinct was selected because of its varied weather characteristics throughout the year, and its high volume of pedestrians due to tourism, business and retail outlets, and local residents who work and study in the CBD. This made it a suitable and interesting study vehicle for investigation. These streets, especially the Queen Street Mall, have high volumes of pedestrians during the day, especially during lunchtime and the morning and afternoon peak periods when people commute to and from work. George and Adelaide Streets also carry heavy traffic throughout the day, while Albert Street and Burnett Lane mainly carry local traffic. Local traffic on Albert Street (off Adelaide Street) ends beyond Burnett Lane, where the Underground Busway starts. Queen Street Mall is for pedestrians only and thus has no traffic.

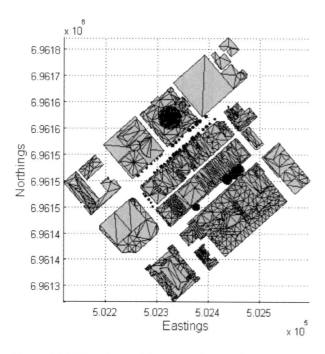

Figure 14.1 Plan views of the selected area of Brisbane CBD.

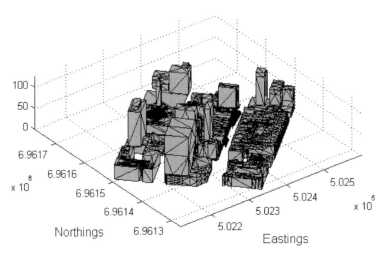

Figure 14.2 Perspective views of the selected area of Brisbane CBD.

The buildings vary significantly in height, but their geometry is usually prismatic. Retail outlets and entrances into buildings all have air conditioning, which can play a significant role in changing the immediate outdoor climate at ground level. Representations of trees and awnings have also been included as part of the model. However, miscellaneous structures, bus stops, seats, signage, etc., which may also affect the microclimate at different locations within the test site, have not been accounted for in the building models.

The project had four related problems that needed to be solved:

1   establishing the degree of correlation between meteorological data recorded at the Bureau of Meteorology sites and the chosen area in the CBD;
2   defining a representation of the buildings, infrastructure and surfaces in the CBD;
3   choosing methods of calculating the required environmental parameters;
4   developing the software to enable a convergence of design and environmental assessment.

Figure 14.3 shows the Queen Street Mall and Figure 14.4 shows Burnett Lane.

Assessing local meteorological data

Bureau of Meteorology data is available for three locations near the chosen site (BOM 2008): Archerfield (040211), Brisbane Airport (040842)

Figure 14.3 Queen Street Mall.

and Brisbane (040913), with the Brisbane station located on the Kangaroo Point Cliffs on the opposite side of the Brisbane River from the test site (approximately 3 km from the centre of the test site).

A weather station was mounted on the rooftop of the Brisbane City Council's Administration Centre (BAC), adjacent to the test site (marked 'X' in Figure 14.5). A cart fitted out with measuring equipment was taken around the path shown in Figure 14.5 at regular intervals. Both the BAC weather station and the cart recorded:

- humidity (%);
- pressure (hPa);
- light (Lux);
- solar radiation (kW/m²);
- temperature (°C);
- rainfall (mm);
- wind direction (deg.);
- speed (m/s).

There were several issues with the data recorded during the project. The weather station was reliable for measuring all factors except wind

Figure 14.4 Burnett Lane.

direction and speed. Being mounted on the roof of a tall building meant that the wind readings were incorrect through wind shear effects around the upper surfaces. The cart was used to record data at street level. This meant that it was not possible to have simultaneous readings at all locations within the test site, but the site was not considered secure

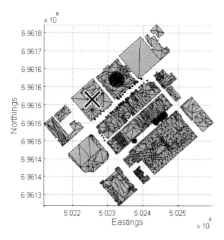

Figure 14.5 Weather station position (X) and trolley path.

enough to leave expensive measuring equipment within easy reach throughout the test period.

Analysis of the microclimate within the test site posed some problems due to the time disparity in sampling at each stop location. However, assuming that the overall climate between successive measurement locations was similar, analyses between locations were possible. Time-series plots proved to be a valuable visual tool in determining trends between the different stop locations. The following summarizes the findings observed from the micro-climate measurements.

- *Humidity.* The maximum difference between successive stop locations for each lap of the test site ranged from as low as 2 to 6 per cent. This is significantly higher than that exhibited by BOM weather stations, which record a maximum difference of less than 1 per cent.
- *Light.* High incident light tended to occur at street intersections due to the larger area of visible sky and reduced shading from awnings.
- *Pressure.* Owing to the test site being on a slope, with George Street at the highest point and Albert Street at the lowest point, small spatial variations in pressure were noticeable. Peak pressure measurements occurred at the lowest points in the test site, while the lowest measurements corresponded to the highest point.
- *Rainfall.* No meaningful results were obtained, as Brisbane was subject to a drought during the study period.
- *Solar radiation.* The microclimatic findings from light were the same as that for solar radiation. For the data collected, there was an almost perfect correlation between light and solar radiation.

- *Temperature*. Temperature measurements showed localized peaks and valleys throughout the test site. The highest readings tended to occur in Burnett Lane. This was unexpected, since it did not correlate with light and solar radiation measurements. Building-related factors are more likely the cause of this temperature increase, and are thought to be the heat sources from air conditioning units, etc., released from buildings on either side of the street. The lowest temperature readings occurred near the George Street ends of Queen and Adelaide Streets. Shop awnings, cooler air escaping from retail outlet air conditioning and proximity to the Brisbane River are possible reasons for this.
- *Wind direction*. No reliable wind data were gathered. The issues with the BAC weather station are discussed above. At street level, the readings at the cart were too chaotic to be analysed.
- *Wind speed*. In general, higher wind speeds were recorded closer to George Street and the Brisbane River.

The results from the cart measurements were very much as expected, with indications of significant variations across the test site. These measurements in an urban environment correlated well with reports in Gokhale (2002) which were taken in suburban areas.

Representing an urban environment

This project aimed at modelling buildings within an urban setting at the early stages of design. Calculations of the microclimatic variables involved the following information:

- definition of the buildings as prisms with outside faces (facades);
- location of each building on its site;
- a geometric model of the area, with the buildings all placed accurately with respect to each other.

A simple massing model, as shown in Figure 14.6, is adequate for this purpose. Besides the geometric data as seen in the figure, the thermal properties of the building surfaces are also required. Figure 14.7 shows one of the existing buildings in the test location. When entering the data for existing buildings, the total value for each facade is averaged over materials, colour and texture.

The urban precinct was modelled using the IFC 2×2 specification (IAI 2007). A number of issues had to be resolved, because the Industry Foundation Class (IFC) specification:

- is aimed at single building projects;
- is not currently aimed at urban infrastructure;

Figure 14.6 Massing model of building.

- does not treat a single facade as a unit;
- uses a Cartesian coordinate system.

During the project, only one computer program was available that could generate multi-building IFC models. This was Autodesk Architectural Desktop 2006 with the Inopso IFC export plug-in (<www.inopso. com/>).

Definition of the urban model was done using the standard IFC structure, under a single project. Separate sites were established for each allotment in the model, with an additional site to contain the roads and footpaths. Public space within the bounds of a city block, such as King George Square, was modelled as a standard site with no buildings on it. *IfcSlabs* were used to model the roads, footpaths and the ground surfaces around buildings, with appropriate entries in a text field to distinguish between them and slabs within a building.

The buildings were modelled as a single storey to the full height of the building. This allowed each facade element to be modelled as one piece. This simplified adding properties to the facade element. It would have been possible to slice each building into storeys and then group the individual facade elements on each face, but this would have required significant extra effort for no significant benefit, even if the number of storeys in each building had been known.

Figure 14.7 Facade of one of the existing buildings in the test location.

The selected area of the Brisbane CBD was small enough that the use of Cartesian coordinates did not cause significant problems. However, some mapping to geospatial coordinates would have been necessary for more extensive models. This issue will be resolved when the IFC 2×4 specification is released.

Calculating environmental parameters

The goal was to quantify and model the potential microclimatic influences and impacts of a building within a precinct of buildings through assessment of a 3D CAD model. This would be used by planners, developers and designers to model this interaction at a conceptual level, and potentially provide local authorities Australia-wide with a tool capable of rapidly quantifying impacts of building designs within a precinct of buildings. The level of detail required in these 3D models was at a 'massing' level rather than requiring detailed inputs. This is reflected in the input data, algorithms and calculation methods used.

The relatively short timeframe for this project (12 months) meant that it was not possible to implement software that covered all of the environmental parameters for which data were available. It was necessary to concentrate on the most critical parameters influencing human comfort – that is, the building design and HVAC systems, which are the air temperature, radiant temperatures, humidity and air flow. Since the initial driver for the project was design conditions for HVAC systems, it was decided that calculating the air temperatures around the buildings was the critical issue. Other parameters that were necessary inputs to this were also calculated. Since the resulting software was also intended as a design tool to support the exploration of building form, the speed and accuracy of results was traded off in favour of speed.

Since internal heat gains in the buildings were not of interest within this project and evaporative gains were considered minuscule, the outdoor air pocket temperatures could be calculated from a simple 'pocket thermal balance' equation (Gokhale 2002):

$$Qcp + Qvp + Qsp = 0$$

This equation can be expanded to:

$$(Qcp_i + Qcp_g) + (Qvp_{nb} + Qvp_{pp'} + Qvp_{ip}) + (Qsp_{gl} + Qsp_s) = 0$$

where

Qcp_i = Conductive heat flow from and into surrounding surfaces
Qcp_g = Conductive heat flow between ground surface and the pocket
Qvp_{nb} = Ventilative heat exchange between neighbourhood and pocket
$Qvp_{pp'}$ = Ventilative heat exchange between two adjacent pockets
Qvp_{ip} = Ventilative heat exchange between indoor spaces and the pocket
Qsp_{gl} = Solar radiation heat flow to pocket from bounding glass surfaces
Qsp_s = Solar radiation heat flow to pocket from bounding opaque surfaces.

Notes:

Qsp_{gl} and Qsp_s include reflected, re-radiated and convected heat flows off the surface.

Qsp_{gl} and Qsp_s are combined within the data as each surface in the data model is defined as an average value.

Adopting the admittance procedure; the above is resolved to give the pocket temperature at any given time 't' as:

$$Tp_t = \frac{mTp_t + sQp_t}{(qap + qcp_{gs} + qvp)}$$

where

$$
\begin{aligned}
sQp_t &= sQvp_{nb} + sQcp_{gs} + sQsp \\
sQv_{nb} &= 0.33 \bullet V \bullet N_{nb} \bullet (Tnb_t - mTnb) \\
sQcp_{gs} &= \Sigma A_{gs} \bullet h_{gs} \bullet (Tgs_t - mTgs) \\
sQsp_{ref} &= A_s \bullet ref \bullet (G_t - mG) \\
sQsp_{gl} &= A_{gl} \bullet (1 - sgf) \bullet (G_t - mG) \\
sQsp_{cnv+rrd} &= A_s \bullet abs \bullet (1 - U/h) \bullet (G_t - mG) \\
mTp &= \{(\Sigma_{\text{surrounding surfaces}} A \bullet U \bullet T_{\text{indoor}} + \Sigma_{\text{ground surfaces}} A_{gs} \bullet h_{gs} \bullet Tgs) + \\
&\quad 0.33 \bullet V_{\text{pocket}} \bullet [(N_{nb} \bullet Tnb + \Sigma_{\text{adjacent pockets}} N_{p'p} \bullet Tp' + \Sigma_{\text{indoor spaces}} \\
&\quad N_{ip} \bullet T_{\text{indoor}}) + \Sigma_{\text{glass surfaces}} A_{gl} \bullet G \bullet (1 - sgf) + \Sigma_{\text{opaque elements}} \\
&\quad G \bullet A_s \bullet ref + \Sigma_{\text{opaque elements}} A_s \bullet h \bullet [(G \bullet abs + U \bullet T_{\text{indoor}})/ \\
&\quad (h + U)]\}/\{\Sigma_{\text{surrounding surfaces}} A \bullet U + \Sigma_{\text{ground surfaces}} A_{gs} \bullet h_{gs} + \\
&\quad 0.33 \bullet V_{\text{pocket}} (N_{nb} + \Sigma_{\text{adjacent pockets}} N_{p'p} + \Sigma_{\text{opaque elements}} N_{ip}) + \\
&\quad \Sigma_{\text{opaque elements}} [A_s \bullet h \bullet U/(h + U)]\}
\end{aligned}
$$

Solar and ventilation (wind) data are needed to calculate the input values. Solar radiation data are available from standard meteorological sources and do not vary significantly over the area between the meteorological stations. Wind data are subject to significant local variations.

The necessary values for temperature and wind effects were derived through the use of computation fluid dynamics (CFD) software. A model was predefined for the entire Brisbane City location, with some additional areas to reduce boundary effects. A global analysis was then undertaken for the area (Figures 14.8 and 14.9). This process required considerable effort to build the underlying virtual model, and then several hours to run each analysis. This made it a highly time-consuming task. Processing time was perceived as being too long to be suitable for this project.

Processing was speeded up by writing a separate computer program that calculates the wind velocities around a single building, using the CFD-generated data to provide the boundary conditions for an analysis of the single building within its context. As the height or width of the building

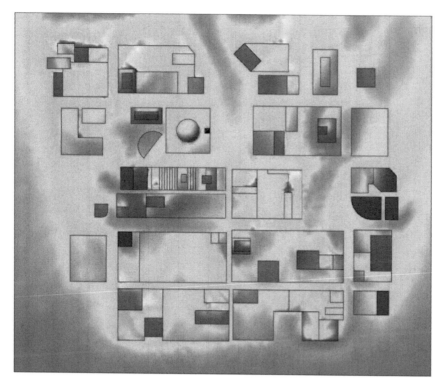

Figure 14.8 CFD-derived temperature effects at ground level.

increases, so does the radius of the area that is used in defining the boundary conditions.

Once the base wind data were generated it was a simple matter to add calculations of the precipitation (rainfall) that would impinge on the various faces of the building (Figure 14.10).

The prediction of other environmental data through the use of neural nets was undertaken on the basis of the local weather data gathered from the BAC weather station and the cart. However, there was inadequate time to build this into the final software deliverable.

Microclimate software

The Microclimate software was developed as a 'perspective' within the Design View platform (see Drogemuller *et al.*, Chapter 7 in this volume). There were no significant issues involved in using the platform as the basis for working with multi-building IFC models. This was as expected, since the underlying information architecture of Design View was built around the IFC 2×2 specification. The user interface for calculating the

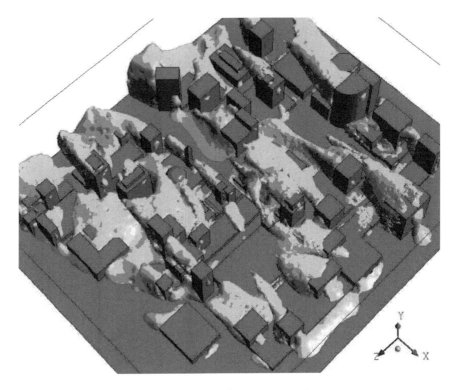

Figure 14.9 Global wind-speed analysis showing iso-surfaces.

pocket temperatures was fully integrated with Design View. The wind and precipitation interface was not fully integrated due to the short timescale of the project.

All the definitions of the micro-environment and the analyses were done in the Microclimates perspective. As shown in Figure 14.11, a 3D viewer is part of this perspective to facilitate the definition of air pockets between buildings.

As described previously, the term 'air pockets' is used to describe the volumes of air in the spaces between buildings (Gokhale 2002). The aim of the air pocket component was to estimate the temperature of air spaces between buildings.

Once the IntelligentBuildingModel has been validated by importing the IFC file and checking that the materials of all surfaces are defined, the Micro-environment Model can be created, based on the building information from the IntelligentBuildingModel. This new model contains specific data about the micro-environment: the zones representing the air pockets are defined, and their surrounding building elements, ground surfaces, other air pockets and neighbourhood are linked to them. The

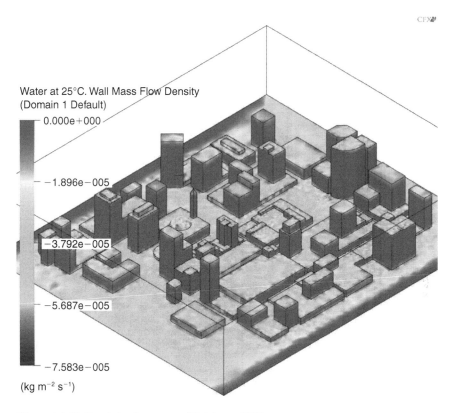

Figure 14.10 Precipitation map of Brisbane CBD.

association of the building and ground elements to a pocket is done by selecting these elements in the viewer and attaching them to a newly defined air pocket.

The thermal properties of materials for each surface are linked to the previously defined materials within the software material library, and the weather files are associated to a specific analysis.

The analysis can then be run and the temperature of the pocket visualized over time as shown in Figure 14.11.

The wind and precipitation analyses required a simpler rectangular prism to fit within the constraints of the CFD algorithm. The dimensions (width, height, length), position (x and y coordinates) and angle from North must be entered. The results appear in separate windows as the underlying software module (written in Fortran) displays its own results (refer again to Figures 14.9 and 14.10). Obviously there is potential here for better integration with the Microclimate perspective and the Design View platform

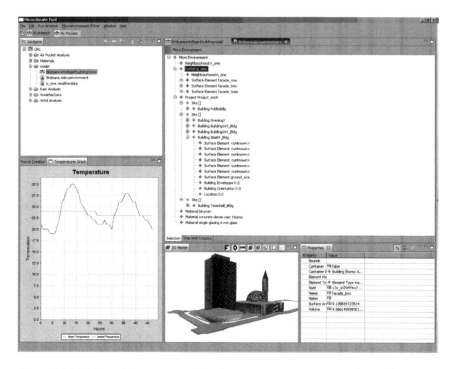

Figure 14.11 Microclimate perspective showing various tree views for working with the data, the 3D view for visualization and definition of the pockets and the resulting graph of temperatures within the defined pocket.

Industry trials

The software was tested by the industry partners to determine whether it could potentially meet their needs. The trial was successful in demonstrating that wind speed and pressure, rainfall precipitation on surfaces and air pocket temperature could be calculated for a highly developed area of the Brisbane CBD. This was a significant achievement from a scientific perspective in that there does not appear to be any equivalent software available worldwide.

The microclimatic influences of solar radiation and access, day lighting and humidity of the Microclimate software were not completed in time to be included in the trial, as originally sought in the project scope. These influences are critical measures to consider in coming closer to a comprehensive analysis of a CBD microclimate. Similarly, assessment of the impact of a proposed design on operational energy performance was not complete for the trial, and the software had not integrated predictive analysis learnings from earlier project research to gain alternative results over the more traditional simulation models

A further constraint was that even modern personal computers were not powerful enough to perform the required set of analyses for a group of buildings. Simplifying assumptions had to be made in order to provide results in real time. For example, the buildings were treated as rectangular prisms for the wind-speed analyses, with the grid spacings set at a relatively coarse spacing to keep computation times within acceptable bounds.

While the admittance procedure-based calculation methodology and algorithms are considered logically robust and adequate for calculations at a massing level, better accuracy of thermal analysis would be best achieved using the heat flow calculations based on a pocket flow analysis (potentially similar to the one used in terms of the development pattern) developed from Fourier equations, since the dynamic heat flow can only be depicted by Fourier equations (Gokhale 2002).

A gap in the knowledge such as this significantly affects the ability to perform accurate analyses of temperatures between buildings and the reliability of the results of the analyses incorporated in the software, even though the best available algorithms and data were used.

It is expected that the availability of software such as the Microclimates software will act as an impetus for the necessary fundamental research that is required to improve the availability of data and algorithms for these analyses.

Conclusion

The Microclimates project was successful in demonstrating that wind, rainfall and air-pocket temperatures could be calculated from an IFC model of an urban area. While the full range of intended environmental parameters were not implemented, this should be seen as an extension of the existing work now that the basic infrastructure has been implemented. Better integration of the CFD modules would improve user perceptions, and automatic recognition of pockets between buildings would improve usability.

Based on the results of this project, it was recommended that:

- more accurate and extensive data be gathered;
- fundamental work on calculation methods be undertaken to improve accuracy of results;
- the existing software be completed to support all of the initially expected parameters;
- further discussion be undertaken to define how the user interface could be modified to improve usability and presentation of results.

Acknowledgements

This project was funded by the Cooperative Research Centre for Construction Innovation. The partners included the Brisbane City Council,

Queensland University of Technology, CSIRO and Queensland Department of Public Works.

Bibliography

Arnfield, A.J. (2003) 'Two decades of urban climate research: a review of turbulence, exchanges of energy and water, and the urban heat island', *International Journal of Climatology*, 23 (1): 1–26.

BOM (2008) *Brisbane Area Weather Observation Stations*, Australian Government, Bureau of Meteorology. Online. Available at HTTP: <www.bom.gov.au/weather/qld/brisbane-observations-map.shtml> (accessed 14 July 2008).

Erell, E. and Williamson, T.J. (2006) 'Simulating air temperature in an urban street canyon in all weather conditions using measured data at a reference meteorological station', *International Journal of Climatology*, 26 (12): 1671–94.

Gokhale, M. (1997) 'Between the building and the macroclimate', paper presented at the Australian and New Zealand Architectural Science Association 31st Annual Conference, University of Queensland, Brisbane, 29 September–3 October.

—— (1999) 'Structure and fluidity of macroclimate', paper presented at the 16th International Conference on Passive and Low Energy Architecture, Brisbane, 22–24 September.

—— (2002) 'Questioning the envelope concept: thermal simulation for urban and suburban built environments', PhD thesis, Brisbane: School of Geography, Planning and Architecture, University of Queensland.

Gokhale-Madan, M. (1994) 'Limitations of the envelope concept: the first threshold', paper presented at the Australian and New Zealand Architectural Science Association 28th Annual Conference, Deakin University, Geelong, September

Hennessy, K. (2008) 'Climate change', in P.W. Newton (ed.) *Transitions: Pathways Towards Sustainable Urban Development in Australia*, Melbourne: CSIRO and Dordrecht: Springer.

Howden, M. and Crimp, S. (2008) 'Drought and high temperatures', in P.W. Newton (ed.) *Transitions: Pathways Towards Sustainable Urban Development in Australia*, Melbourne: CSIRO and Dordrecht: Springer.

IAI (2007) Online. Available at HTTP: <www.iai-international.org/> (accessed 20 June 2007).

Szokolay, S.V. and Gokhale, M. (1998) 'The limitations of simulation', paper presented at the 15th International Conference on Passive and Low Energy Architecture, Lisbon, 31 May–3 June.

Thomas, P.C. and Moller, S. (2007) *HVAC System Size: Getting It Right*, Brisbane: CRC for Construction Innovation. Online. Available at HTTP: <www. construction-innovation.info/images/HVAC_system_size.pdf> (accessed 13 July 2008).

15 Technological innovation in the provision of sustainable urban water services

Steven Kenway and Grace Tjandraatmadja

This chapter focuses on urban water. Specifically, it considers available technological options and innovations capable of improving the sustainability of urban water services. It addresses decentralized development options, at scales from single household through to subdivision scale, designed to source and treat water near the point of use. The challenges of selecting the 'right' option for a particular site and successfully integrating these sustainably are considered.

The chapter uses four Australian cities as case studies. These cities have common challenges: that in the preceding five years water storage facilities have been at unprecedented lows and usage restrictions at unprecedented highs. It is projected that climate change will reduce rainfall and increase evaporation for many of Australia's major cities. The need to address energy use and greenhouse gas emissions is forcing a re-evaluation of traditional water management practices. The chapter also considers the coupling of these decentralized options with the existing centralized water systems.

While technology is a valuable tool in addressing the management of water resources, the extent to which it contributes to the pursuit of urban sustainability is strongly linked to the context in which it is applied and how it is used.

New models of thinking, planning, regulation and system design are clearly necessary to ensure that water supply solutions for cities also contribute cost-effectively to the simultaneous reduction of greenhouse gas emissions and the overall goal of achieving sustainable cities.

This chapter draws on and adds to work undertaken for the CRC for Construction Innovation on water technologies in sustainable subdivisions (Diaper *et al.* 2007). It aims to provide an overview of existing and emerging trends in technologies available for water service provision in Australia, and of technology characteristics and their application to the urban context, and to explore some major challenges that remain to be addressed to achieve greater sustainability of urban water systems.

The challenge of sustainable urban water in Australia

Australia is a highly urbanized country, with 64 per cent of the population (20.1 million in 2004) (ABS 2006) living in ten major urban areas (WSAA 2005a), and the trend of increased urbanization and higher urban densities is continuing (Newton 2008).

Residential demand comprises around 60 per cent of total urban water use (WSAA 2005a). In most major cities, residential use of centrally supplied water typically varies between 75 and 110 kl/cap/year (Figure 15.1).

Historically, most major Australian cities (excepting Perth) have relied on surface water storage facilities. Increasingly, the influence of global warming and climate change is suggesting a reduction in the reliability of these sources (Howe *et al.* 2005; IPCC 2001; WSAA 2005b). In 2007, the supply reliability (yield without restrictions) of many traditional surface and groundwater resources has been reduced on this basis. Within cities, increased restrictions on water use and the associated changes in usage patterns also create potential for reduced and more concentrated wastewater flows. Climate change can also increase sewer overflow events, particularly in combined sewer and stormwater systems (through more intense storms), and contribute to increased failure of water and wastewater networks (as soils dry and swell at higher than traditional levels) (Chan *et al.* 2007).

The pressure on surface water storage facilities has led to a call for strategies of 'diversity for security' and for 'climate-proof' supply solutions. In responding to the increasingly apparent challenges, most state agencies have developed strategies to address water supplies for the next 25 years.

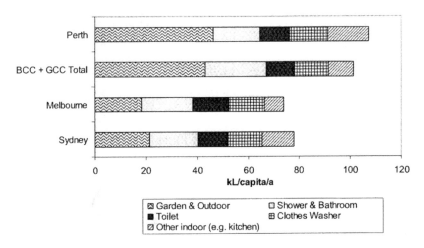

Figure 15.1 Categories of urban water use 2004–05 (source: WSAA (2005a) and water utility public documents).

Note
The use of water varies substantially in each city and this affects the viability of particular technologies.

Nationally, the anticipated growth in desalination capacity by 2030 to supply centralized systems is expected to be over 300 GL (Figure 15.2). Additional private and industry-based desalination schemes are also being developed. However, many question the environmental ethics of investing large amounts in an energy-intensive water source, such as desalination, to address a problem arguably caused by excessive use of fossil fuels. Anticipated reuse of wastewater and stormwater is increasing, with over 150 GL of wastewater reuse schemes and several stormwater harvesting schemes proposed and/or under construction. Simultaneous with the growth of city-scale integration of wastewater and stormwater into the centralized water supply system, many traditional technologies have regained popularity. For example, in South East Queensland (SEQ), 148,000 rainwater tanks were installed in the 18-month period to August 2007 (Ted Gardner, Department of Natural Resources and Water, personal communication).

Ideally, integrated urban water management (IUWM) aims to promote the integration of multiple sources (drinking water, wastewater, groundwater, surface water and stormwater), maximizing the benefit achieved. However, the initiatives to improve household water efficiency and increase reuse have not always considered 'system-wide' implications, such as the compatibility of technologies integrated within a household or development and integration with the centralized system. For example, as water restrictions continue and as households convert to water-efficient appliances and divert greywater for garden irrigation, flow volumes at wastewater treatment plants (WWTPs) have been observed to fall by around 20 to 30 per cent (Taylor *et al.* 2007), thereby reducing the efficiency of reuse schemes and, potentially,

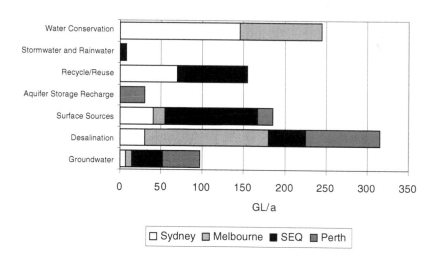

Figure 15.2 Proposed water source strategies by city (source: extracted from city water strategy documents (2007)).

the functioning of the cities' main sewerage network. Furthermore, a shift from centralized to decentralized systems requires the development of different management, regulation and enforcement models to mitigate public health and environmental risks.

The shift away from the traditional paradigm of managing water, wastewater and stormwater as separate resources to an integrated approach for providing urban water services presents a new set of challenges. In some cases, innovative supply solutions have been necessary to enable continued urban development, given that centralized systems have faltered in their ability to deliver unrestricted water supplies.

Technology as a tool for sustainable cities

Newton (2007) sees the future challenges to sustainable cities as a series of successive planning and innovation development horizons which require navigation. He describes Horizon 1 as technologies and solutions that are able to be implemented now, Horizon 2 as technologies that require some challenging extensions of existing technology and modified policy, governance and regulation, and Horizon 3 solutions as being based on technologies and planning concepts which are radically different to those currently operating and require major barriers to be overcome.

Water-efficient appliances, rainwater and greywater reuse for outdoor irrigation are examples of Horizon 1 innovations now being implemented. In South East Queensland, the Western Corridor recycled water scheme proposes to use around 80 GL/year for industrial purposes as well as indirect potable reuse (Gardner and Dennien 2007), representing a Horizon 2 innovation. Large-scale reuse of stormwater is an example of a Horizon 3 challenge. Continued development of centralized urban water infrastructure is requiring significant national investment. This investment, combined with enhanced awareness of the risk of failure of centralized systems, is contributing to a keen search for alternatives, and is fuelling innovation.

New technologies have emerged rapidly to meet a growing need for rain and stormwater harvesting, water reuse and increased water use efficiency. Small-scale and medium-scale technologies are increasingly seen as part of the solution, and certainly as ones that an individual householder or developer can adopt.

Technology options for the exploration of alternative water sources

Conventional urban water infrastructure relies on engineering solutions to implement and operate large-scale centralized water and wastewater treatment plants, and transport water and waste over long distances. While such systems are typically more efficient at larger scales, they require extensive use of material and energy resources to ultimately release treated wastewater at a concentrated location. The ideal of sustainable water use is to achieve maximum economic, social and environmental benefits from water resources, while minimizing adverse impacts. This implies a consideration of the total water cycle when assessing and utilizing water options at local, regional and wider scales (Diaper *et al.* 2007).

Mass balance analysis of four Australian cities in 2005 indicates that wastewater and stormwater flows could theoretically meet between 65 and 170 per cent of urban needs (Kenway *et al.* 2008), provided local issues, including treatment and storage needs, can be met (Figure 15.3). The magnitude of the opportunity is heavily dependent again on local factors, including the nature of existing centralized infrastructure. Technological, policy and management step-change will be necessary to enable harvesting of this resource.

Emerging systems can be characterized by a mix of innovation and re-engineering of existing systems. Examples of systems currently in use in Australia and which could be used to explore the diversity of water resources available at various scales can be found in Table 15.1. Information on the selection and adoption of decentralized systems is further explored in Diaper *et al.* (2007).

The range of technology available for alternative water and wastewater services is increasing at rapid pace; for instance, rainwater tanks and accessories are now available in a huge range of sizes, shapes and materials. Greywater reuse technologies too are booming, and include diverter systems (e.g., Waterwise) as well as more complex treatment systems based on

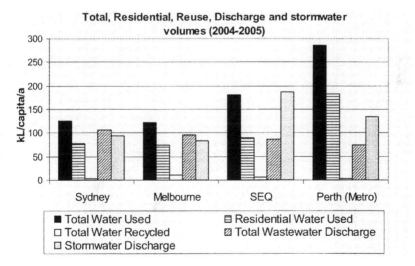

Figure 15.3 Urban water balance for Australia's major cities.

Note
There is a significant quantity of wastewater and stormwater lost from each city. Local factors are important and influence the quantity.

Table 15.1 Alternative water harvesting: applications and technologies

Case study site	Raintanks	Stormwater	Greywater	Wastewater
Pimpama Coomera	Household	Cluster WSUD, ponds for irrigation	Discharged with wastewater	Subdivision reduced infiltration gravity sewers and regional wastewater treatment to class A and recycling via 3rd pipe for non-potable use
Payne Road	Household and communal	Communal collection and WSUD for run-off treatment	Household for garden irrigation	Communal collection and timed discharged into sewer
The Currumbin Ecovillage	Household	WSUD	Discharged with wastewater	Mix of cluster and household treatment for reuse
CH2	Rooftop collection for building use	Not applicable	Discharged with wastewater	On-site treatment and sewer mining for non-potable use
Newington (Sydney Olympic Park)	Rooftop collection from public facilities	WSUD treatment including wetlands. After treatment stormwater is collected for reuse	Discharged with wastewater	Tertiary treatment plant and water reclamation plant with Reverse Osmosis/MF treats all wastewater from commercial and residential areas, recycling via 3rd pipe for non-potable use
Atherton Gardens	Rooftop collection from one building for garden irrigation	WSUD treatment prior to discharge via raingardens, wetlands	Communal laundry greywater treatment for use in garden irrigation	Discharged to sewer
Sustainable house	Household	Household On-site mini-wetland	Discharged with wastewater	Household on-site treatment for non-potable use
60L	Communal	Not applicable	Discharged with wastewater	Communal collection and on-site treatment
City of Salisbury (Parafield Stormwater harvesting facility)	Up to individual household	Large-scale stormwater collection, treatment via reed beds and storage in underground Aquifer Storage and Recovery. Treated stormwater is sold to local businesses; stormwater is also mixed with treated effluent from a municipal WWTP for non-potable use at Mawson Lakes residential development	Up to individual household	Discharged to sewer

Source: Adapted from Diaper et al. (2007).

physical sedimentation and adsorption (e.g., Perpetual Water), biological treatment and ultraviolet (UV) disinfection (e.g., Pontos Aquacycle), biological treatment and membrane filtration (e.g., Nubian) and physico-chemical systems (e.g., New Water Aqua Reviva). Decentralized wastewater treatment technologies are also available for a range of scales (for allotment, cluster and larger-size developments) and complexity, with technologies including textile, peat or membrane filtration, membrane bioreactors and aerobic biological treatment (Landcom 2004).

The boom in technologies presents a significant challenge to the development of uniform performance evaluation standards and regulation. It is also a challenge to the government and authorities who need to safeguard public health, and to the community who are encouraged to adopt and sometimes operate some of these new technologies. An overview of existing technologies is provided here to better instruct the reader about the current possibilities, based on water source type.

Rainwater

Rainwater harvesting is an example of a Horizon 1 water system. Rainwater systems comprise collection, storage, treatment and distribution technologies.

Collection systems are comprised of roof screening devices, aimed at reducing contaminant ingress from the roof during the first rain wash-off, and transfer systems (i.e., pipework from roof to tank). Their effectiveness in treating collected rainwater is not well understood, and at present there are no statistically relevant data showing the relative contaminant-removal efficiency or improvement in stored water quality of the different systems (Diaper *et al.* 2007).

Storage systems for collected rainwater include underground tanks, bladder systems placed underneath houses or decks, and external tanks. Designs in the market range in size, shape and fabrication material, with the storage needing to be appropriate for the lot size, dwelling type, existing infrastructure and end-user needs. However, in urban areas, aesthetics (shape and size) of a rainwater tank are often the main factors considered during technology selection, to the detriment of reliability of supply.

The degree of treatment required for rainwater is dictated by the intended application. Rainwater for garden irrigation or non-potable applications (toilet flushing and laundry) can be used untreated, but for potable indoor uses it requires treatment. The options available can provide effective disinfection, but do not provide residual disinfection, and include filtration, thermal disinfection and UV treatment. Filtration (micro- and/or ultrafiltration) provides a barrier to micro-organisms. Thermal disinfection using the hot water service is currently being investigated. All have associated maintenance and operation issues (Diaper *et al.* 2007).

Distribution of collected rainwater is usually via a pump, although in certain situations end uses can be gravity fed. In cases of need for mains back-up, an automatic diverter device is preferred over a tank top-up system, as it offers greater energy efficiency (Gardner 2006, in Diaper 2007). Additionally, an alarm or float system is suggested to alert the householder when mains water is being used, so that consumption can be adjusted accordingly. There are a number of innovative float systems available to monitor tank water level, and some can be retrofitted to existing tanks.

Major issues for consideration when adopting rainwater for water supply include the following:

- The reliability of supply is influenced by tank capacity, roof collection area, climate (rainfall distribution) and water end use, as these impact on the match between rainwater collection and demand during the year. Roof collection area has the single greatest impact on total yield from a raintank (Marsden Jacob Associates 2007).
- Improved water savings can be achieved when rainwater is supplied to non-seasonal uses, such as toilet, laundry and bathroom, rather than seasonal uses such as irrigation (ISWR 2006), particularly in high-density developments (Gray 2004).
- The energy requirement of rainwater tanks may be lower compared to large-scale water resource options such as desalination. However, local issues, including configuration of particular options and uses of the water, have considerable impact.

Benefits of rainwater harvesting include reduction of stormwater flows and, if designed accordingly, reduced mains water peak flows (Marsden Jacob Associates 2007). An additional benefit is that maintenance requirements are typically simple and minimal, particularly for the collection and disinfection systems.

Stormwater

Stormwater consists of run-off from all pervious and impervious surfaces, and traditionally is diverted to pipe mains maintained by municipal councils. Stormwater technologies can be adopted at allotment, cluster or sub-divisional scales. A review of Australian practices in integrated stormwater treatment and harvesting found that reuse was largely restricted to smaller-scale sites and that treatment was still based on systems designed for environmental protection (control of pollutant transfer and attenuation of peak flows), not domestic use (Hatt *et al.* 2004).

As scale increases to subdivisional level, there are a number of options for collection, treatment, storage and redistribution. A range of technical documents and guidelines for design and implementation of stormwater systems aimed at environmental protection are available (City of Melbourne 2004; Wong 2006; Brisbane City Council 2005). More recently, the potential for large-scale harvesting of urban stormwater has been explored in projects across Australia (Urban Stormwater Initiative Executive Group 2004; Government of Victoria 2004; Stormwater Industry Association Queensland 2006). For example, the Parafield stormwater recovery scheme in Salisbury, South Australia, collects, treats and sells stormwater (Salisbury City Council 2008). Interestingly, city-scale reuse of treated wastewater may open up stormwater reuse options as well, by temporarily storing stormwater and directing it (off-peak) to wastewater treatment plants for reuse. Capturing and using this stormwater via the wastewater system provides many benefits, including a source of low-salinity water. Interestingly, no clear methodology exists to identify the optimum scale at which to harvest stormwater, given existing infrastructure configurations.

Research on the quality of stormwater and its potential reuse applications other than non-potable is underway. Further work is required to develop reliable, robust techniques and technologies to provide water of a quality suitable for drinking-water substitution. A recent document providing technical guidance for stormwater treatment and harvesting (ISWR 2006) suggests UV as a preferred disinfection technique, non-seasonal end uses to minimize storage requirement, and a requirement for closed storage for minimization of evaporation.

Major considerations in the use of this technology include the following:

- Reliability of supply is dependent on storage capacity and seasonality. Design considerations also need to account for quality and quantity variability (Diaper *et al.* 2007).
- The traditional focus of stormwater technologies has been pollution control and downstream protection; however, it can have potential as a water source at cluster and subdivision development scale.
- Storage capacity – the area required and cost of this area – must be taken into account.

Innovative designs for subsurface storage that also provide some stormwater treatment are available, with systems having been used in single dwellings, residential aged care facilities, parks and schools (Atlantis Water Management 2007). Alternatively, if geological conditions are favourable, adoption of aquifer storage and recovery (ASR) at cluster or subdivisional scale offers the advantage of storage capacity and minimal infrastructure restrictions. Examples of test sites for stormwater ASR can be found in South Australia (Martin and Dillon 2002), New South Wales (Argue and Argue 1998) and Victoria (Dillon *et al.* 2006), and opportunities have been explored or are planned in other states, e.g. Pimpama Coomera in Queensland. In addition to the general advantages of stormwater use, ASR also has the benefits of potential reduction in groundwater salinity (depending on salinity levels), reduced storage costs and storage which has a small spatial footprint.

Reuse of stormwater for irrigation is a Horizon 1 to 2 goal; however, potable reuse is Horizon 3.

Greywater

Greywater encompasses any wastewater stream originating from the bathroom, laundry and kitchen, excluding toilet wastewater. Most commonly, bathroom and laundry water is diverted, whilst kitchen wastewater is often excluded due to oil and grease and the risk of higher pathogen content. Greywater contains higher levels of human pathogens, dissolved salts, nutrients and biodegradable matter than do rainwater and mains water.

Greywater, however, offers the advantage over rainwater systems of a continuous supply available during dry periods. The potential water savings associated with greywater treatment and use are well documented, and will be primarily dependent on end use (Gardner and Millar 2003; Priest *et al.* 2003; Diaper *et al.* 2003). Depending on local circumstances, greater than 20 per cent saving can be made with a 1,000-L tank collecting bathroom and laundry greywater and supplying both garden uses and toilet flushing (Gray 2004). Collection technologies range from direct diversion of untreated greywater for use in the garden and subsurface irrigation, to systems for collection, treatment, storage and distribution which produce higher-quality water, allowing application for spray irrigation, toilets and, potentially, laundries. Treatment and costs vary widely, and are influenced by house design, space restrictions and piping requirements (Diaper *et al.* 2007).

Important considerations for adoption of greywater as a water source (if allowed under local regulations) include the following:

- End use – greywater components can have detrimental environmental impacts on groundwater and soil structure. However, toilet flushing has less impact on local environment.
- Householder involvement and awareness is more necessary and noticeable with greywater due to the scale of its reuse (allotment) and impact (e.g., the type of laundry detergent used has a strong influence on the quality of the greywater, and householders often can see the impact on the garden).
- Greywater systems also require a certain level of general maintenance (e.g., cleaning of lint filters and declogging).
- Retrofitting for non-seasonal uses (toilet flushing, laundry) may be limited in established developments, depending on access to existing pipe-work infrastructure.
- Storage and reuse of greywater has the potential to reduce wastewater flows.

While studies have gauged acceptance of wastewater reuse at larger scales (Po *et al.* 2003, 2005; Hurlimann and McKay 2003, 2004), none reviewed have examined public perceptions of large-scale greywater reuse (Diaper *et al.* 2007).

Wastewater

Wastewater is any used water leaving a dwelling, including greywater and blackwater from the toilet. Domestic wastewater contains a range of contaminants, including organic matter, organic and inorganic contaminants and pathogens. In urban areas, restrictions apply to the use of on-site wastewater management systems at the domestic allotment scale. However, a number of projects are exploring its potential for medium- and high-density developments, e.g., Inkerman Oasis (a residential complex) and 60 Leicester Street (a commercial building), both in Melbourne.

On-site wastewater management can occur either by reducing the volume of water used for transport of waste, due to water demand management and low flow appliances, or by on-site treatment and reuse of treated effluent. Initiatives in sewer mining (harvesting of wastewater from sewers) and non-potable reuse have also become more prevalent, with trials conducted at commercial and public developments. More radical technologies are available that eliminate or reduce water used for solid transport, such as dry toilets and urine-separating toilets, but have not yet been adopted for use in urban Australia, although individual trials have been conducted for solid–liquid separation toilets at the University of New South Wales and dry toilets have been used at remote public parks (Tjandraatmadja and Burn 2004).

At the household or dwelling scale, wastewater tends to have a lower level of persistent contaminants and pollutants than from stormwater run-off (hydrocarbon residues, pesticides) and is less diluted, so simpler treatment methods can be employed. Additionally, separation of individual streams (greywater and blackwater) provides opportunity for different treatment levels.

Treatment and reuse of wastewater at allotment scale has been implemented in a number of sites; however, technologies for on-site wastewater management in urban areas often focus on cluster scale developments (e.g., high-density residential and commercial). There are also some examples of larger-scale reuse (e.g., Rouse Hill). Treated effluent can be reused in non-potable applications including garden irrigation and, after disinfection, for toilet flushing. Sludge is usually removed by a contractor or disposed of to the sewer – as, for example, in the City of Melbourne's Council House 2 (CH2).

At subdivision scale, treated effluent from a centralized treatment plant is generally the technology of choice. Treated effluent is sold for golf-course irrigation, landscaping, and to a growing number of residential developments for non-potable applications (e.g., Pimpama Coomera, Rouse Hill and Newington).

Reviews of a range of technologies on advanced wastewater treatment technologies available for on-site treatment have been conducted by Geolink (2005) and Landcom (2004), and include a variety of systems,

Table 15.2 Technology options for alternative water sources

Technology	Rainwater	Stormwater	Greywater	Wastewater
Collection	Gutter guards First flush diverters Inlet filters	Gross pollutant traps Stormwater drains Permeable paving	Greywater diverter Bin collection and pump systems	Solid–liquid separation toilets Gravity sewer Vacuum sewer Pressure sewer Tap into wastewater sewer (sewer mining)
Storage	External rainwater tank Rainwater bladder or tank (underneath house)	Lagoons or lakes Aquifer storage and recovery (ASR)	Bladder or greywater tanks (max. hold time 24 h untreated) with disposal to sewer	After treatment
Treatment	UV disinfection Filtration Thermal disinfection	WSUD Particularly bio swales, infiltration trenches, raingardens, wetlands (surface and subsurface)	Untreated systems (diversion) Treatment systems: biological (e.g., Pontos GMBH), physico-chemical (e.g., New Water)	Allotment (Advanced treatment systems) with subsurface irrigation or non-potable use. Cluster (advanced treatment systems) (often coupled with RO/UF and disinfection) Wastewater treatment plant and water purification plant for piped effluent
Distribution	Gravity Pump systems with or without an automatic diverter	Pump systems	Subsurface irrigation Dedicated pipework	Dedicated pipework (purple pipe or lilac network)
Reliability	Seasonal Mains water top up	Seasonal	Constant, but dependent on household occupancy Potential salinity issues	Constant
Other	Tank level float			
Scale	Household Cluster	Cluster Large development	Household	Household (less common) Cluster Sub-development Mixed systems
Uses	Irrigation Potable water use Non-potable uses	Irrigation Non-potable uses Also mixed with other alternative water sources, e.g., treated effluent	Irrigation (non-edible crops) Non-potable uses	Non-potable uses
Quality	Low nutrient content Major contaminant sources would include residues from airborne pollution, leaves, animal droppings	Can contain surface run-off residues, including animal faeces, pollutants, oil and grease	Has a higher salt content than storm and rainwater, and micro-organisms from human activities	Contains majority of the micro-organisms and pathogens

such as Biolytix (peat filters), Orenco systems (textile filter systems) and combined systems where treatment stages are interspersed between the household and a centralized facility.

At the cluster or subdivisional scale, collection system options range from the use of vacuum and low-pressure sewerage with low-diameter flexible pipes and shallow burial depths, through to 'smart pipes' of flexible material with fused joints and inspection points rather than manholes, which minimizes infiltration. However, their application is dependent on local constraints such as terrain (Diaper *et al.* 2008).

Technologies for wastewater treatment and recycling can be based on biological, chemical and physical processes, or combinations of these. Some are appropriate for both greywater and wastewater. Appropriate scale, end uses, physical footprint and capital and operating costs need to be considered (Landcom 2004). The level of wastewater and greywater treatment should be guided by quality requirements based on intended usage, such as irrigation of public places, or indoor non-potable use. Advances in monitoring technology mean that remote real-time monitoring of such systems can be adopted.

Although a range of technologies and their effectiveness have been evaluated in the literature (Landcom 2004; Geolink 2005; Diaper 2004), assessing the suitability of each system and the development of options for their application to different sources is a major challenge that requires consideration of the context in which the technology is applied, guided by factors such as scale, and local geological and climatic conditions (Diaper *et al.* 2007).

Many of the decentralized wastewater technologies have simple maintenance requirements. However, compliance with maintenance schedules requires strict observance due to implications for public health, and is preferably assigned to an organization with the proper expertise and capacity to service and monitor such systems, particularly as the system increases in size.

Adoption of wastewater as a non-potable source is a Horizon 2 target. Piped effluent systems are being developed across Australia, for example at Pimpama Coomera, Rouse Hill, Aurora, Lynbrook and Mawson Lakes. For non-potable applications, its acceptance and proper use requires further development of strategies for social engagement, community education, regulation and management.

Challenges of technology selection and integration

The design and integration of different technological options and water sources poses a range of challenges. The previous section provided an overview of the technologies available for individual water streams that offer a range of options in regards to treatment, volume, quality, cost, maintenance, usage, etc. The major challenge in technology selection is in the configuration of a system and the integration of the various technology options to achieve the optimal solution for a site, development and region.

The selection of technology needs to consider the characteristics of each site (geological, climatic and built environment), its integration into the overall water cycle, and the potential for future growth or upgrade. The range of factors to consider is discussed below.

The management and maintenance liability of decentralized wastewater systems is one of the major issues for their adoption. Public health was the major driver for the development of centralized water and wastewater systems in the nineteenth century. Contamination of drinking water with wastewater resulted in major disease outbreaks. The same concerns are still valid today. Although advances in technology and management practices offer decentralized systems better opportunities and options, the need remains for strong definition of roles and responsibilities that allocates duty of care and maintenance in decentralized systems. This requires regulation through appropriate legislation.

Management of decentralized systems represents a challenge to institutions and regulatory frameworks that have evolved to support centralized wastewater treatment. Australian and overseas experience with septic systems maintenance by householders (Sarac *et al.* 2001) has highlighted the need for the adoption of centralized management in the maintenance of on-site systems due to the high risk to public health. For example, a body corporate or a centralized service provider can offer adequate expertise and take responsibility for the assets. This model has been adopted in Australia for commercial buildings, including CH2 and 60L, and also in developments such as Currumbim Ecovillage and Sydney Olympic Park Aquatic Centre, where the local water utility (Sydney Water) is responsible for the treatment plant and treated water distribution. Under such arrangements, the onus of infrastructure maintenance is removed from individual householders, and the governing authority can exert a greater degree of control on upkeep of the infrastructure system. The involvement of a centralized service provider offers economies of scale, with treatment facilities shared across multiple dwellings.

Matching demand and water supply availability throughout the year determines the efficiency of a system, particularly for rainfall-dependent sources (Grant *et al.* 2006). Designing adequate storage is important for systems that depend on seasonal activities, when demand and supply do not always match. Integration of multiple sources (e.g., rainwater tanks with mains water top-up) is another strategy adopted to ensure reliability (Diaper *et al.* 2007). As scale shifts from household to cluster to larger developments, different opportunities and challenges arise. As scale increases, the strength of association that householders make between the water source and their behaviour decreases. However, the communal supply and storage can improve reliability and allow pooling of resources for investment in capital works – for example, excess rainwater directed to a communal water tank for fire fighting at Payne Road, and treatment of communal laundry greywater and storage in Atherton Gardens (Diaper *et al.* 2007).

The level of treatment necessary is determined by the intended end use. This allows different technologies to be adopted, based on the risk that the use of water poses to the public and the environment. For example, indoor use of greywater requires tertiary-level treatment and disinfection, whilst untreated greywater is allowed for subsurface irrigation within 24 hours of generation without treatment. As the source composition includes more greywater and wastewater, the treatment train complexity increases, as does the level of technical expertise and maintenance requirements.

Technologies are also having to respond to altered qualities of wastewater flow in the sewerage system. In CH2, for example, the sewer mining plant had to deal with higher concentrations of waste paper in the sewer as water use progressively decreased. Consequently, the plant had to be upgraded to include fibre removal. Additionally, traditional treatment methods are currently not designed to remove salts. This may become important for greywater systems which discharge wastewater (and salt) into the environment or garden, and can only currently be dealt with by encouraging households to use low-salt detergent formulations.

Many technologies are currently being tested at demonstration sites and implemented in small- to large-scale developments. Applications and uses for existing technologies are growing as new issues arise and the number of sites where wastewater technology is applied increases. As the technology matures the knowledge base will increase, but at present there is still a lack of understanding of some of the risks involved in many applications, and both industry and regulators are still unfamiliar with many of the new systems. Furthermore, the increase in uptake of decentralized technologies, demand management and integrated water management has the potential to change the quality and flows of water and wastewater streams transported across stormwater and wastewater networks and city catchments. A better understanding of the potential that such changes can have on existing collection infrastructure, chemical and energy use, and biosolids generation is required to better optimize the operation and integration of existing and innovative systems.

Adoption of alternative water sources requires a higher degree of involvement and education of householders and plumbers as water management practices change. Likewise education, adequate legislation and enforcement are also needed to ensure proper use of water sources – for example, piped effluent and avoidance of cross-connections. Existing infrastructure and property configurations can restrict the technologies that can be adopted in certain sites. For instance, in the Sustainable House, the size of the property limited the selection of tanks that could be used, requiring a custom-made tank (Mobbs 1998).

Insufficient data currently exist regarding the impact of water conservation on wastewater flows. It is, however, a positive that stormwater source control can assist to mitigate flood events and also prevent the transport of pollutants into receiving waters. However, some concern is emerging regarding the potential ecological implications of widespread stormwater harvesting.

A key question revolves around whether centralized, decentralized or hybrid systems offer the highest overall sustainability benefits. A recent analysis by Yarra Valley Water in Melbourne was conducted using Life Cycle Assessment (LCA) of new developments in areas requiring the removal of septic tanks. The results indicated that decentralized services can deliver improved environmental and cost results (Kelly 2008). Comparison of decentralized and centralized systems needs to be conducted using sustainability criteria that include wider system implications (i.e., a full complement of externalities). The benefits and costs of centralized and decentralized options are likely to vary for each individual development, so need to be evaluated on a case by case basis.

Implications of policy and technology change for management strategies

To date, only limited analysis of the energy implications of water strategies has been undertaken. Australia became a signatory to the Kyoto Protocol in early 2008, and therefore has committed to achieving reductions in emissions of greenhouse gases. However, a price for carbon is yet to be set. Transfer of these policy changes into practice presents a significant challenge for the sustainable management of urban water in Australia.

Issues associated with linkages between climate, energy and water will become more critical in future. The need to address urban water cycle issues while reducing energy use (or greenhouse gas emissions) represents a clear Horizon 3 challenge, in that it will require radically different planning concepts and technologies coordinated across both the water and energy cycles.

Use of energy in the water cycle

Urban water service provision includes the planning and delivery of supplies for residential, commercial and industrial uses. The collection, treatment and disposal of wastewater and the management of stormwater all use energy (Gleick 1994). The investment of energy in the water cycle varies significantly from city to city, depending on local factors including topography, pumping requirements and degree of treatment.

Energy is also consumed in homes and businesses when water is heated, pumped, filtered or treated. This energy use is substantial in Australia, with approximately 4,900 MJ/capita per year used in the heating of water for residential purposes such as showers, washing machines and dishwashers (Belsham 2005). This equates to approximately ten times the energy of current water and wastewater services provision (461 MJ/capita per year) (Kenway et al. 2007).

Energy use in water services provision is increasing. The energy requirements of the water industry world-wide have increased with city expansion and increased treatment standards (Chartres 2005). As cities grow, water often has to be moved from more distant locations and additional treatment undertaken to meet current standards (Zakkour *et al.* 2002). The majority of proposed centralized new water sources, such as desalination and reuse, are more energy intensive when compared with traditional sources (Medeazza and Moreau 2007). Even rainwater tanks can use a relatively large amount of energy when pumped using small and inefficient pumps (Gardner *et al.* 2006).

The increased use of energy creates a real dilemma for planners, because it contributes further greenhouse gas emissions. Consequently, the solution to climate-change induced urban water stress can compound the problem, through increased burning of fossil fuels. In addition, energy generation itself requires a constant source of water which is in competition with other uses and can place further pressure on urban water supplies.

Urban metabolism and how it gives context to water supply choices

Urban metabolism is a systems perspective on the resource inputs and outputs necessary to sustain cities. Metabolism analysis is a means of quantifying the overall fluxes of energy, water, material and wastes into and out of an urban region (Sahely *et al.* 2003). Undertaking a mass-balance of a city 'entity' was first proposed by Wolman (1965) as a means of resolving 'shortages of water and pollution of water and air'. This model is viewed by some as having the potential to add significant value to the analysis of sustainability of urban water systems (Pamminger and Kenway 2008; Tambo 2002) (see Figure 15.4).

This is because it will help us find solutions which simultaneously reduce the mass flow of water and energy through our urban systems. It will also give a context for the movement of other materials through our cities, notably nitrogen, phosphorus and carbon, which have significant flows in the wastewater network. It is possible that multi-criteria modelling based on metabolic models could begin to displace traditional 'triple bottom line' modelling, and/or that the relative weighting applied to parameters will change. Operationalizing the metabolic framework could be assisted by the relatively rapid emergence of carbon and water markets. Similarly, some nutrient trading markets (e.g., for wastewater discharges) are emerging and will contribute to the drive for a metabolic framework.

Substantial work will be necessary to implement the urban metabolism model and its application at cluster, subdivisional and city scales into the planning process. To date, the model has been applied at city scale (Kennedy *et al.* 2007). However, it could potentially offer guidance at development and individual household scales. For example, how much energy should a development or individual property owner invest in

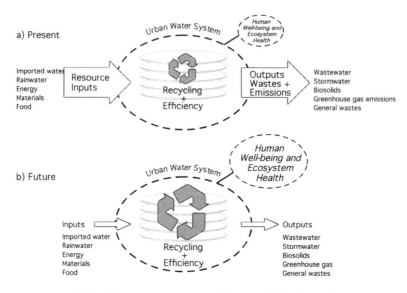

Figure 15.4 The urban metabolism model as applied to the urban water system.

resolving water issues? Can subdivisional development provide water and wastewater services at lower energy cost than centralized systems? To ensure that sustainable future pathways are chosen, it will be important that the overall urban planning processes make sure that new development applications and the technologies that they propose lower the overall metabolic rate (water, energy and materials throughput) of the city.

Conclusions and recommendations

Integrated water systems are here to stay. Further attention is warranted at the subdivisional and household scales to ensure the most efficient solutions are identified and that the challenges of integration are overcome.

Most new developments have not yet thought through integration at the development scale, let alone with the centralized system. Examples of integration issues include the following:

- while individual technologies are understood, the integration of the technologies and impacts on other systems is less clear;
- technologies can be applied at a range of scales from allotment, to cluster to subdivision, often simultaneously, with associated benefits and challenges;
- improved knowledge regarding the energy use, management and maintenance and householder acceptance of new systems and technologies represents an important area for future research;

- guidance on managing risks in relation to treating water to acceptable standards to meet technical, health and environmental safety criteria, including the management of waste and greenhouse gas emissions from those systems, is necessary.

Selecting the most appropriate technologies relies on many site-specific factors. To fully reap the benefits, clear understanding is required of proposed uses of the water and the quality necessary for each use, system requirements including design areas (e.g., for wastewater use), and technology or project characteristics and limitations. Additionally, in assessing decentralized systems, greater understanding is required regarding the implications of adoption of new technologies and systems in the wider city/society context in relation to increased sustainability benefits (environmental, social, future economic).

Planning and management approaches need to be adapted to meet the needs of decentralized systems. A strategic approach is needed for our infrastructure to allow flexibility in design and to ensure an outcome of reducing the total water, energy and materials throughput of cities. Suggested research needs to support the adoption of diverse appropriate technologies, including performance monitoring of alternative systems over the life cycle. This will require a deeper understanding of the impact of alternative technologies (e.g., greywater reuse), and needs to be supported with social and economic analysis.

While our technology options are increasing dramatically, technology alone will not provide the solution; this requires appropriate and simultaneous supporting social, institutional and economic factors to be addressed. Development of technology selection frameworks and guidance for any particular water harvesting or reuse option is necessary. It is here that broader impact assessment frameworks such as urban metabolism can play a significant role in helping to transition to urban developments which meet defined service needs, while minimizing energy, water and nutrient use.

Acknowledgements

The authors would like to thank Claire Diaper and Steve Cook for detailed reviews of drafts of this chapter, and the CRC for Construction Innovation and CSIRO Water for a Healthy Country Flagship for funding the project on sustainable subdivisions with a water focus.

Bibliography

ABS (2006) *Australian Demographic Statistics*, Cat. no. 3101.0, Canberra: Australian Bureau of Statistics.

Argue, J.J. and Argue, J.R. (1998) 'Total stormwater management at Figtree Place, Newcastle, New South Wales', in *Proceedings of the 3rd Novatech Conference on Innovative Technologies in Urban Storm Drainage*, Lyons: GRAIE.

Atlantis Water Management (2007) Home page. Online. Available at HTTP: <www.atlantiscorp.com.au/projects/storm_water_filtration_&_reuse_system>.

Belsham, T. (2005) 'Running hot! National residential hot water strategy: strategic options and issues', paper presented at National Appliance and Equipment Energy Efficiency Committee forum, Canberra, 15 September.

Brisbane City Council (2005) *Water for Today and Tomorrow: A Proposed Integrated Water Management Strategy for Brisbane*, Consultation Draft, Brisbane.

Chan, D., Kodikara, J.K., Gould, S., Ranjith, P.G., Choi, X.S.K. and Davis, P. (2007) 'Data analysis and laboratory investigation of the behaviour of pipes buried in reactive clay', in *Common Ground: Proceedings of the 10th Australia New Zealand Conference on Geomechanics*, Brisbane. Online. Available at HTTP: <http://civil.eng.monash.edu.au/expertise/geomechanics/mapps/pubs/expertise/geomechanics/mapps/PB1.pdf>.

Chartres, C. (2005) *Water Scarcity Impacts and Policy and Management Responses: Examples from Australia*, Canberra: National Water Commission.

City of Melbourne (2004) *Water Sensitive Urban Design Guidelines*. Online. Available at HTTP: <www.melbourne.vic.gov.au/rsrc/PDFs/Water/WSUDGuidelines.rtf>.

Diaper, C. (2004) *Innovation in On-Site Domestic Water Management Systems in Australia: A Review of Rainwater, Greywater, Stormwater and Wastewater Utilisation Techniques*, MIT Technical Report 2004–073, Melbourne: CSIRO.

Diaper, C., Sharma, A., Gray, S., Mitchell, G. and Howe, C. (2003) *Technologies for the Provision of Infrastructure to Urban Developments: Canberra Case Study*, MIT Technical Report 2003–183, Melbourne: CSIRO.

Diaper, C., Sharma, A. and Tjandraatmadja, G. (2008) 'Decentralised water and wastewater systems', in P.W. Newton (ed.) *Transitions: Pathways Towards Sustainable Urban Development in Australia*, Melbourne: CSIRO and Dordrecht: Springer.

Diaper, C., Tjandraatmadja, G. and Kenway, S.J. (2007) *Sustainable Subdivisions: Review of Technologies for Integrated Water Services*, Brisbane: CRC for Construction Innovation.

Dillon, P.J., Pavelic, P., Molloy, R. and Dudding, M. (2006) 'Developing aquifer storage and recovery opportunities in Melbourne', in *Proceedings of the VicWater Conference*, Geelong.

Gardner, T. and Dennien, B. (2007) 'Why has SEQ decided to drink purified recycled water?', paper presented at the Water Reuse and Recycling Conference, University of New South Wales, Sydney, 16–18 July.

Gardner, T. and Millar, G. (2003) 'The performance of a greywater system at the Healthy Home in South East Queensland: three years of data', in R.A. Patterson and M.J. Jones (eds) *Proceedings of On-Site '03 Conference: Future Directions for On-Site Systems: Best Management Practice*, Armidale: Lanfax Laboratories.

Gardner, T., Millar, G., Christiansen, C., Vieritz, A. and Chapman, H. (2006) 'Urban metabolism of an ecosensitive subdivision in Brisbane, Australia', paper presented at the Enviro 06 Conference, Melbourne, 9–11 May.

Geolink (2005) *Clunes Wastewater: Assessment of Onsite and Community-Scale Wastewater Management Options*. Online. Available at HTTP: <www.geolink.net.au/infocentre/index.html>.

Gleick, P. (1994) 'Water and energy', *Annual Review of Energy and the Environment*, 19: 267–99.

Government of Victoria (2004) *$12 Million Boost for Stormwater Use and Water Savers*, media release, Melbourne, 15 September. Online. Available at HTTP: <www.dpc.vic.gov.au/domino/Web_Notes/newmedia.nsf/798c8b072d117a01ca25 6c8c0019bb01/60a4e3dd69854c04ca256f10007aab35!OpenDocument> (accessed 7 May 2007).

Grant, A., Sharma, A., Mitchell, V.G., Grant, T. and Pamminger, F. (2006) 'Designing for sustainable water and nutrient outcomes in urban developments in Melbourne', *Australian Journal of Water Resources*, 10 (3): 251–60.

Gray, S. (2004) 'Reducing potable water use: a comparison of on-site greywater treatment systems and raintanks in urban environments', paper presented at the 6th Specialist Conference on Small Water and Wastewater Systems, Murdoch University, Fremantle, 11–13 February.

Hatt, B., Deletic, A. and Fletcher, T. (2004) *Integrated Stormwater Treatment and Reuse Systems: Inventory of Australian Practice*, Industry Report, Melbourne: CRC for Catchment Hydrology.

Howe, C., Jones, R., Maheepala, S. and Rhodes, B. (2005) *Melbourne Water Climate Change Study: Implications of Potential Climate Change for Melbourne's Water Resources*, Melbourne: CSIRO.

Hurlimann, A. and McKay, J. (2003) 'Community attitudes to an innovative dual water supply system at Mawson Lakes, South Australia', in *Proceedings of the Ozwater 2003 Convention*, Perth.

—— (2004) 'Attitudes to reclaimed water for domestic use: part 2: trust', *Water: Journal of the Australian Water Association*, 31 (5): 40–5.

IPCC (2001) *Climate Change 2001: Impacts, Adaptions and Vulnerability*, Geneva: Intergovernmental Panel on Climate Change.

ISWR (2006) *Integrated Stormwater Treatment and Harvesting*, Technical Guidance Report 06/05, Melbourne: Institute for Sustainable Water Resources, Monash University.

Kelly, T. (2008) 'Transitioning to sustainable water and sewerage infrastructure: a Melbourne perspective', paper presented to Australia-German Workshop, Melbourne, 1 April.

Kennedy, C., Cuddihy, J. and Engel-Yan, J. (2007) 'The changing metabolism of cities', *Journal of Industrial Ecology*, 11 (2): 43–59.

Kenway, S.J., Pamminger, F., Gregory, A., Speers, A. and Priestley, A. (2008) 'Urban metabolism can help find sustainable water solutions: lessons from four Australian cities', paper presented at the IWA World Water Congress, Vienna.

Kenway, S.J., Priestley, A. and McMahon, J.M. (2007) 'Water, wastewater, energy and greenhouse gasses in Australia's major urban systems', paper presented at the Water Reuse and Recycling Conference, University of New South Wales, Sydney, 16–18 July.

Landcom (2004) *Best Planning and Management Practices*, Water Sensitive Urban Design Strategy, Book 2, Parramatta: Landcom. Online. Available at HTTP: <www.landcom.com.au/downloads/file/WSUD_Book2_FactSheet.pdf>.

Marsden Jacob Associates (2007) *The Economics of Rainwater Tanks and Alternative Water Supply Options*, Melbourne: report prepared for Australian Conservation Foundation, Nature Conservation Council NSW and Environment Victoria. Online. Available at HTTP: <www.acfonline.org.au/uploads/res_rainwater_tanks.pdf> (accessed 9 May 2007).

Martin, R.R. and Dillon, P.J. (2002) 'Aquifer storage and recovery in South Australia', *Water: Journal of the Australian Water Association*, 29 (2): 28–30.

Medeazza, G. and Moreau, V. (2007) 'Modelling of water-energy systems: the case of desalination', *Energy*, 32: 1024–31.

Mobbs, M. (1998) *Sustainable House*, Sydney: Choice Books.

Newton, P.W. (2007) 'Horizon 3 planning: meshing liveability with sustainability', *Environment and Planning B: Planning and Design*, 34: 571–75.

—— (ed.) (2008) *Transitions: Pathways Towards Sustainable Urban Development in Australia*, Melbourne: CSIRO and Dordrecht: Springer.

Pamminger, F. and Kenway, S.J. (2008) 'Urban metabolism: a concept to improve the sustainability of the urban water sector', *Water: Journal of the Australian Water Association* (forthcoming).

Po, M., Kaercher, J.D. and Nancarrow, B.E. (2003) *Literature Review of Factors Influencing Public Perceptions of Water Reuse*, Perth: CSIRO Land and Water.

Po, M., Nancarrow, B.E., Leviston, Z., Poter, N.B., Syme, G.J. and Kaercher, J.D. (2005) *Predicting Community Behaviour in Relation to Wastewater Reuse: What Drives Decisions to Accept or Reject?*, Perth: CSIRO Land and Water.

Priest, G.M., Anda, M., Mathew, K. and Ho, G. (2003) 'Domestic greywater reuse as part of a total urban water management strategy', in *Proceedings of the Ozwater 2003 Convention*, Perth.

Sahely, H.R., Dudding, S. and Kennedy, C.A. (2003) 'Estimating the urban metabolism of Canadian cities: Greater Toronto Area case study', *Canadian Journal of Civil Engineering*, 30 (2): 468–83.

Salisbury City Council (2008) *Parafield Stormwater Harvesting Facility*, Salisbury. Online. Available at HTTP: <http://cweb.salisbury.sa.gov.au/manifest/servlet/page?pg=8424> (accessed 18 March 2008).

Sarac, K., Kohlenberg, T., Davison, L., Bruce, J.J. and White, S. (2001) 'Septic system performance: a study at Dunoon, Northern NSW', in R.A. Patterson and M.J. Jones (eds) *Proceedings of the On-Site '01 Conference: Advancing On-Site Wastewater Systems*, Armidale: Lanfax Laboratories.

Stormwater Industry Association Queensland (2006) *Comment on SEQ Regional Plan*, Brisbane. Online. Available at HTTP: <www.stormwater.asn.au/qld/SEQ-RegPlanReview-SIAQ2006.pdf> (accessed 09 May 2007).

Tambo, N. (2002) 'A new water metabolic system', *Water: Journal of the Australian Water Association*, 21: 67–8.

Taylor, B., Gardner, E.A. and Kenway, S. (2007) 'South East Queensland recycled water aspects and soil impacts', paper presented at the Australian Water Association Regional Conference, Sunshine Coast, 9–11 November.

Tjandraatmadja, G.T. and Burn, S. (2004) *A Review of Technologies for Wastewater and Bioresource Recovery in Urban Centres*, MIT Technical Report 2004–271, Melbourne: CSIRO.

Urban Stormwater Initiative Executive Group (2004) *Urban Stormwater Management Policy for South Australia*, Draft 18, Adelaide. Online. Available at HTTP: <www.lga.sa.gov.au/webdata/resources/files/Draft_Urban_Stormwater_Management_Policy_for_SA.pdf> (accessed 7 May 2007).

Wolman, A. (1965) 'The metabolism of cities', *Scientific American*, 213: 179–90.

Wong, T.H.F. (ed.) (2006) *Australian Runoff Quality: A Guide to Water Sensitive Urban Design*, Sydney: Institution of Engineers.

WSAA (2005a) 'The Australian urban water industry', in *WSAAfacts 2005*, Melbourne: Water Services Association of Australia.

—— (2005b) *Testing the Water: Urban Water in Our Growing Cities: The Risks, Challenges, Innovation and Planning*, Melbourne: Water Services Association of Australia.

Zakkour, P., Gaterell, M., Griffin, P., Gochin, R. and Lester, J. (2002) 'Developing a sustainable energy strategy for a water utility, part I: a review of the UK legislative framework', *Journal of Environmental Management*, 66: 105–14.

Part IV

Construction

16 Virtual design and construction

Martin Fischer and Robin Drogemuller

The concept of 'build virtually then build actually' has been a key driver in research and process innovation within the AECO industry for the last 20 years, and was the vision that led to the formation of the Center for Integrated Facility Engineering (CIFE) at Stanford University in 1988. After many years of research, testing and incremental uptake, this concept is now being realized under the banner of virtual design and construction (VDC). VDC is the use of multidisciplinary performance models of design-construction projects, including their products (the facility), organization and work processes, to meet the business objectives across the stakeholders. Design and construction are intimately linked in the process, as the uptake of this technology enables changes in the role and responsibilities of designers and constructors. There will also be implications for facilities managers (see Ding *et al.*, Chapter 20 in this volume).

The major motivation for using VDC is the perceived underperforming of design and construction firms. For example, in the analysis of a representative *design* firm, it was found that 6 per cent of the total time was spent on planning design processes, 36 per cent on designing and 58 per cent on managing information (Flager and Haymaker 2007). The high proportion of time spent on managing information is essentially 'lost' time, and the firm has now proceeded to reduce this.

On the *construction* side, a study over the period 2001–05 (Figure 16.1) by SPS found that of all the factors involved in on-site construction delays, unforeseen site conditions were responsible for 1 per cent and inclement weather for 10 per cent. The remaining 89 per cent were caused by factors that were under some form of control by the contractor. These high levels of delays could easily be reduced by earlier recognition of the potential problems.

Perceptions of VDC

A range of 12 general beliefs comparing the impacts of 3D/4D modelling were identified in the literature by Gao and Fischer (2008). Thirty-two case studies were analysed to assess the accuracy of these general beliefs.

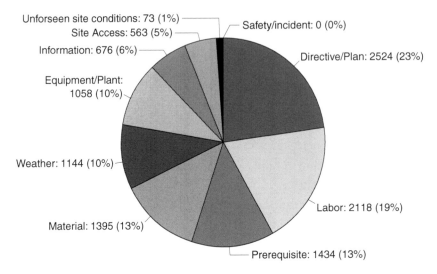

Figure 16.1 Factors causing on-site delays on construction projects (source: courtesy of Strategic Project Solutions, Inc., San Francisco, CA, USA).

The functions for which the 3D/4D models were used varied depending on the business drivers of the project participants, where in the project procurement process the modelling occurred, and the challenges faced by the teams. The major uses of the models were for communication with non-professionals, construction planning, production of drawings and design coordination and clash detection. They were also beginning to be used for the analysis of a wider set of design options, management of the supply chain, estimating cost, and definition of owner requirements.

When 3D models were developed in the earlier design phases, benefits were felt both immediately through improved processes and organization, but also over the long term as the benefits flowed to the later stages of project procurement and assisted in improving the performance of the completed project. When 4D processes were used there were immediate benefits to the users, but these did not flow to as many other team members. Not surprisingly, the maximum benefits to the individual team members and the team as a whole were gained when all members used the 3D/4D technologies.

The 3D/4D models needed to be created at the appropriate level of detail for the current stage of the project to maximize the benefits. This depends on the uses to which the model is put and the amount of information available at that time.

When using 4D models, the first three stages of collecting data, modifying the original schedule and creating or modifying the 3D model can all be performed on standard software, and are often performed as a matter of course. Hence the major effort is 'for free', as it is the result of other processes. The linking of schedule and model is a straight technical application of 4D software, while the final steps of reviewing and then updating the 4D models rely on the 4D software, data exchange issues and organizational alignment.

Gao and Fischer (2008) found that 3D models were used for:

- establishing the owner's requirements;
- interacting with non-professional stakeholders;
- analysing design options;
- checking multidisciplinary system clashes and constructability issues;
- producing construction documents;
- supporting cost estimating;
- managing supply chains;
- planning for construction execution;
- managing facility operations.

The 4D modelling was used for:

- strategic project planning;
- developing contractor's proposals;
- comparing proposals between the general contractor and subcontractors;
- permitting;
- master scheduling;
- constructability review;
- analysis of site operations.

The general beliefs (GBs) were broken up into five groups: model use, timing of 3D/4D modelling, key stakeholders involved in the 3D/4D modelling and review process, level of detail in 3D/4D models, and effort required for the workflow of the 3D/4D modelling. They are presented below (from Gao and Fischer 2008) with the results of the analysis.

Model use

GB1: A 3D model is useful for a wide range of purposes, such as cost estimating, construction planning, analysis, automated fabrication, and project control applications; but for now 3D/4D models are primarily used as a visualization and marketing tool.

The first part of this belief was accurate, but the second part was becoming less accurate as organizations became more comfortable with 3D/4D and learned new ways to exploit them. Half of the sample projects used the models for more than one purpose, with interaction with non-professionals, construction planning, drawing production and design checking being the most popular.

GB2: The use and benefits of three-dimensional design in residential and commercial buildings have not been shown.

All of the sample projects used 3D models. Ten projects were residential, nine were institutional, eight were commercial, three were industrial and two were transportation facilities. While the sample projects could not be considered representative of the proportion of projects undertaken by industry, it can be claimed that 3D models are commonly used across residential, institutional and commercial projects. 4D CAD had significant use on commercial (seven) and institutional (nine) projects.

GB3: Lack of design-build or other collaborative contractual models should not be viewed as a reason to avoid 3D/4D modelling practices.

In the sample, the number of uses of 3D/4D across the different contractual models are shown in Table 16.1. These results support the idea that non-collaborative contractual models also benefit from 3D and 4D technologies.

GB4: 3D/4D models are more applicable for use on certain projects which involve characteristics such as a complex design, fast-paced project delivery, tight budget, high-tech facilities, etc.

The results from the surveys supported this belief. The drivers for 3D and 4D use both varied depending on whether owners, developers, architects or construction managers/general contractors were promoting the use of the technology.

GB5: There are many 3D/4D models developed for many different uses.

This belief was supported by the results. Only five of the projects used the model for a single purpose. Additionally, there was strong evidence that,

Table 16.1 Use of 3D/4D across contractual models

Contractual model	Number of 3D uses	Number of 4D uses
Design-bid-build	16	12
Construction management/ general contractor	8	8
Design-build	8	2

Source: Fischer (CIFE).

the more uses made of a model, the higher the number of benefits that were obtained.

Timing of 3D/4D modelling

GB6: There is a time lag between corresponding benefits.

This belief was reformulated to capture the findings: 'Among all the benefits attainable from one particular 3D model use, some benefits come along immediately with that use of 3D models while other benefits occur later.' In general, the earlier the use of 3D/4D models the wider the time span and the number of downstream participants over which the benefits were obtained.

GB7: It is essential to capitalize on project opportunities early on to make 3D/4D models have a lasting and positive effect on the facility over its total lifespan.

The results confirmed this belief. Three of the projects used 3D models in the early planning and design stages. This allowed the setting of aggressive but realistic goals for energy, cost and environmental performance through the early exploration of building form and layout. One of the projects quantified the benefits as a 10 to 15 per cent saving in upfront cost, and potential 5 to 25 per cent life cycle cost savings.

GB8: Designers benefit directly from building detailed 3D models. A design in 3D costs less than a design in 2D for an architect.

The contention by some designers that the client should contribute to model development costs because there were not sufficient direct benefits to designers was not supported by the results, supporting the belief above. Many of the projects used 3D to support client briefing or construction document generation. A smaller but significant proportion used the models for design analysis or checking with respect to coordination between systems. Benefits were experienced by team members downstream, but there were also significant benefits experienced by the designers, including:

- reduction in errors and inconsistencies;
- reduction in late design changes due to delays in identifying problems – early identification reduced the cost to the designer of fixing the problem;
- 3D modelling does not significantly reduce total design effort, but it does change the distribution of effort between design development and contract document preparation. The productivity improvements from the reduction in contract document preparation were used strategically

to improve design quality and hence increase the likelihood of repeat business.

Key stakeholders involved in the 3D/4D modelling and review process

GB9: The more stakeholders are involved in implementing 3D/4D modelling, the more benefits will accrue to them as a whole and to each stakeholder individually.

Analysis of the results strongly supported this belief, independently of who was motivating the use of 3D/4D. The benefits not only accrued to the team as a whole, but also to each individual organization. As one of the subcontractors commented, the more groups involved in the modelling, the higher the reliance that can be placed on the information. This leads to increased levels of off-site manufacture through the increase in confidence.

Level of detail in 3D/4D models

GB10: Creating 3D and 4D models at the appropriate level of detail is instrumental in reaping their benefits.

A hierarchy of level of detail was defined to allow analysis of this belief:

- project (building/site);
- system;
- sub-system;
- component;
- part.

The appropriateness of the level of detail was dependent on the requirements of proposed use of the model and on the information available at that stage of the project. The results showed that the creation of the models 'just in time' and at the appropriate level of detail were critical in maximizing the benefits of their use.

Effort required for the workflow of the 3D/4D modelling

GB 11: The limitations of 3D/4D modelling software tools and issues pertinent to data exchange and organizational alignment are the main stumbling blocks to an efficient modelling process.

The findings supported this claim. Most identified issues related to information availability and interoperability. Information availability can be addressed through changes in procurement methods or revisions to responsibilities in current procurement methods.

Limitations of 4D modelling tools that were identified include:

1   lack of naming conventions that support both 3D components and activities;
2   lack of support for automatic linking of components and activities;
3   lack of support for hierarchical views of schedules at multiple levels of detail;
4   poor support for distributed work environments;
5   limitations of current hardware and software in displaying large models;
6   no support for simultaneous synchronization of updates to the 3D data and schedule;
7   lack of support for automatic link updates between objects and activities.

The issues involved in addressing points 2, 6 and 7 are discussed later in this chapter.

VDC technologies

The discussion above illustrates that while there are a wide range of potential VDC technologies, these also vary significantly in readiness, usability and reliability. The study indicates that the VDC technologies that can be used confidently under current industry practice include:

*   visualization;
*   capturing existing conditions;
*   coordination in 3D;
*   fabrication from 3D models;
*   3D model-based Bill of Materials extraction;
*   coordinating construction with 4D simulations/models.

Visualization is possibly the easiest and most influential change enabled by VDC. Visualization techniques are normally part of projects. Significant improvements in the delivery of projects can be achieved by thinking about how additional value can be obtained from visualization material that is already available. An area where significant gains can be made is in improving communication between designers and clients/stakeholders. If accurate visualizations of proposed designs rather than 2D drawings are used in the early stages of design (Figure 16.2), then the clients are more likely to understand what is being proposed and to comment in an informed manner. This is likely to increase the accuracy and usefulness of client requirement documents and to decrease the need for variations during construction caused by clients not understanding spatial issues until construction is underway.

The capture of existing conditions is especially important for projects in congested areas and for refurbishment projects. The use of new technologies

Figure 16.2 Use of VR technologies for stakeholder assessment of design (source: Walt Disney Imagineering).

Figure 16.3 Point cloud-defined using laser scanning (source: Andrew Peterman, Stanford University).

such as laser scanning can significantly improve the amount and accuracy of information. Laser scanning is also an extremely fast process compared with traditional measurement systems. While laser-scanned models only contain masses of points which require human interpretation (Figure 16.3), laser-scanned data can also be used as the basis for the development of 3D CAD models of the existing facility.

A significant proportion of problems during construction are due to interference between different building elements. The use of 3D spatial coordination tools that perform clash detection has the potential to minimize this problem. There are a number of issues that can result in clashes occurring:

- use of 2D drawings that 'hide' clashes by not including a third dimension;
- lack of coordination between groups, such as structural and HVAC where air conditioning ducts are designed to pass through areas where beams or slabs exist;
- lack of coordination in planning where building components are placed outside the required sequence, such as installation of a ceiling grid before the HVAC ductwork;
- conflict between temporary works and final construction;
- conflict between temporary works for different trades through poor time coordination.

Software tools that support clash detection comprehensively will also need to handle the range of file formats used by the project participants.

Conflicts between building elements on site are a major cause of requests for information (RFIs) and rework during construction. 3D models can be used to coordinate the information from different disciplines to identify and resolve conflicts (Figure 16.4). This is applicable during design (i.e., conflicts between structural and HVAC elements) and during construction (i.e., interference of temporary works with permanent elements).

Figure 16.4 Fulton Street Transit Center, New York: design conflict (source: Hartmann *et al.* (2007)).

There is an increasing range of software available that supports the direct download of 3D CAD data to numerically controlled machinery. This allows the direct or partial fabrication of components. One example is the CadDuct software. This supports the design of HVAC ductwork in the software with download of the data to numerically controlled machinery. Compatible machinery can then automatically cut and fold the duct component below a particular size or fold the sheet for larger ducts for later completion by hand (Figure 16.5).

The major 3D CAD systems support simple counting functions, capable of counting objects in the system much faster and more reliably than by human effort. However, this information is not normally structured in a way that is directly useful in an estimating system. Quantity take-off and estimating software that supports the take-off of quantities is now available. The early versions of these systems required the user to manually link the items in the bill of quantities (BOQ) with the relevant components. Recent versions of these computer programs provide various levels of support to semi-automate the process of linking building elements with the relevant item(s) in the BOQ. One current research effort (Automated Estimator) to take this further is discussed by Drogemuller *et al.* (Chapter 7 in this volume). The user interface is shown in Figure 16.6. The use of 3D models to support BOQ generation and estimation is leading to reported savings of 50 per cent on estimation time.

4D simulation models can be considered as adding the time dimension to clash detection in 3D. As mentioned above, static clashes can be identified using straight 3D. Clashes that occur in time, such as two subcontractors wanting to use the same space simultaneously, or overhead lifting processes occurring over the top of workers, need the time dimension for identification and resolution. 4D tools will normally show a list of the current activities together with a visualization of the current planned state of the project (Figure 16.7).

Figure 16.5 Duct work as designed in CAD and as produced on numerically controlled machine (source: courtesy Triple 'M' Mechanical Services (Qld) Pty Ltd).

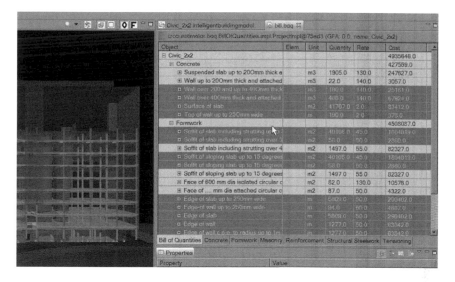

Figure 16.6 User interface to Automated Estimator, automated quantity take-off software.

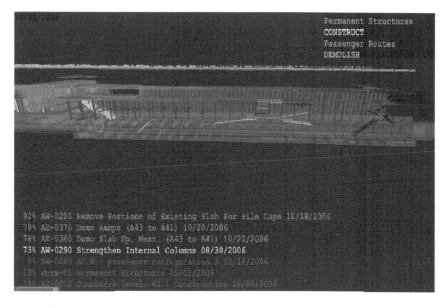

Figure 16.7 Snapshot from a 4D model built by Parsons Brinkerhoff on the Fulton Transit Center Project (Hartmann *et al.* 2007) to support coordination of the design and construction of the subway and the building renovation.

While many construction personnel can interpret a Gantt chart and drawings to build a picture of operations in their head, these are typically senior personnel with extensive experience. The use of 4D simulations allows relatively junior staff to perform at a similar level. They are also extremely useful in communicating with the public. One instance was the identification by hospital staff of cranes impinging on emergency helicopter airspace when viewing 4D simulations. Under standard scenarios, it is extremely unlikely that this would have been detected until construction started.

Technical issues underlying VDC

This section describes a joint research project undertaken by the CIFE and the CRC for Construction Innovation – the development of the Automated Scheduler software. The goal was to examine the issues involved in automating the generation of initial construction schedules and then automatically simulating the operations in a 4D environment. A number of constraints were placed on the project to maintain a feasible scope:

1   The method was tested using multi-storey buildings. This meant that a starting point could be automatically determined (at the foundation of the building) and the dependencies between building elements could be determined by identifying which elements support others. In contrast, civil engineering projects often have multiple potential starting points and multiple potential sequences thereafter.
2   The IFC model was used as the input building data. The model's internal structure supports automatic identification of building elements, and contains sufficient information to identify which elements support which other ones.
3   The scope of the building elements was limited to foundation elements, slabs, columns, beams and wall.
4   Common Point was used as the 4D software, as this had been developed at Stanford University and both groups were comfortable with it. Among the options for data exchange supported by Common Point, a format supported by Microsoft Project was used to exchange the construction program information and a subset of VRML was used to import geometry.

The goal was to automatically generate a 'first cut' construction program, linked to the geometry and ready to import into Common Point for analysis. The missing ingredient was the knowledge required to build the construction plan. This knowledge was captured using the Component/Action/Resource/Sequencing constraint method (Aalami *et al.* 1998) developed at CIFE.

When a user runs the software, it analyses the IFC model and adds 'supports' relationships to the model. The underlying assumption is that a building element that supports another must be completed before the supported element is started. This defines a partial order on the building elements. Those of the same type (column, beam, etc.) on the same floor are grouped. Recipes for the construction of each of the known component/material types are then accessed. The durations, activities and resources required to construct the group of elements are then defined, and the elements attached to them. The precedence relationships are then added, based on the 'supports' relationship. This completes the required analysis of the building model to generate the construction program. These data are then exported to a database. The last step is to generate the VRML geometry which is exported to a file.

There was no point in replicating the functionality of existing commercial software such as Microsoft Project. The list of activities and the building elements that they operate on are imported directly into Microsoft Project. The scheduling algorithm automatically generates the Gantt chart for the activities, together with the start and end dates of each activity based on the precedence relationships between them.

The construction program can then be imported from Microsoft Project, the VRML from the file and the construction simulation can be run (Figure 16.8).

Facility life cycle coverage

With some exceptions, the first 15 years of VDC research at CIFE focused mostly on VDC support for the design and construction phases of facilities. In the last few years CIFE researchers, with support and guidance

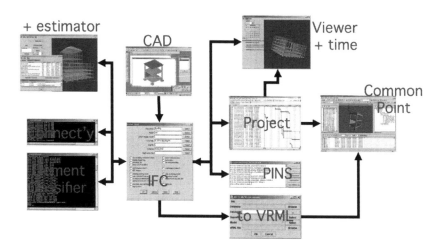

Figure 16.8 Software architecture and information flows in Automated Scheduler.

from its industry members, have expanded this focus to address opportunities in the very early generative project phases and in the facility operations and renovation phases. Through a few illustrative examples of ongoing projects, this section gives a short synopsis of these efforts (Figure 16.9).

Master planning

CIFE researchers developed a software tool that takes a simple 3D model of a planned development and a development schedule as input and then calculates the economic, environmental and social impact over 20 or 30 years (or longer if desired). The goal is to provide a holistic, visual and analytical perspective on the scope, evolution and impacts of large-scale developments that are understandable by all key project stakeholders, and that can help with understanding of the desired and less desired impacts so that appropriate goals can be set and tracked. Such models and simulations can be built for greenfield developments and for renovations and extensions of existing developments. This tool also illustrates a strong trend to centre computer-based analyses of engineering projects on 4D models instead of static 3D models as is more common today.

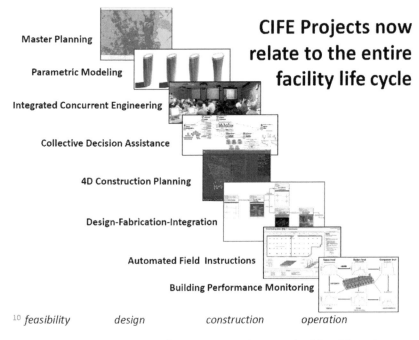

Figure 16.9 CIFE's VDC research projects over the entire building life cycle.

Parametric modelling

Designers are increasingly leveraging parametric modelling tools to understand the architectural and structural options for a building more quickly and fully. For example, in a few hours or days they can explore hundreds of layout options for a new building. Issues studied include how to best use such parametric models in the design process, how to communicate the insights gained to the other project team members, and how to carry out multidisciplinary parametric modelling.

Integrated concurrent engineering

Design is a social process. The integrated concurrent engineering research at CIFE addresses the integration of VDC tools into the multidisciplinary design process. The main goal is to reduce the latency to answer questions so that good design ideas can be developed quickly with multidisciplinary perspectives, and poor ideas occupy as little design time as possible. The VDC tools and a multiscreen display environment are helpful in supporting rapid analysis, visualization and documentation of solutions. However, most VDC tools are designed for single-user operation and do not support multi-user operation. This is inspired by the Jet Propulsion Laboratory's highly successful use of a process called 'extreme collaboration' for conceptual design, where about 20 experts from all the key disciplines work in one room for one week to develop the conceptual design of a space mission. This method has shortened the conceptual design schedule by about 10×, reduced costs for conceptual design by about 3× and increased the number of design options and criteria considered by about 5×.

Collective decision assistance

Understanding the project goals for all the important project stakeholders, relating them to each other, prioritizing them and showing how a design solution addresses each goal is a challenge on every project. CIFE researchers have developed two methods to address this: MACDADI (Multi-Attribute Collective Decision Assistance for Design Integration) and the Decision Dashboard with an underlying Decision Breakdown Structure. These make the setting and management of project goals and design solutions much more effective and transparent than today's methods allow. They relate the information in VDC models to the strategic and tactical decisions that must be made frequently on projects.

4D construction planning

Research and application in 4D (time plus 3D) construction planning at CIFE spans 15 years, with applications of 4D models to all types of projects. Increasingly, 4D modelling is used not only for construction schedules but also for the modelling and analysis of schedules in or for other life cycle phases. The production of 4D models is also being automated as much as possible. Furthermore, they are increasingly supporting computer-based analyses of schedules in their spatial context.

Design–fabrication integration

The main purpose of VDC models is not only to make the design of facilities more productive and to design better facilities, but also to make their physical assembly more productive and safer. Consideration of fabrication constraints during design and the hand-off of digital design data to support the precise and timely fabrication of parts, sub-systems and modules are critical elements in the VDC-enabled digital and physical production of facilities. Furthermore, a VDC-enabled design process often allows the identification of project elements that can be prefabricated so that the faster, safer and more precise fabrication methods available in factories can be leveraged wherever possible. The better coordinated design and documentation processes and the more accurately modelled facility components are both key to increased prefabrication. Issues addressed in CIFE research include formalization of the fabrication knowledge and constraints to enable timely (i.e., as late as possible but as early as needed) design decisions that consider the manufacturability of the designed systems and components. Of course, the hand-off format and timing of design information to fabrication needs to be well specified and synchronized between the two functions.

Automated field instructions

VDC methods have largely been used in offices and sometimes in site trailers. They rarely reach the workers – the final actors in the production of a facility. CIFE researchers have formalized a method that leverages detailed, company-specific, but project-independent process models of how work gets done in the field (e.g., how the form for a parking deck gets erected) and project-specific schedules and 3D models to produce daily field instructions for workers that include all the information needed for construction (including accurate and specific dimensions for a location, bills of materials, tool lists and the necessary design details). Attempting to produce such instructions every day for every worker on a construction site using existing methods demonstrated that it is extremely difficult, costly and time-consuming, and therefore practically infeasible. The field instructions method developed at CIFE automates as much of this process as possible. This project also illustrates how the digital project models and the actual construction (its physical and spatial context and the organizations and processes employed) are getting increasingly closer in scope, scale and timing.

Building performance monitoring

Even though 3D-model-based building energy performance simulations are becoming increasingly common, and sensors to measure all kinds of internal and external environmental data and system performance are increasingly widespread, there is as yet no practice and method of closing the loop between design assumptions and solutions and the actual operation of buildings. Therefore, we do not learn whether and how the design and operation of buildings can be improved (see Reffatt and Gero, Chapter 21 in this volume). CIFE research is developing and testing a method to relate building simulation and measured performance data to each other, and to understand the relationships between the spaces (where the user expects particular environmental conditions such as temperature, indoor air quality, etc.), the building systems (that should condition the spaces as needed) and the components of the building systems (where building operators can adjust the performance of the building automation systems).

An example of VDC: One Island East, Hong Kong

One Island East is a 70-storey office tower recently completed in Quarry Bay, Hong Kong. Swire Properties Limited, the project owner, established a building information modelling (BIM) group near the construction site. The project team, including the owner, BIM consultant (Gehry Technologies), main contractor (Gammon Construction Limited) and a number of subcontractors, worked together on building a project BIM. The owner initiated the use of BIM to improve processes on this project and, eventually, the Hong Kong construction industry. The improvements that were realized included (from Riese 2008):

- comprehensive 3D geometric coordination of all building elements prior to tender;
- enhanced quantity take-off from the BIM to improve the speed and accuracy of the preparation of the BOQ in Hong Kong Institute of Surveyors format prior to tender;
- lower, more accurate tender pricing resulting from the contractors' unknowns and risks being reduced;
- automation and interoperability of 2D documents with 3D BIM data;
- creation of a reusable catalogue of intelligent parametric building parts (knowledge capture);
- management of construction sequence and process modelling using the BIM elements;
- reduction of waste in the construction throughout the entire process;
- reduction in contractor RFIs;
- reduction of claims on site resulting from incomplete design coordination;

- quicker construction;
- lower construction cost;
- standardization of the construction supply chain and regulatory authorities;
- enhanced site safety;
- better build quality.

Additionally it is expected that the BIM elements will be used for facilities maintenance (see Ding *et al.*, Chapter 20 in this volume).

Figure 16.10 shows the use of clash detection on one of the levels, while Figure 16.11 shows the 3D model together with 2D views that are generated directly from the model to ensure full, accurate coordination.

An example of VDC: Granlund

Granlund is a Finland-based building services design company that developed a strategy for the use of VDC in the early 1990s (Figure 16.12). This enabled the company to move from the traditional position of such companies as a pure design consultant to engage in relevant activities further up and down the supply chain (Figure 16.13). The strategy has also supported Granlund's move into the international arena.

Figure 16.10 One Island East: clash detection in operation (source: courtesy Swire Properties and Gehry Technologies Asia).

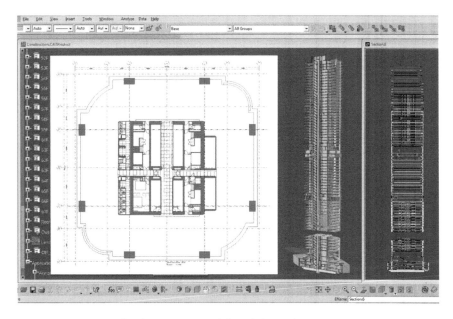

Figure 16.11 One Island East: 3D model and dependent 2D views (source: courtesy Swire Properties and Gehry Technologies Asia).

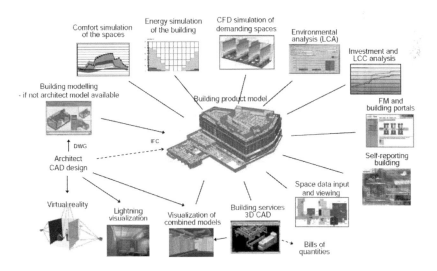

Figure 16.12 Granlund strategy for using virtual design and construction tools (source: used with permission, Olof Granlund Oy).

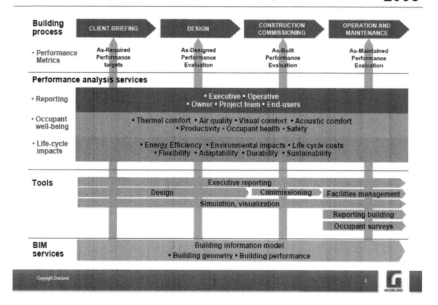

Figure 16.13 Range of services currently offered by Granlund (source: used with permission, Olof Granlund Oy).

A key part of the strategy was to establish an internal software development group that provided support for expansion of activities and has also provided a basis for a sustainable competitive advantage. Granlund now work in the areas of information management, maintenance management and integrated building performance assessment in addition to design.

Return on investment of VDC

One of the first questions which business asks when discussing the concept of VDC is 'What is the return on investment?' Table 16.2 shows the results for a number of projects.

The use of VDC on the first three projects was led by the general contractor. The cost of using VDC is an actual cost against the project budget. The savings are estimated, but are reliable. The return on investment varies significantly, but is worthwhile in each case. Project 4 was led by the owner and VDC technologies were not used; hence the actual cost is known, as the money has been expended. The VDC model cost and the savings are estimated, as they were not realized.

Actual projects which used VDC realized savings in a number of ways:

Table 16.2 Measured/estimated return on investment (ROI) from VDC (CIFE)

	Type of project, organization	Project cost	VDC model cost	Savings from using VDC model	ROI
1	Music centre	$250 M	$100,000	$500,000	5:1
2	Office complex	$200 M	$50,000	$3,000,000	60:1
3	Retail complex	$100 M	$40,000	$575,000	14:1
4	Large campus with complex facilities	$250 M	$400,000	$16,800,000	42:1

Notes
1, 2 and 3: General contractor, VDC used, costs and savings were realized
4: Owner, VDC not used, cost of VDC not incurred but estimated, potential savings, costs incurred

- no interferences in the field through the use of 3D visualization and clash detection;
- 90 per cent fewer RFIs and change orders;
- engagement of all stakeholders before finalizing design to ensure full needs capture and expectation management;
- 20 to 30 per cent higher productivity in the field through reductions in waste time by ensuring people and resources were available when needed;
- 100 per cent prefabrication, reducing on-site labour significantly;
- exploration of multiple design options to ensure full exploration of alternatives and close conformance to stakeholders' needs;
- 50 per cent faster quantity take-off through the use of automation.

There are a number of issues that must be resolved when considering the use of VDC. First, the cost is a financial cost against the budget. Is there a budget allowance? Are the savings credited to this budget? Is the use of VDC providing a cash benefit to the body that is meeting the cash cost? Are the benefits being realized by groups that are not contributing to the cost?

A significant issue if VDC is being proposed to clients is the implicit suggestion that they will not be receiving the best 'service' without its use. Most clients assume that the consultants working for them are already using the best available methods and technologies.

The above discussion has treated VDC as a tactical rather than as a strategic investment. The strategic value is difficult to quantify, as the benefits flow well outside that of the initiating organization. Additionally, its use will have impacts on future projects that use the skills and techniques developed in the early projects.

Finally, the system boundaries that are used to consider the impacts of VDC are not firm. Different results will be obtained if life cycle cost is considered, rather than just first cost. If costs and benefits are spread across environmental and social costs in addition to financial costs, the results will also vary. An additional consideration is the difference between cost, which is quantifiable, and value, which is often subjective.

Adoption of VDC

Adoption of VDC is a three-stage process. First, breakthrough goals need to be identified and listed, together with timelines for achievement. Examples are given in Table 16.3. Second, a set of incremental goals, as in Table 16.4, need to be defined that provide stepping stones towards these goals. Finally, a set of predictable performance objectives needs to be identified that will lead towards achieving the incremental and breakthrough goals. Some examples of these are given in Table 16.5.

Future work

Figure 16.14 shows the focus of most construction research in the first half-century of its academic existence on first cost aspects of the design and

Table 16.3 Examples of breakthrough goals (CIFE)

	Practice: 2002	*Goal: 2015*
Schedule	1–6 year design ~18 months construct Variance 5–100%	1 year design <6 months construct Variance <5%
Cost	Variance 5–30%	Variance <5%
Function/scope	Poor	25% better than 2002
Safety	Good	Better (e.g., lost time incidents per 200,000 hours)
Sustainability	Poor	25% better than 2002

Table 16.4 Examples of incremental goals (CIFE)

Business objectives of CIFE members

2006	*2010*
Operate with a strategic plan to implement VDC incrementally	Strategic plan to implement VDC broadly
	Manage by public and explicit metrics
Use first stage (visualization) of VDC confidently	Serve ≥ business purposes on ≥10 major projects p.a.
	Automate >30% of routine design and construction on >2 pilots
Staff each project with four VDC-trained engineers	Staff each project with four VDC-trained engineers

Table 16.5 (Multiple) predictable performance objectives (CIFE)

Controllable	Process	Outcome
Product, organization, process designs	Decision latency (promptness): mean <1; 95% within 2 working days	Safety: 0 lost hours
Coordination activity: planned, explicit, public, informed >90%	Response latency: mean ≤1; 95% within 2 working days	Schedule: 95% on-time performance
Facility-managed scope: 100% of items with >2% of value, time, cost or energy	Field-generated requests for information: 0	Cost: ≥95% of budgeted items within 2% of budgeted cost
Prediction basis: >80% of predictions founded	Rework volume: 0 (for field construction work) Goal = 10–20% (virtual work)	Delivered scope: 100% satisfaction
Design versions: 2 or more ≥80%	Meeting effectiveness: >90% participation Meeting efficiency: >70% prediction, evaluation	Sustainability: 25% better than 2002

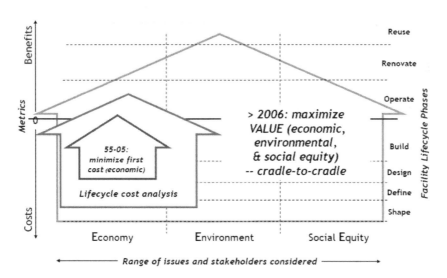

Figure 16.14 History of focus of decision-making in development (CIFE) (source: CIFE).

construction phases of the life cycle. Recently, life cycle cost analysis has extended the range of phases considered and included some environmental considerations. However, facilities are part of the physical, economic, environmental and social context of communities and regions. Therefore, they should be managed with a triple bottom line perspective to consider all their cost and benefit values over the entire life cycle. Significant research is needed to formalize the metrics and models and methods of analysis and visualization to support the consistent and rapid triple bottom line evaluation of projects.

Future research needs to address the current limitations of VDC methods and expand the application areas for VDC.

Limitations of VDC methods

Leveraging knowledge from multiple disciplines

For many disciplines, the design and life cycle knowledge has not been formalized so that it can be leveraged in VDC model-based analysis and simulations of facility projects. This limits the applicability of parametric and generative computational methods, leading to too much manual work, which is difficult to do with the large amounts of data required for 'full' project models.

Providing multi-user interaction functionality

Much progress has been made to separate data from software applications to support more effective and efficient sharing of digital project data. However, most applications are designed for single-user operation, with user interfaces designed for authoring the data and models that are the focus of a particular application. Such interfaces do not work well in a multi-user environment because authoring interfaces tend to be complex, which means that an expert user for each application needs to be present to leverage each application in collaborative design sessions. Research is needed to separate the user control – at least for basic interaction and information consumption – from the application and the data. This will make it much more likely that the VDC data can be leveraged for all important project decisions and by all important stakeholders.

Comparison of models over time and across criteria

VDC applications are becoming increasingly powerful in order to model, simulate and visualize design solutions for the purposes and perspectives for which they are made (e.g., structural engineering applications, energy simulations). However, little functionality exists to compare such models across levels of detail, over time (to truly understand changes in context, requirements or design), across disciplines and across projects. Hence, our ability to learn from these models and past performances and to shape the design of a project's physical form and organization and process with well-founded decisions is very limited. Research should develop methods and intuitive user interfaces to allow project teams to compare VDC models much more quickly and richly than possible today.

Application of VDC models

As Figure 16.14 shows, the responsibility of engineers involved in facility design, construction and operation should broaden significantly to consider the cost *and* value of facilities over their *entire* life cycle for economic, environmental *and* social concerns. It is unimaginable that this could be done without formal VDC methods to model, document, visualize, simulate and analyse design options. The development of such methods will require an incredible research effort, since in many cases we lack the metrics, data and methods to carry out the analysis for a particular concern. This is *the* major challenge and opportunity for all parties involved in making and operating buildings for the next decade or two.

Conclusion

The development of the concept of virtual design and construction to its current stage of understanding has taken 20 years. As described above, a suite of tools is now available that enables the project team to exploit its capabilities on live projects to gain identifiable benefits. Examples of the use of these technologies in live projects were given. Their use within the Center for Integrated Facility Engineering at Stanford University was also described. This provided the basis for a discussion of the gradual expansion of focus over time.

Two examples of the use of VDC technology within organizations were discussed. The One Island East project provides an example of significant benefits within a single multi-storey office development. The use of ICT to support strategic positioning of Granlund as HVAC consultants and to provide a sustainable commercial advantage was discussed.

The measurement of return on investment on the use of VDC technologies was also presented. These were assessed against tactical (within project) criteria. The difficulty of identifying the strategic benefits was noted, while acknowledging that these do exist.

Significant levels of research and development are required for VDC to reach its full potential. It was argued that VDC is essential if the full range of environmental considerations is to be met in the future. However, this is not just an issue of technology; it will also require a significant effort in gathering data and defining algorithms and metrics.

The major point that should be made is that while VDC tools are not perfect, they provide measurable benefit now. The sooner they are used, and the more widespread their use, the sooner more advanced tools will become available.

Bibliography

Aalami, F.B., Fischer, M.A. and Kunz, J.C. (1998) *AEC 4D Production Model: Definition and Automated Generation*, Working Paper 52, Stanford, CA: Center for Integrated Facility Engineering, Stanford University.

Fischer, M. and Hamburg, S. (1999) 'Construction planning and management using 3D CAD models', in *Proceedings of the Computers for Contractors '99 and A/E/C Systems Fall Conference*, Chicago, 8–11 November.

Flager, F. and Haymaker, J. (2007) 'A comparison of multidisciplinary design, analysis and optimization processes in the building construction and aerospace industries', in D. Rebolj (ed.) *Proceedings of the 24th W78 Conference, Maribor 2007*.

Gao, J. and Fischer, M. (2008) *Framework and Case Studies Comparing Implementations and Impacts of 3D/4D Modelling Across Projects*, Technical Report TR172, Stanford, CA: Center for Integrated Facility Engineering, Stanford University.

Hartmann, T., Goodrich, W.E., Fischer, M. and Eberhard, D. (2007) *Fulton Street Transit Center Project: 3D/4D Model Application Report*, Technical Report TR170, Stanford, CA: Center for Integrated Facility Engineering, Stanford University.

Riese, M. (2008) 'One Island East, Hong Kong: a case study in construction virtual prototyping', in P.S. Brandon and T. Kocatürk (eds) *Virtual Futures for Design, Construction & Procurement*, Oxford: Blackwell.

17 Internet-based construction project management

Achi Weippert and Stephen Kajewski

In an attempt to gain a better understanding of the current state of play of the international construction industry's use of innovative internet-based solutions – from both a technological and end-user (culture/social) perspective – the authors directed an extensive international literature review, as well as identifying, testing and trialling various innovative internet-based electronic tender (e-tender) systems and handheld devices. This provided the basis for developing industry guidelines designed to benefit both the public and private sectors in three specific areas:

1   *Electronic tender*: guidelines to help improve the identification, implementation and application of:

- features, functionality and capabilities of an internet-based e-tender system;
- risks that industry professionals should consider when choosing an internet-based e-tender process or system;
- innovative training and education requirements;
- legal issues.

2   *Handheld technology*: guidelines to help improve the implementation and application of innovative handheld technology solutions.
3   *Culture change*: guidelines that highlight sustainable methods of promoting a culture change philosophy that can assist construction industry stakeholders overcome their inherent resistance towards changing traditional work habits.

Revolution or evolution?

The construction industry is classed as being an information-intensive industry, where efficient information processing is continuously being challenged by the extreme fragmentation of the industry's demand and supply chain. Today's internet-based information and communication technologies (ICT) can help improve project communications by:

- reducing the dissemination of large volumes of paper-based documents and drawings to and from project team professionals;
- improving on the traditional paper-based document management and archiving systems;
- providing faster, cheaper and more accurate communication flows.

The Internet has arguably revolutionized the way in which information is stored, exchanged and viewed, opening new avenues for business which only two decades ago were almost inconceivable (DCITA 1998; Information Industries Bureau 2002). Even with a conservative uptake, the construction industry and its participating organizations are making concerted efforts (generally with positive results) in taking up innovative forms of doing business via the Internet (Anumba and Ruikar 2002; ITCBP 2003). Furthermore, governments, who are often key clients of the construction industry with an increased tendency to transact their business electronically, will undoubtedly have a continued influence on how various private industry consultants, contractors and suppliers do business (Murray 2003).

Today's construction industry is also faced with many new challenges, including the need to:

- change current work practices;
- become more client orientated;
- increase competitiveness and productivity.

These challenges are attributable to many factors that affect the current working environments of all industry stakeholders, including:

- globalization of the economy;
- greater performance expectations from the clients;
- increased competition between local and international contractors;
- continued restructuring of work practices and industrial relations.

Furthermore, when considering the implementation or adoption of innovative internet-based ICTs into long-established organizational arrangements and multiple work structures, one cannot assume that once an organization or project team is electronically and simultaneously linked it will automatically ensure an increased sense of community, improved ability to collaborate or improved understanding of co-workers (Graham 1996). On the contrary, internet-based ICT users are faced with increased burdens of regular and mostly unintended project communication misinterpretations, errors and inaccuracies.

Therefore construction industry leaders need to realize that investing in innovative internet-based ICT solutions is no longer about simply buying a piece of hardware or software. They need to recognize that the purchase of an internet-based ICT solution has evolved into being more of a long-term investment that needs to be fully embraced by all its end-users to ensure sought-after increases in efficiency and productivity.

E-tender: friend or foe?

Doing business electronically is found to have a profound impact on the way construction businesses operate. Streamlining existing processes and the growth in innovative tools such as e-tender offer the construction industry new responsibilities and opportunities for all parties involved (ITCBP 2003). It is therefore important that these opportunities should be accessible to as many businesses as possible.

Historically, there is a considerable exchange of information between various parties during a tendering process, where accuracy and efficiency of documentation is critical. This is either paper-based (involving large volumes of supporting documentation) or via stand-alone, non-compatible computer systems, usually costly to both client and contractor. As such, having a standard electronic exchange format that allows all parties involved to access only one system via the Internet saves both time and money, eliminates transcription errors and increases the speed of bid analysis.

Increased knowledge, awareness and successful implementation of innovative systems and processes raise great expectations regarding their contribution towards stimulating the globalization of electronic procurement activities, and improving overall business and project performances throughout the construction industry sectors and overall marketplace. Yet achieving the successful integration of an innovative e-tender solution with an existing/traditional process can be complex and, if not done correctly, can lead to failure. In an attempt to help confront these ongoing complexities, and based on an in-depth investigation into the construction industry and government's current state of play on e-tendering applications, CRC for Construction Innovation (CRCCI) researchers identified five main guidelines (Kajewski *et al.* 2001a) (Table 17.1).

Handheld technology: the future is in your hands

It is well documented that the quality of communication and document management has great impact on the outcome of construction projects (Kajewski *et al.* 2001b; Kajewski and Weippert 2003a). Mobile workforces have been provided with new technologies to help improve their communication and document management performance. Ordinary laptop computers, for example, provide an electronic mobile device capable of carrying large amounts of documentation, and have the ability to access back-office systems through the phone-line and, more recently, through various wireless technologies. Yet this addresses only a small portion of the workforce that is typical for site-office based personnel.

Table 17.1 E-tender guidelines

#	Outline		Recommendation
ET1	**Basic features**	1	Distribute all tender documentation via a secure internet-based tender system, thereby avoiding the need for collating paperwork and couriers.
	Research identified 11 main e-tender system recommendations in relation to its basic features, functionality and capabilities	2	The client/purchaser should be able to upload a notice and/or invitation to tender onto the system.
		3	Notifications are sent out electronically (usually via email) for suppliers to download the information and return their responses electronically (online).
		4	Updates and queries are exchanged through the same e-tender system during the tender period.
		5	The client/purchaser should only be able to access the tenders after the deadline has passed.
		6	Hold all tender-related information in a central database, which should be easily searchable and fully audited, with all activities recorded.
		7	It is essential that no tender documents can be read or submitted by unauthorized parties.
		8	Users of the e-tender system are to be properly identified and registered via controlled access. In simple terms, security has to be as good as if not better than a manual tender process. Data is to be encrypted and users authenticated by means such as digital signatures, electronic certificates or smartcards.
		9	Assure all parties that no 'undetected' alterations can be made to any tender.
		10	The tenderer should be able to amend the bid right up to the deadline, whilst the client/purchaser cannot obtain access until the deadline has passed.
		11	The e-tender system may also include features such as a database of service providers with spreadsheet-based pricing schedules which can make it easier for a potential tenderer to electronically prepare and analyse a tender.
ET2	**Risks**	1	Where tender information is simply posted on the Internet as 'pure information'. Recommendation 1: Although exposed to minimum levels of risk, attention must be given to its contents, that is, truth, accuracy, not misleading or defamatory, etc.
	When industry professionals choose an e-tender process or system, the following three potential risks in its use are identified as being directly proportional to the increasing levels of electronic interaction	2	Where the e-tender internet site claims to have tender-related information that tenderers need to rely on and perhaps download.

Recommendation 2: In this case, owners or managers of the e-tender system are to spend more time ensuring that what is on their site is complete, accurate and true. The inclusion of a 'non-reliance' exclusion clause may also be necessary. Ensure the tender documentation can in fact be successfully downloaded in its entirety, especially if tenderers are asked to reply in hard copy format.

3 At the top end of the risk scale is having a fully interactive internet-based e-tender system where tenderers both receive an invitation to tender and reply with a bid electronically, that is, with no option of obtaining a paper copy of the tender documentation (except by printing out the contents of the internet site).

Recommendation 3: In this case, security of information and integrity of the system is of paramount importance. Here, legally binding and enforceable contracts are being formed electronically, leaving little room for error in receiving, sending, or storing the information. Furthermore, owners or managers of the e-tender system are unable to simply exclude all liability for what could happen during the process, and will likely have to assume some of the unforeseen risks (especially when an electronic reply is the only option) (Worthington 2002).

ET3 Improved training and education	1 In an attempt to help increase construction industry participants' uptake of innovative technologies, systems and processes (such as e-tendering), it is strongly recommended that construction organizations become learning organizations.
	2 Due to the increasing electronic integration of construction processes, industry participants have no choice but to acquire a complete range of new skill sets and to rethink the way current construction education is organized in delivering these skills, thereby implying a need for cross-disciplinary education (Foresight 2000).
	3 There is also a significant role for tertiary education to develop and support the understanding of how to accept, evaluate and implement technological change and innovation. This provision is required both in undergraduate and postgraduate courses to create a more receptive and able cadre of construction professionals (CRISP 2000).

Continued

Table 17.1 Continued

#	Outline	Recommendation
ET4	**Improved implementation** Research identifies seven basic recommendations when it comes to the implementation of an e-tender system	1 Having an extremely robust and secure e-tender system – by having an enhanced security policy in place and by carrying out regular security 'health checks' on the system itself and its users. 2 Ensuring confidential information cannot get into the wrong hands, for instance: – whilst many aspects of an e-tender process are similar to traditional arrangements, there are certain legal issues (possibly contractually binding issues) that need special consideration, for instance, people often let work colleagues check their email inbox, allowing 'unrestricted' access to dedicated e-tender system usernames and passwords. 3 Clarification of certain 'grey areas' regarding timing of electronic tender documents – that is, the need for an e-tender system to automatically generate and archive dispatch and receipt times of electronically distributed/submitted tender documents. 4 Providing access to advanced capabilities within the system, for instance: – allowing one to compare data from project to project in order to view relative prices and timely decision making. 5 Allowing the reuse of standard information of regular tenderers, for instance: – storing the pre-qualification documents and information of a regular pool of tenderers. 6 Tender terms, conditions, application forms and software installation procedures (if applicable) are to be uncomplicated – to help persuade certain contractors, consultants, suppliers, etc., to participate in an e-tendering process. 7 Additional e-tender implementation issues that require consideration include: – liability for lost or corrupted data; – ensuring that the servers are well protected, that is, having 'fallback' plans and procedures in place for when the service is unavailable (off-line); – ensuring that firewalls do not restrict the dissemination of supporting tender-related documentation (ITCBP 2003).
ET5	**Legal issues**	1 The successful implementation of an

e-tendering process within the industry is susceptible to the current legal status regarding electronic transmissions, use of electronic signatures, etc.

2 Commitment by both government and industry sectors is required to help develop more innovative strategies to build a stronger and more competitive construction industry.

3 Ongoing legal investigations aimed at strengthening organizational and individual use of electronic communications on projects must continue, by providing better management of communication risks such as:

– authenticity: This concerns the source of the communication – does it come from the apparent author?

– integrity: Whether or not the communication received is the same as that sent – has it been altered either in transmission or in storage?

– confidentiality: Controlling the disclosure of and access to the information contained in the communication.

– matters of evidence: This concerns e-communications meeting current evidentiary requirements in a court of law, for example, a handwritten signature.

– matters of jurisdiction: The electronic environment has no physical boundaries, unlike those of an individual state or country. This means that it may be uncertain which state's or country's laws will govern legal disputes about information placed on the Internet, or about commercial transactions made over the Internet.

In general, anecdotal evidence suggests that handheld devices provide improved productivity through the reduction in doubling of efforts. Recently, mobile workforces at the coalface of projects have been given handheld computing devices, providing a much greater proportion of the construction team with the ability to record data and communicate between site office and head office systems. This new trend in applying handheld communication devices on remote construction sites potentially results in improved project communication and document management quality, and hence increased project efficiencies (bottom line).

As there is an array of software applications available that cover the full range of construction-related activities, there are a number of issues that need to be considered pertaining to, for example, the selection of compatible handheld devices and how they will be accessed, and what the RAM requirements are. For instance:

- if the handheld device is accessed through an internet browser (ASP or Client/Server), then it will require Wireless Application Protocol (WAP) capabilities;
- if the software is loaded on the handheld device, then the required RAM must be provided either as a standard inclusion or as an add-on.

Although the current crop of handheld devices is found to provide improved productivity over traditional/paper-based systems and processes, they are usually faced with three common technological limitations:

- small screen size;
- cumbersome data entry;
- limited or no navigation functionality.

Emerging technologies such as orientation-driven navigation and near-eye displays are continually being developed and improved on to help overcome these limitations, with increased research and development efforts focusing on areas such as:

- power sources (fuel cells, photovoltaics, micro-engines, silver polymer batteries);
- keyboards (virtual laser image, holographic);
- printing (random movement technology);
- navigation (flexible handhelds).

To help the construction industry successfully utilize the many benefits handheld technologies have to offer, CRCCI research generated four main recommendations (Table 17.2) to assist construction industry stakeholders to:

- successfully select suitable handheld devices that facilitate more productive and sustainable construction projects;
- identify individual construction activities/tasks best suited for handheld devices;
- overcome emerging technology limitations and promote positive opportunities;
- identify and develop enhanced handheld devices, models, processes and technologies for future applications (Kajewski *et al.* 2001b).

Culture change: need or want?

In an increasingly competitive and ever-changing construction industry environment, there are no quick fixes that will truly deliver long-term excellence within individual organizations, groups and project teams (Kajewski and Weippert 2003b). This is due to the pace, size and complexity of

Table 17.2 Handheld technology guidelines

#	Outline	Recommendation
HT1	**Selecting suitable handheld devices**	When selecting a suitable handheld device for construction applications, the main things to consider include:

1. Individuals working within a team/organization and selection must consider its existing frameworks and systems, such as connectivity and platform interoperability.
2. Operating system: different operating systems have different characteristics and are more suited to specific tasks and customization, including:
   - task management
   - power management
   - user interface
   - memory management
   - security
   - memory protection
   - supported processors
   - typical handheld usage.
3. Processor speed: generally 'the faster the better' – specified processor speeds between devices with different operating systems may not be directly comparable. Some operating systems require less processing than others, and therefore, the net speed may be similar. (Typical ranges are 33–200 MHz for Palm OS devices and 200–400 MHz for Windows OS devices.)
4. Read only memory (ROM): generally 'the larger the better' – presently typical for high-end devices is 32–48 MB of ROM.
5. Random access memory (RAM): generally 'the larger the better' – one needs to check the required level software applications to be used (RAM expansion is available through add-ons).
6. Connectivity options: need to consider the type of construction jobs – i.e., horizontal or vertical – in relation to their distance from available site/centralized network office locations or individual devices to connect:
   - For field workers who only require synchronization of data at start and end of shift, USB hard-wired cradle is suitable;
   - For connectivity to provide data transfer within a range of 1m and in line of-sight, Infrared (Irda) is suitable;

Continued

Table 17.2 Continued

#	Outline	Recommendation
		– For connectivity to provide data transfer and PIM synchronization type applications within a range of 10 m (100 m with amplifier), Bluetooth technology is suitable;
		– For full network connectivity within a range of 200 m, Wi-Fi is suitable. One needs to consider the cost of setting up the Wi-Fi enabled server. not just the individual devices;
		– For full network connectivity greater than 200 m (within a range limited by service only), WAN communication systems are suitable. Available systems include GSM (2G), CDMA (2G), GPRS (2.5G), EDGE (3G), WCDMA (3G) and UMTS (3G). Cost and data download rate needs to be considered. These are currently respectively slow and expensive, but are rapidly improving.
7		Required ruggedness: One needs to consider the type of applications and environmental conditions (dust, water) to which the device is likely to be subjected. PDT and PPT devices are inherently 'ruggedized'. The traditional PDAs can be 'ruggedized' through add-ons.
8		Data collection: Automatic data collection is available through various technologies, either built-in or as an add-on, including:
		– Speech (speech recognition, text to speech and interactive voice response)
		– Barcode reading
		– Radio frequency identification (RFID)
		– Fingerprint sensing technology.
9		Add-on/expansion capability: As mentioned previously, the capability of a handheld device can be greatly enhanced and/or brought to speed through the use of expansion slots. The type of slots currently available include:
		– Secure Digital (SD)/Multimedia Card (MMC) slot
		– Compact Flash Slot (CF)/Micro-drive slot
		– Springboard Slot (Handspring)
		– PC Slot
		– Memory Stick slot.
10		Available software applications that cover construction-related activities. The things to consider in relation to the selection of a

	handheld device and the available software are how it will be accessed and what is the RAM requirements: – If it is to be accessed through an internet browser (ASP or Client/Server), then the handheld device will require Wireless Application Protocol (WAP) capabilities. – If the software is loaded on the device, then the required RAM must be considered provided either as a standard inclusion or as an add-on.
HT2 Construction activities/ tasks suited to handheld devices	Tasks that require: 1 Access to large amounts of text information; 2 Viewing a small detail of a document; 3 Entry of binary data; 4 Entry of data into a form; 5 Instant transfer of small amounts of information to and from a network (Saidi *et al.* 2002).
HT3 Overcoming screen size, cumbersome data entry and navigation limitations of emerging technologies	To overcome these limitations: 1 Emerging technologies such as orientation-driven navigation and near-eye displays have been and are being developed. 2 Other emerging technologies that are looking to improve current handheld devices include: – Power sources (fuel cells, photovoltaics, micro-engines, silver polymer batteries) – Keyboards (virtual laser image, holographic) – Printing (random movement technology) – Navigation (flexible handhelds).
HT4 Future applications	Looking to the future: 1 Traditional handheld computing devices will be competing with wearable computers. 2 Wearable computers provide more potential benefit to the mobile worker through greater use of automated processes. 3 Handheld devices of the present are a testing ground for some of the technologies slated for wearable computers in the more distant future.

change being greater than ever before, overwhelming many of those who face it. Changes include:

• transforming a business that has succeeded for years by focusing on customer service alone but must now focus on its technical proficiency to keep up with increased domestic and global competitors;

- redesigning and adapting existing roles to incorporate these new and never before used technologies;
- transforming the current culture and subcultures of an organization from, for example, a cautious and reactive culture (follower) into tomorrow's highly efficient and 'first mover' culture (industry leader) (Black and Gregersen 2002).

One of the last available mechanisms left for organizations to improve their competitive position within the construction industry is by considering its people (culture) along with its technology. In other words, if one wants to make construction companies, consortia and project teams more efficient and effective, then one needs to better understand the role that culture plays within them (Schein 1997).

> If people fail to see the need for change ... whether threat or opportunity drives it ... they will not change.
>
> (Black and Gregersen 2002: 20)

The term 'culture' is identified as one of the most difficult and complex approaches to understand and change. This complexity in trying to change or understand culture is mainly based on it being defined in so many different and sometimes conflicting ways (Pepper 1995). For example, culture...:

> ... begins to form wherever a group has enough common experience ... [which in turn becomes] the property of that group ... [and] ... is a pattern of shared basic assumptions that has been learnt whilst solving problems, that has worked well enough to be considered valid and, therefore, to be taught to new members as the correct way to perceive, think, and feel in relation to those problems.
>
> (Schein 1997: 12)

> ... is influenced by traditions, myths, history and heritage ... [and is] the sum of how we do things around here.
>
> (Hensey 2001: 49)

> ... pervades the decision-making and problem-solving process of the organization by influencing the goals, means and manner of actions ... [and] ... is a source of motivation, de-motivation, satisfaction and dissatisfaction ... thereby underlining much of the human activity within an organization.
>
> (Williams *et al.* 1993: 15)

The reasons why the study and understanding of an organization's, group's or team's culture is so important to its survival can be summarized as follows:

- Culture focuses on communication at all levels of an organization and project team hierarchy, where individuals identify who they are in relation to one another and their environment and where shared understandings form identifiable subgroups/subcultures.
- By focusing on culture, one inevitably focuses on the daily routine and 'sense-making' process of building identities, developing shared beliefs and forming perceived realities among members.
- A cultural approach focuses on largely ignored issues, such as assumptions, and brings underlying values and motives to the surface.
- The understanding of culture offers a better insight to the managers and leaders – not in order for them to better shape the culture, but so that they can better understand and participate in the sense-making activities of members.
- Finally, culture is pervasive, being not simply a variable that affects the organization, group or team but indistinguishable from it (Pepper 1995).

Change, on the other hand, has always been and remains difficult, and when attempts are made to change culture, it is inevitably a slow process where the all too common phrase 'You can't change culture overnight' becomes a major excuse for not changing it at all. Research defines culture as being complex, multi-levelled and deeply rooted – a concept that must be observed and analysed at its every level before it can be fully understood or successfully changed and managed. In many cases organizations have attempted to change their culture, resulting in employees only learning the basics of this 'new' culture without fundamentally altering their old culture (beliefs, values, attitudes, etc.). An organization will change only as far and as fast as its collective individuals are willing to change, because people are and always will be instinctively programmed to resist any form of change. Therefore, to strategically and successfully change any organization, research recommends one must first attempt to change the individual beliefs, values and attitudes (culture) within the organization before the organization as a whole can benefit from the overall change initiative.

> When we know what culture is, we know what needs to be changed for culture to change ... only once we appreciate its true nature can we understand how it might be changed ... and when we know its role, only then can we comprehend its importance.
>
> (Williams *et al.* 1993: 11)

The industry also has to realize that the cost of delaying any technologically driven changes is in many cases not only inconvenient but also often catastrophic. When the implementation of a new ICT solution or process drives the change in an organization's culture, its leaders have to realize from the outset that hierarchically imposed solutions usually do not work well when subcultural differences and conflicting assumptions are involved. Instead, new intercultural processes need to be developed, permitting better communication between the subgroups and allowing the strengths of each to interact to form an integrative and new implementation solution. If this process is not undertaken and managed correctly, then the old (traditional) and new (ICT) practices will only superficially and temporarily coexist, resulting in the organization's original 'way of doing things' eventually resurfacing.

Promoting a more cohesive, transparent and intercultural philosophy that will assist all industry stakeholders in better managing and maintaining a more sustainable solution towards innovative change, and incorporating lessons learned from other leading international industry sectors, is essential to Australia's construction industry's survival in tomorrow's dynamic and highly competitive arena. This unique attitude towards recognizing the potential value and need for innovative change within the industry forms the foundation of the 24 general culture-change guidelines (Table 17.3), which in turn encapsulate the following four critical aspects of changing culture:

1   harmonizing the attitudes, values and behaviours of all industry stakeholders;
2   understanding the role and importance of culture itself;
3   identifying the need for change and for change incentives (motivation and rewards);
4   identifying a suitable culture change strategy by aligning innovative solutions and business processes with the industry's most valuable resource – people.

Future research: where to from here?

Due to the nature of the Australian construction industry, involving large numbers of geographically dispersed projects, organizations and professionals, communication activities are inevitably complex. Ongoing research efforts in determining ways to improve traditional (paper-based) methods of communicating, and through the implementation and application of commercially available ICT tools and systems, will accentuate increased recognition of the opportunities and benefits these innovative technologies have to offer (Anumba and Ruikar 2002).

Debatably, today's businesses and personal worlds are being increasingly dominated by a wide range of ICT- and internet-based initiatives. The ICT revolution and its dramatic impact on communication practices will inevitably continue transforming the way the industry operates. Yet, even though these innovative ICT tools and systems have a great deal to offer, and despite increased international research and development (R&D) activities, the construction industry's ever-present resistance to change, and the need to improve current and traditional communication and information processes, still dominates.

Table 17.3 Culture change guidelines

#	Outline	Recommendation
CC1	Harmonize attitudes, values and behaviour	If members' attitudes, values and behaviour are in harmony, then a stronger and effective culture is likely to result where members are committed to the overall change, goals and methods of the organization, group or team.
CC2	Understand the role of culture	To make industry organizations, groups and project teams more efficient and effective, one must better understand the role that culture plays within them.
CC3	Culture is never singular, always plural	Attempts to change the whole culture of an organization must be abandoned, because every culture is made up of a whole range of mentalities and subcultures, all of them different, and at different stages of development.
CC4	Identify the need for change	It is important for organizations, groups and teams to realize and create a need for change, before the act of change can take place.
CC5	Motivate people	People are motivated to change when they are confronted with real or perceived threats and/or opportunities.
CC6	Suitable change strategy	To ensure successful change in culture, a suitable change strategy needs to be identified and properly implemented and managed, which in turn can promote a new business strategy.
CC7	People and places	People in key positions may need to be changed, moved or rotated to ensure successful change in culture within an organization.
CC8	People's beliefs, attitudes and values	To ensure successful change in culture, individual beliefs, attitudes and values may need to be altered by applying one or more suitable change methods.
CC9	Structures, systems and technology	An organization's existing communication network may entail restructuring and require the implementation of improved reward, appraisal, monitoring, budgeting and/or control systems to ensure successful and sustainable change in culture.
CC10	Corporate image	Promote an improved corporate image to help develop positive attitudes between both customers and staff, which in turn will enhance overall commitment towards the organization.
CC11	Invest in people	To ensure successful change in culture, organizations need to improve their attitude and performance towards respecting and recruiting their people in order to retain their best talent.

Continued

Table 17.3 Continued

#	Outline	Recommendation
CC12	Create a feeling of shared ownership	Employee participation is essential to ensure increased commitment and feeling of ownership towards the implementation of a culture change process.
CC13	Suitable culture change process	To ensure successful change in culture within an organization, a suitable change process needs to be identified, properly implemented and managed.
CC14	Timing of change	Timing and cost-effectiveness of implementing a change process or method in an organization determines the success or failure of change.
CC15	Align technology with people	Understanding the interconnections between technology and people (culture) is essential during the implementation of a technology-driven culture change process. This can be achieved by designing the technology to fit the organization's current structure and culture, or by reshaping the organizational structure (processes) and its culture (people) to fit the demands of the new technology.
CC16	Promote an electronic culture	Organizations need to investigate and implement a suitable transition strategy to help ensure a technology-driven culture change, i.e., assisting an organization in its transition from existing/traditional business operations and processes, to industry required (electronic) operations and processes.
CC17	ICT champion	An organization pursuing technology-driven advancement or change requires strong support from an ICT champion (preferably senior management) to undertake and lead the difficult task of managing its impact upon organization structures and cultures.
CC18	Three cornerstones of successfully implementing innovative ICT	Industry organizations are to consider three success factors when implementing ICT: a Vision: a durable vision of the change process is required to ensure progress, shared with top management, construction managers, developers and ICT staff. b Commitment: obtain overall commitment from top management, construction and ICT managers (re-allocation of financial and human resources). c Possibilities: apply a 'migration strategy' that enables ICT staff to balance the ICT strategy with the company's business needs, thereby underwriting the success of the change process.

CC19	ICT implementation strategies	Construction industry executives and management need to consider various ICT implementation strategies, and select the one that best serves the needs of the application and its users.
CC20	Overcome fear	The construction industry is to lessen and ultimately remove the fear of 'exploitation' arising from technology-led innovation and change.
CC21	Camouflaging change	It is important for implementers of, for example, an innovative ICT tool or system not to 'camouflage' the true nature of this change or new way of 'doing things' prior to its implementation, i.e., not to portray the change as less dramatic and positively beneficial to the employees and the company.
CC22	Promote ICT adoption benefits	The construction industry will increase and strengthen the rate of technological adoption by promoting its benefits, developing and running short courses, establishing industry-wide awards for ICT best practice and taking relevant action.
CC23	Continued training and education – a must	Construction industry organizations need to become learning organizations, attuned to absorbing and using knowledge and strive for lifelong learning.
CC24	Enhanced tertiary training and education	Tertiary education (both undergraduate and postgraduate) is to further develop and support the understanding of how to evaluate and implement technological and cultural change and innovation within construction industry organizations.

Potential ICT benefits and competitive advantages experienced through the electronic distribution of project-related documentation and information within a virtual team will undoubtedly continue to transform the way the industry operates. More importantly, researchers agree that the success of any profession means going beyond simply exchanging electronic information – that determining new and improved ways of doing business is dependent on the innovation of the user, not only the technology itself – requiring careful consideration and a greater emphasis on the 'human touch' (Gore 1999; Ahmad 2000; Claver *et al.* 2001).

Yet, for one of the oldest industries to change its traditional ways of doing things (culture) and to embrace innovative technologies and processes will be an ongoing challenge. Unless one can uncover and expose the fundamentals of these inherent restraining forces, there is little hope of successfully implementing permanent cultural change (Black and Gregersen 2002). As Furst *et al.* (1999) have commented: 'We know a great deal about the technical aspects of being "virtual" – we now need to know more about making the human and teams more virtuous.'

There is also a perceived fear within the industry that technology-led innovations may result in exploitation. In an attempt to lessen and ultimately remove this fear, construction industry researchers are to continue efforts in determining ways to:

- create a common understanding that would enable the industry to take positive action;
- provide appropriate and easily accessible information on risk evaluation and implementation;
- provide both cultural and contractual changes to remove the constant fear of liability and the concern to assign blame to individuals and organizations, creating an environment that is receptive to ideas, challenges and opportunities;
- investigate historic (past) projects to provide lessons for the future;
- lessen constraints imposed by regulations, codes and standards that tend to oppose novel and innovative solutions (CRISP 2000).

When it comes to researching and developing innovative technologies, the construction industry is said to be lagging when compared to others (Michel 1998). Yet, the level of ICT adoption by the Australian construction industry appears to be neither more nor less advanced than that of our international competitors. Still, current R&D efforts need to be increased in order to manage ongoing industry implications and the inevitability of ICT-driven change, including its effect on organizational cultures (Black and Edwards 2000). If Australian organizations continue to explore the competitive dynamics of the industry without realizing the current and future trends and benefits of adopting innovative ICT solutions, it will limit their opportunity to internationalize.

The need for continued R&D efforts in identifying, developing and implementing best practice methods, processes and systems that will assist Australia's construction industry sector to overcome its inherent resistance towards innovative change is essential in order for it to:

- generate and maintain a competitive advantage in the international arena;
- contribute towards major social and technological change that will fuel the overall knowledge, awareness and skills of all construction industry stakeholders;
- revolutionize its traditional methods of integration in a unique, distinctive and never before experienced way.

Summary

In this uncertain and ever-changing world, the construction industry and its participants need to be creative, alert to opportunities, responsive to external stimulus, have a good grasp of the changing environment, and increase existing levels of confidence in its ability to adapt. It is over

40 years since the introduction of ICT tools and systems into the industry, yet organizations are still unable to obtain the full benefits of ICT investment, many years after the initial expenditures have been incurred. As such, the industry has inevitably been identified as relatively slow in embracing innovative ICT tools and systems.

Culture is difficult to change and manage. This is due to it essentially representing the accumulated beliefs, attitudes and values of usually a large number of individuals within an organization, group or project team. Therefore, having a better understanding of the overall effects a change initiative has on the subcultures of an organization, group or project team will in turn help leaders to better understand the reasons behind experiencing any resistance during the implementation of a change process, and provide a more realistic approach on how to better manage it.

The three sets of guidelines summarized in this chapter will assist in:

- modifying traditional work habits;
- improving responses to technical and social challenges;
- encouraging the use of innovative ICT and internet-based solutions.

Acknowledgements

The authors of this chapter would like to thank the following Australian industry-, government- and university-based project partners for their unwavering support and invaluable contributions during this exhaustive research undertaking: Queensland University of Technology, Commonwealth Scientific and Industrial Research Organisation, University of Newcastle, Queensland Department of Public Works and Queensland Department of Main Roads.

Bibliography

Ahmad, I. (2000) 'Success in the wake of the IT revolution', *Journal of Management in Engineering*, 16 (1): 28.

Anumba, C.J. and Ruikar, K. (2002) 'Electronic commerce in construction: trends and prospects', *Automation in Construction*, 11 (3): 265–75.

Black, J.A. and Edwards, S. (2000) 'Emergence of virtual or network organizations: fad or feature', *Journal of Organizational Change Management*, 13 (6): 567–76.

Black, J.S. and Gregersen, H.B. (2002) *Leading Strategic Change: Breaking Through the Brain Barrier*, Upper Saddle River, NJ: Pearson Education.

Claver, E., Llopis, J., González, M. and Gasco, J. (2001) 'The performance of information systems through organizational culture', *Information Technology and People*, 14 (3): 247–60.

CRISP (2000) *Report on a Workshop: Technological Change and Rethinking Construction*, London: Construction Research and Innovation Strategy Panel.

DCITA (1998) *Where to Go? How to Get There*, Canberra: Department of Communications, Information Technology and the Arts.

Foresight (2000) *Constructing the Future: Making the Future Work for You*, London: Construction Associate Programme, Built Environment and Transport Panel, Foresight Programme.

Furst, S., Blackburn, R. and Rosen, B. (1999) 'Virtual team effectiveness: a proposed research agenda', *Information Systems Journal*, 9 (4): 249–69.

Gore, E.W. (1999) 'Organizational culture, TQM, and business process reengineering: an empirical comparison', *Team Performance Management*, 5 (5): 164–70.

Graham, M.B.W. (1996) 'Changes in information technology, changes in work', *Technology in Society*, 18 (3): 373–85.

Hensey, M. (2001) *Collective Excellence: Building Effective Teams*, Reston, VA: ASCE Press.

Information Industries Bureau (2002) *e-Business & m-Commerce*, Brisbane: Department of Innovation and Information Economy. Online. Available at HTTP: <www.iib.qld.gov.au/guide> (accessed May 2002).

ITCBP (2003) *e-Construction, e-Procurement and e-Tendering*, London: IT Construction Best Practice.

Kajewski, S.L. and Weippert, A. (2003a) *A Brief Synopsis in the Use of ICT and ICPM in the Construction Industry*, Report 2001–008-C-01, Brisbane: CRC for Construction Innovation.

—— (2003b) *Industry Culture: A Need for Change*, Report 2001–008-C-05, Brisbane: CRC for Construction Innovation.

Kajewski, S.L., Tilley, P.A., Crawford, J., Remmers, T.R., Chen, S.-E., Lenard, D., Brewer, G., Gameson, R., Martins, R., Sher, W., Kolomy, R., Weippert, A., Caldwell, G. and Haug, M. (2001a) *Electronic Tendering: An Industry Perspective*, Report 2001–008-C-07, Brisbane: CRC for Construction Innovation.

Kajewski, S.L., Tilley, P.A., Crawford, J., Remmers, T.R., Chen, S.-E., Lenard, D., Brewer, G., Weippert, A., Caldwell, G. and Haug, M. (2001b) *Handheld Technology Review*, Report 2001–008-C, Brisbane: CRC for Construction Innovation.

Michel, H.L. (1998) 'The next 25 years: the future of the construction industry communications', *Journal of Management in Engineering*, 14 (5): 26–31.

Murray, M.B.D. (2003) 'The development of e-commerce within the global construction industry', in B.O. Uwakweh and I.A. Minkarah (eds) *Construction Innovation and Global Competitiveness: Conference Proceedings of the 10th International Symposium*, Abingdon: CRC Press.

Pepper, G.L. (1995) *Communicating in Organizations: A Cultural Approach*, New York, NY: McGraw-Hill.

Saidi, K.S., Haas, C.T. and Balli, N. (2002) 'The value of handheld computers in construction', *Proceedings of 19th International Symposium on Automation and Robotics in Construction*, Gaithersburg, MD: National Institute of Standards and Technology.

Schein, E.H. (1997) *Organizational Culture and Leadership*, San Francisco, CA: Jossey-Bass.

Williams, A., Dobson, P. and Walters, M. (1993) *Changing Culture: New Organizational Approaches*, London: Institute of Personnel Management.

Worthington, R.C. (2002) 'The devil is in the details: e-tendering poses potential legal risk', *Summit: Canada's Magazine on Public Sector Purchasing*, Autumn: 3. Online. Available at HTTP: <www.summitconnects.com/Articles_Columns/PDF_Documents/02it2_01.pdf>.

18 Project diagnostics

A toolkit for measuring project health

Achi Weippert

Despite the availability of a large number of published reports, reviews and research reports providing guidance to successful project execution, many high-profile and publicly funded projects still attract adverse publicity through failing to meet predetermined objectives and intended outcomes. As a result, the CRC for Construction Innovation (CRCCI) sought to explore and better understand the myriad reasons why construction projects still fail to achieve their intended outcomes. This chapter reports on the development of a Project Diagnostics Toolkit protocol which includes metrics to:

- indicate the areas where projects are going wrong;
- diagnose why projects may be failing;
- suggest means of returning projects to 'better health' (with a direct link to better business outcomes).

International snapshot: project success or failure?

Project success or failure means different things to different people. Each stakeholder can have a different definition of success and/or failure that is consistent with their perception and interests in relation to the project outcome. In order to develop common measures that broadly represent the interests of all stakeholders, the subject of project success or failure has been one of the main areas of focus for a number of international researchers. Success and failure measures were first introduced by Rubin and Seeling (1967). By basing their research on investigating the impact of a project manager's experience on project success or failure, and by using technical performance as a measure of success, it was concluded that the project manager's previous experience had minimal impact. A theoretical study by Avots (1967) concluded that the wrong choice of project manager, unplanned project termination and unsupportive top management were the three main reasons for project failure.

In the early 1990s, predictive models were developed to focus on explaining failure factors at a project level. These models use financial ratios derived by statistical analysis through a number of plausible financial indicators (Russell and Jaselskis 1992; Abidali and Harris 1995; Kangari 1992). More recent investigations suggest that the use of financial ratios may not be very reliable, as they can only highlight certain project failure 'symptoms'. Furthermore, the sole use of financial indicators is questionable, as they may be relying on data that have been created in an attempt to hide the poor (actual) condition of a project (Arditi 2000). In some cases, researchers reportedly tried to link project success to different stages, such as delivery and post-delivery. Similarly, project success was linked to individual perspectives of the stakeholders by, for example, employing various methodologies in an attempt to identify a robust set of success and failure measures, ranging from unstructured interviews that ask respondents to list measures that are important to project success or contribute to project failure, to structured interviews that require them to rank a list of measures that affect project performance.

Reports (Pinto and Pinto 1991; Pinto and Slevin 1987; Morris and Hough 1987) further suggest that project success measures should also include psychosocial outcomes such as safety, litigation and other factors that relate to interpersonal relationships within a project team, identifying the following 'likely candidates' for measuring project success or failure:

- communication;
- environmental events;
- community involvement;
- team-member conflict;
- lack of negotiation and arbitration;
- legal disputes;
- management inability to understand site staff;
- stakeholder value.

These further need to be combined with outcomes of research efforts that suggest time, cost and quality (the 'iron triangle' of measures) were the basic criteria of project success (Belassi and Tukel 1996; Skitmore 1997; Shenhar and Levy 1997; Atkinson 1999). This notion is echoed by Songer and Molenaar (1997), who consider a project to be successful if it:

- is completed within budget (cost);
- is on schedule (time);
- conforms to user expectations (quality);
- meets specifications (quality);
- attains quality of workmanship (quality);
- minimizes construction aggravation (time, cost and quality).

Examples of cost overrun and quality overrun measures identified during this extensive literature investigation are presented in Tables 18.1 and 18.2.

Table 18.1 Project cost overrun measures

More common	Less common/ frequently reviewed
• Poor estimating	• Lack of contractor project type experience
• Inclement weather	• Contractor's lack of familiarity with local regulations
• Insufficient and untimely cash flow	• Complexity of project
• Communication gap between project parties	• Inflation
• Inaccurate prediction of production output	• Lack of supply of plant, equipment and materials
• Design changes	• Site storage problems
• Safety issues	• Geographic location (restricted site access, etc.)
• Industrial action	• Production of design drawings
• Skill shortages	

Table 18.2 Project quality overrun measures

More common	Less common/ frequently reviewed
• Reluctance to adopt quality systems	• Lack of control of inspection, measuring and testing equipment
• Inadequate quality assurance and control systems	• Lack of control of nonconforming product
• Lack of product identification and traceability	• Poor data control
• Lack of internal and external audits	• Lack of employee conscientiousness
• Infrequent inspections	• Lack of encouraging specialization in construction work
• Insufficient training	

Finally, more recent literature investigations undertaken by the CRCCI Project Diagnostic Research Team indicate that the majority of today's construction industry clients and stakeholders take the issue of quality conformance more seriously, believing that the issue of resuscitating failing projects, due to, for example, poor quality of documentation or workmanship, is a vitally important topic that needs to be further investigated and better understood to help secure a healthier and vibrant industry future.

Industry need: what's in it for me?

A review of publications indicates that a comprehensive tool to assist construction industry stakeholders in assessing the health of a project is not readily available. Availability of such a tool would significantly enhance the opportunity for an underperforming project to be appraised and then corrected in a focused and systematic way. The construction industry therefore will benefit greatly from the development and application of an innovative Project Diagnostic Toolkit that can:

- appraise the current condition of a project;
- identify performance against industry benchmarks for the key success factors;
- ensure further analysis of any underperforming area is carried out, enabling the probable root causes of poor performance to be captured;
- provide a prognosis for the success of a project;
- point the way to remedial actions that could be taken;
- be applied repeatedly at subsequent stages of a project, to monitor the effectiveness of remedial action.

The evolution of measuring project health

Human physical health can broadly be thought of as the condition of the body whereby its performance or quality of life can be compromised when physical health is poor. Poor physical health often has associated symptoms that can be used to help pinpoint the cause quickly and accurately. Once the cause has been identified, a remedy can be implemented to assist the body in returning to good health. If symptoms are left unchecked or untreated, they can worsen and develop into a more critical situation (Humphreys *et al.* 2004; Mian *et al.* 2004).

In many ways, the health of a construction project is arguably analogous to human physical health. That is, if a project or any particular aspect of a project is not performing as expected, it would be perceived as unhealthy or failing. On the other hand, if it is fulfilling the expectations of its various stakeholders, it would be perceived as healthy or successful. The various parallels between construction project health and human physical health tend to include the following:

- state of health influences performance;
- health often has associated symptoms;
- symptoms can be used as a starting point to quickly assess health;
- symptoms of poor health are not always present or obvious;
- state of health can be assessed by measuring key areas and comparing these values to established norms;
- health can change over time;
- remedies can often be prescribed to return to good health;
- correct, accurate and timely diagnosis can avoid small problems becoming large.

Consequently, a broad but useful 'poor health classification' strategy can be based on two variables: whether the state of poor health is known, or whether the reason for poor health is known. These in turn allow construction project health to be characterized as four distinct states (Table 18.3).

- States 1 and 2: where poor health is either known or unknown and the reasons are known:

  - see that all relevant health problems are identified;
  - help provide clarity on complex health issues;
  - allow the freedom to bypass the initial health check stage (saving time and effort).

- States 3 and 4: where the reasons for poor health are *not* known:

  - an initial health check could provide direction for the more detailed investigation.

These four fundamental health states of a construction project are accommodated in Stage A of the Construction Project Health Model (Figure 18.1), derived from the Continuous Improvement Management Cycle (Deming 1986).

Although the broad concept of the Construction Project Health Model satisfies the initial requirement for rapid, accurate diagnosis of a construction

Table 18.3 Four states of construction project health

State	Poor health	Reason for poor health	Characteristic or state
1	Known	Known	• Characterized by a project that is well monitored, and sufficient analysis, understanding or experience is available to allow diagnosis of the underlying health problem
2	Unknown	Known	• Although not so straightforward, it could be characteristic of a project that is not well monitored, yet experience, observation or inside knowledge suggests that there are some underlying problems
3	Known	Unknown	• Could be characterized by a project that is well monitored, but management lacks sufficient experience or analysis to accurately diagnose an underlying problem
4	Unknown	Unknown	• Potentially the worst-case scenario, characterized by poor monitoring and lack of experience, understanding or analysis

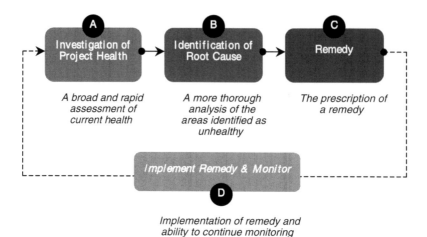

Figure 18.1 Construction Project Health Model (source: adapted from Deming (1986)).

project's health, a more detailed investigation is required and a comprehensive methodology/model needs to be developed to help identify:

- factors contributing to a project's poor health;
- performance indicators (to assess the state of these contributing factors);
- prescribed remedial actions (to improve project performance), based on the condition of the contributing factors investigated.

Research approach: the project diagnostics roadmap

It would be of great benefit to stakeholders to have direct and ready access to an evaluation system that informs them of how a project is progressing at any stage of its life cycle. The chances of bringing a failing project to recovery are far greater if a problem is identified as early as possible. If stakeholders do not have an efficient (early enough) warning system in place, or if the warning signs of a problem (real or perceived) are viewed with scepticism, then there is a strong possibility that it would be near impossible to recover from the potential damage it can cause. This in turn may cost stakeholders much more than they had originally committed to, and further tie up valuable resources. On the other hand, costs associated with being able to identify and remedy a problem early are, in most cases, much less. This situation is again analogous to human health, where the chances of remedying a disease or illness are far better if it is diagnosed at an earlier stage.

In an attempt to meet the above investigative requirements, the Project Diagnostics Research Team further examined and developed the Construction Project Health Model (Figure 18.1), including all associated measures and methodologies used to assess these measures, in order to:

- ensure its suitability/compatibility with a typical construction project scenario;
- improve the potential for a project to achieve the outcomes expected.

From this the team developed the Project Diagnostic Model (Figure 18.2), which:

- allows the immediate and detailed assessment/evaluation of current key project health symptoms;
- facilitates a detailed investigation to identify the root causes of poor project performance;
- suggests a remedy and means of returning the project to better health.

The cyclic process of the Project Diagnostic Model was used to assess the state of health of the three case-study projects identified by industry partners in the Project Diagnostics Research Project. Projects were valued at more than A$10 million and covered a variety of procurement methods, including design and build, lump sum and schedule of rates. Following is a summation of the Project Diagnostic Model process, identifying the key elements that form part of the cyclic mechanism that repeats the investigation

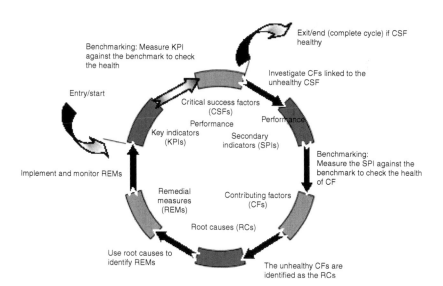

Figure 18.2 Project Diagnostic Model.

of a project until the problem(s) are remedied – that is, bringing the project back to good health.

Critical success factors (CSFs)

The task of assessing a project's health was achieved by recognizing the numerous performance measures identified by the case-study project stakeholders. In order to make the extensive list of over 120 different yet highly relevant sources of stakeholder success and failure measures more manageable, they were grouped under the following seven key measures of success: cost, time, quality, relationships, environment, safety and stakeholder value.

As is the case with human physical health, the seven measures of success – referred to here as critical success factors (CSFs) – ultimately represent seven critical areas or themes that facilitate a broad evaluation of construction project health, i.e., in terms of specific success factors that are crucial to the interested stakeholders.

Key performance indicators (KPIs)

In order to use the CSFs effectively, they had to be properly assessed and validated by developing an associated series of key performance indicators (KPIs) for each of the seven CSF themes. The purpose of a KPI is essentially to enable the measurement of a construction project's performance, defined here as 'a number or value which can be compared against an internal target or an external target benchmark to give an indication of performance' (Ahmad and Dhafr 2002).

Using KPIs to assess the performance of the CSF themes allows the Project Diagnostic Model to be applied to most (if not all) construction projects using the majority of procurement methods, regardless of whether a performance target was set by an interested party, legislation or by other projects. To achieve this, KPIs had to be calibrated using various benchmark statistics from the UK, USA and Australia (CBPP 2003; Cole 2003; CII and ECI 2003). Benchmarking is defined here as 'a technique of evaluating performance in specific areas when compared to recognized leaders' (Plemmons 1994).

Although a large number of KPIs were identified from the literature investigation, many lacked attributes that would make them applicable, useful, independent and practical enough for the immediate health assessment of ongoing or historic projects. Therefore, in order to undertake a robust, accurate and immediate assessment of a project's current health (in terms of the seven CSFs), six critical characteristics were identified (Table 18.4) and then used to narrow down the extensive KPIs list to a total of only 33.

Due to the nature of some of the performance attributes described in Table 18.4, it was necessary to validate the robustness of the 33 KPIs by

Table 18.4 Six critical KPI characteristics

1	Easily measurable	• Must be able to be measured quickly, directly and accurately with comparatively little effort
2	Broadly applicable	• Must be able to be measured at any stage of a project, or at least a combination of indicators across a CSF should be able to represent all or most stages. The indicators should also be able to represent different procurement methods
3	Assessable	• Once measured, the indicator must be able to be compared to a known value to allow correct judgements of health to be made
4	Independent (not duplicated)	• Independence from other project variables is very desirable to provide clarity in assessment of a specific CSF by avoiding overlaps or dependencies which can give misleading results
5	Reflect reality	• The measured variable must encourage a description of reality rather than 'ideal' or perceived situations
6	Sensitivity	• The indicator must be tuned to project health to allow accurate health assessment

testing them on actual projects. The results of the seven CSFs were then analysed and the overall health of the case-study projects was assessed. If the results indicated a healthy project, the investigative cycle ended. If they indicated an unhealthy project, the use of the Project Diagnostic Model proceeded to the next sequential stage – that of identifying contributing factors (CFs).

Contributing factors (CFs)

The main aim of this stage is to identify the factors that can contribute most to project failure in relation to the seven previously identified CSF themes. A total of 28 interviews were conducted on the case-study projects. The structured questionnaire was designed to identify CFs and to allow these to be ranked in terms of relative importance using a numeric scale. Respondents included clients, consultants, contractors and subcontractors.

As mainly successful case-study projects were evaluated, the initial list of CFs was not considered comprehensive enough – i.e., the data collected were insufficient and did not encompass all the factors that contribute to potential failure. Consequently, augmentation of additional CFs (identified from the extensive international literature investigation) occurred. The Project Diagnostics Research Team added these to the existing list of CFs (obtained from the three successful case study projects) so as to achieve a more comprehensive list.

Industry partner specialists then validated the CFs by committing themselves to two consecutive rounds of a Delphi process. The list of CFs was discussed in a workshop (attended by the same industry partner specialists), where the final list was confirmed. Finally, the overall importance index and rank for each of the seven CSFs was determined (Table 18.5), whereby the ranking of every CF (for each of the seven CSFs) was calculated using an Importance Index equation (Figure 18.3).

Secondary performance indicators (SPIs)

This stage of the Project Diagnostic Model process focuses on a more detailed investigation into the CSFs that were found to be in poor health. This required the cause of each unhealthy CSF to be further analysed and assessed by way of its previously identified list of related CFs, in order to pinpoint the areas that most likely cause poor project health. This was accomplished by identifying a series of secondary performance indicators (SPIs) for each CF. An SPI is defined here as 'an indicator showing the level of performance achieved against an operation that is of secondary importance to the successful completion of the services being provided' (CBPP 2003)

As with KPIs, each SPI was benchmarked using various statistics identified from an extensive investigation of literature and research material from the UK, USA and Australia. In most cases, comparisons were made against industry averages rather than market leaders. This shift in benchmarking was attributed to the nature of the Project Diagnostic Model, and the Project

Table 18.5 Overall index and rank of the seven CSFs

1	Cost		6.68	1
2	Time		3.86	4
3	Quality of	a Documentation (increase in requests for information)	3.20	8
		b Construction (increase in rework)	4.65	2
4	Safety		3.60	5
5	Relationships		4.15	3
6	Environment		3.40	7
7	Stakeholder value		3.43	6

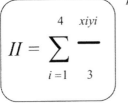

KEY:
- *II* represents *Importance Index*
- *X? is a constant that represents the **weight** of the ?*[th] *response*
  - *Where ? = 1,2,3,4*
- *Y? is a constant that represents the **frequency** of the ?*[th] *response*
  - *Where ? = 1,2,3,4.*
  - *Hence:*
    - *y1 = frequency of the **least important** contributing factor*
    - *y2 = frequency of the **less important** contributing factor*
    - *y3 = frequency of the **important** contributing factor*
    - *y4 = frequency of the **most important** contributing factor*

$$II = \sum_{i=1}^{4} \frac{xiyi}{3}$$

Figure 18.3 Importance Index equation.

Diagnostics Research Team deciding it to be more appropriate to classify a project as being unhealthy if it is not performing better than the industry average, rather than comparing its health against industry best practice.

Although a large number of SPIs were identified from the literature investigation, many lacked attributes that would make them more applicable, useful, independent and practical. Consequently, in an attempt to increase the robustness and usefulness of the Project Diagnostic Model it was decided that, prior to their acceptance, each SPI had to possess at least four characteristics (Table 18.6), similar to those used for selecting KPIs.

Remedial measures (REMs)

Commencing the final stage of the Project Diagnostic Model process relies on the successful completion of the previous two stages: the correct and timely identification of the CFs as well as the accurate assessment of SPIs. This allows the Project Diagnostic Model to confidently prescribe an effective remedy or course of action that will potentially return a project to good health: a remedial measure (REM). The various REMs suggested by the Project Diagnostic Model ensure that project failures are not continually repeated; this is achieved by recognizing the potential effect of a potential failure early in a project, and taking the proactive (suggested) step(s) necessary to avoid any unwanted consequences (unhealthy project).

The practical nature of construction further suggested that a suitable approach for the development of a suite of remedies for a range of project health problems will inevitably have to be based on previous lessons learned and experiences gained by industry partners and specialists, which in turn may raise concerns regarding the validation or relevance of certain REMs.

Limitations to using REMs: when based on lessons learned

There are two basic limitations in using historic experiences or lessons learned (REMs) from previous projects. The first is that the various remedies used at

Table 18.6 Four critical SPI characteristics

1	Easily measurable	• Must be able to be measured quickly, directly and accurately with comparatively little effort
2	Assessable	• Once measured, the indicator must be able to be compared to a known value to allow correct judgements of health to be made
3	Reflect reality	• The measured variable must encourage a description of reality rather than 'ideal' or perceived situations
4	Sensitivity	• The indicator must be tuned to project health to allow accurate health assessment

the time of application tend to be dependent on the individual experience(s) of the person(s) applying the remedy. This means that remedies for any given CF may vary from person to person or from team to team, sometimes in contradictory or conflicting ways. The second limitation is the need to realize that each project is unique, with its own set of distinctive issues and problems that usually require a select suite of remedies or a distinctive course of action(s) to help bring it back to good health. The set of remedies developed for the Project Diagnostic Model therefore had to prove that they were workable, relevant and able to achieve the required results on any given project. It is for this reason that the REMs nominated in the Project Diagnostic Model can be no more than generic and should only be seen as such, unless additional influencing dynamics of a project are confirmed or better understood.

Implementing and monitoring REMs: for continuous improvement

Implementing a remedy is the most important step towards bringing a project back to good health. The implementation of an REM may require the coordination and support of more than one stakeholder and, once implemented, the allocation of a certain amount of time to bring the project back to good health (i.e., it is likely that some time will be required before the effects of the remedy can be realized or even measured). This potential lag in results suggests that ongoing monitoring of a proposed remedy and its desired effect is an essential component of the process. Due to the model being based on the 'continuous improvement loop' by Deming (1986), it allows the required monitoring to take place without the need to re-enter the loop at the start of the process. This is achieved by simply reusing the previously identified CFs and reassessing their performance by way of the various KPIs, saving valuable investigative time, effort and resources.

The Project Diagnostic Toolkit: ensuring successful project outcomes

As stated at the beginning of this chapter, one of the key business imperatives of the Project Diagnostics Research Project is to develop a toolkit that enables potential end-users to:

- investigate the health of a construction project;
- identify the probable root causes of a project's poor health;
- give an indication of possible REMs which could be implemented to improve a project's performance and outcomes.

It is envisaged that the use of the Project Diagnostic Toolkit will be highly cost-effective for project clients and stakeholders alike, potentially reducing associated costs of adverse project impacts, including time overruns, inadequate build quality, poor project relationships, loss of reputation, public clamour and legal disputation.

When to use the Project Diagnostic Toolkit

The Project Diagnostic Toolkit has the potential to be used either when clients or other stakeholders believe that their project is not performing in accordance with initial expectations (cost, time, quality), or at regular intervals during the life of a project to assess/monitor its health and likely success. When used on a regular basis, much of the data for the Project Diagnostic Toolkit can be collected concurrently with the data collection efforts of regular project status reports.

Why use the Project Diagnostic Toolkit?

Following earlier references made in this chapter to the construction industry's need for a comprehensive tool to assist stakeholders in assessing the state of health of construction projects, a further five key reasons as to why the implementation and application of the Project Diagnostic Toolkit is of great benefit to all construction industry stakeholders are presented in Table 18.7.

Table 18.7 Benefits of using the Project Diagnostic Toolkit

Integrated benefits	• The Project Diagnostic Toolkit is a three-in-one package that:
	  • identifies areas of poor project health
	  • pinpoints the root causes
	  • identifies remedial measures

Research based	• The development of the Project Diagnostic Toolkit involved rigorous academic review with literature from industry and research institutes in the UK, USA, Europe and Australia, and is being comprehensively validated using many real-life projects of differing sizes, with various procurement methods and at different stages
Economical	• The cost associated with using the Project Diagnostic Toolkit is very small when compared with those costs related to the adverse impacts of failing projects
Relevant	• The Project Diagnostic Toolkit is based on a cyclic mechanism that repeats the investigation until the problems are remedied • It is dependent on benchmarks for performance evaluation, most of which are based on industry standards • It has the provision for updating these benchmarks as required
Easy to implement	• An independent and objective team is needed to implement the Project Diagnostic Toolkit

Future recommendations: where to from here?

Due to the sensitive nature of the project failure syndrome, a number of clients and stakeholders preferred not to provide access to certain potential case-study projects. Consequently, the Project Diagnostic Research Team was able to gain access to no more than three projects in the allocated timeframe of this research. The future and ultimate success of the Project Diagnostic Toolkit will therefore largely depend on how it is implemented, applied and used on additional projects. The Project Diagnostic Research Team further recommend that the various KPIs, CFs and SPIs and associated linkages identified are further validated on these case-study projects.

Furthermore, in an attempt to help overcome the various limitations of using lessons learned and personal experiences as REMs, and to help target the most appropriate remedial action(s) that will potentially bring a project back to good health, the Project Diagnostic Research Team recommended developing a dedicated REM model that aims to take the traditional judgement-based approach of applying REMs, and transform it into a more scientific decision-support system. That is, reducing client and stakeholder reliance on the traditional decision-making process of individual judgement and human perception will in turn reduce any wasted time, financial cost and additional resources associated with the current 'hit and miss' approach, and improve the overall effectiveness of applying relevant and proven REMs on a project.

The design of the proposed REM model would also allow for its seamless integration into the current Project Diagnostics Toolkit, as proposed by the preliminary research and development concept illustrated in Figure 18.4.

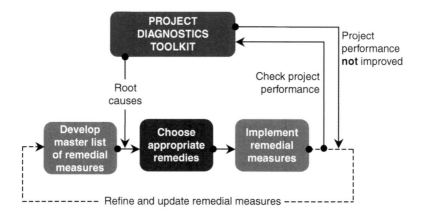

Figure 18.4 Remedial Measures Model: preliminary concept.

Summary

Despite the availability of a large number of international publications providing guidance on successful project execution, the construction industry continues to suffer from projects failing to achieve key deliverables expected by clients and key stakeholders. The development of a construction project-specific 'health check' model that allows for early detection and accurate diagnosis of emerging problems, followed by the recommendation of an appropriate remedy, solution or action, is integral to enhancing and maintaining overall project performance. On the other hand, if the deteriorating health of a project is left unchecked, if key symptoms of project illness are mistreated or if the causes of project failure are ignored, the inevitable result will be an amplification of reduced project performance.

Acknowledgements

The author of this chapter would like to thank the following Australian industry-, government- and university-based Project Team Members for their unwavering support and invaluable contributions during this exhaustive research undertaking: John Tsoukas (project leader) and Sheldon Sherman (Arup), Chris Evans (John Holland), John Collin (Queensland Department of Public Works), Mike Swainston (Queensland Department of Main Roads), Matthew Humphreys, Tony Sidwell and Daniyal Mian (Queensland University of Technology) and Paul Tilley (CSIRO).

Bibliography

Abidali, A.F. and Harris, F. (1995) 'A methodology for predicting company failure in the construction industry', *Construction Management and Economics*, 13 (3): 189–96.

Ahmad, M.M. and Dhafr, N. (2002) 'Establishing and improving manufacturing performance measures', *Robotics and Computer-Integrated Manufacturing*, 18 (3): 171–6.

Arditi, D. (2000) 'Business failures in the construction industry', *Engineering, Construction and Architectural Management*, 7 (2): 120–32.

Atkinson, R. (1999) 'Project management: cost, time and quality, two best guesses and a phenomenon, it's time to accept other success criteria', *International Journal of Project Management*, 17 (6): 337–42.

Avots, I. (1967) 'Why does project management fail?', *California Management Review*, 12 (1): 77–82.

Belassi, W. and Tukel, O.I. (1996) 'A new framework for determining critical success factors/failure factors in projects', *International Journal of Project Management*, 14 (3): 141–51.

CBPP (2003) *Construction Industry Performance Indicators*, Watford: Construction Best Practice Programme.

CII and ECI (2003) *Contractor Questionnaire Version 7.0*, Austin, TX: Construction Industry Institute and European Construction Institute, University of Texas. Online. Available at HTTP: (accessed 2003).

Cole, T.R.H. (2003) *Final Report of the Royal Commission into the Building and Construction Industry*, Canberra: Commonwealth of Australia.

Deming, W.E. (1986) *Out of the Crisis*, Cambridge, MA: MIT Center for Advanced Engineering Study.

Humphreys, M., Mian, D. and Sidwell, A.C. (2004) 'A model for assessing and correcting construction project health', paper presented at the International Symposium of the CIB W92 on Procurement Systems, Chennai, India, 7–10 January.

Kangari, R. (1988) 'Business failure in construction industry', *Journal of Construction Engineering and Management*, 114 (2): 172–90.

Mian, D.M., Sherman, S.R. and Humphreys, M.F. (2004) 'Construction projects immediate health check: a CSF and KPI approach', paper presented at Project Management Australia conference, Melbourne, 12–13 August.

Morris, P.W. and Hough, G.H. (1987) *The Anatomy of Major Projects: A Study of the Reality of Project Management*, New York, NY: John Wiley.

Pinto, J.K. and Slevin, D.P. (1987) 'Critical success factors in successful project implementation', *IEEE Transactions on Engineering Management*, 34 (1): 22–7.

Pinto, M.B. and Pinto, J.K. (1991) 'Determinants of cross-functional cooperation in the project implementation process', *Project Management Journal*, 22 (2): 13–20.

Plemmons, J. (1994) 'Measuring and benchmarking materials management effectiveness', *AACE Transactions 1994*, 2.1–2.9.

Rubin, I. and Seeling, W. (1967) 'Experience as a factor in the selection and performance of project managers', *IEEE Transactions on Engineering Management*, 14 (3): 131–4.

Russell, J. and Jaselskis, E. (1992) 'Quantitative study of contractor evaluation programs and their impact', *Journal of Construction Engineering and Management*, 118 (3): 612–24.

Shenhar, A.J. and Levy, O. (1997) 'Mapping the dimensions of project success', *Project Management Journal*, 28 (2): 5–13.

Skitmore, M. (1997) 'Evaluating contractor pre-qualification data: selection criteria and project success factors', *Construction Management and Economics*, 15 (2): 45–56.

Songer, A.D. and Molenaar, K.R. (1997) 'Project characteristics for successful public sector design and build', *Journal of Construction Engineering and Management*, 123 (1): 34–40.

19 Engineering sustainable solutions through off-site manufacture

Nick Blismas and Ron Wakefield

The construction sector is under the spotlight in relation to its poor 'sustainability' record. Taken very broadly, the built environment is viewed as a major contributor to resource depletion, environmental pollution, and industrial accidents and fatalities. It is further seen as wasteful and inefficient. Many of these accusations are well founded. Australia's construction industry, like many around the globe, is characterized as adversarial and inefficient, and in need of structural and cultural reform (Cole 2003). Similarly, several UK government reports (e.g., Latham 1994; Egan 1998) have called for significant improvement of the industry.

The nature of construction management research understandably attempts to address these challenges individually, concentrating on improvement to efficiency, human relationships and worker safety. Often these are simply 'patches' to larger, more systemic issues. While best practice solutions are very useful, their contributions are incremental rather than revolutionary and therefore offer marginal improvements. Developments in materials, processes and knowledge have not substantially changed construction over the past century, and it is argued that this reductionist approach will not effect any substantial improvements into the future, and cannot hope to provide solutions to construction's 'sustainability' challenges. A fundamental paradigm shift is lacking in these approaches. The industry needs to engineer new and different solutions. This chapter proposes that off-site manufacture (OSM) in construction is one approach which has the potential to overcome many of the challenges facing the construction industry in the twenty-first century.

The chapter briefly outlines some of the common challenges that construction faces, arguing that modern initiatives do not provide a sufficient shift from current practice to be able to effect any lasting change. This is followed by an outline of the benefits offered by OSM, before suggesting how these can fulfil all major dimensions of sustainable construction. The chapter ends with some speculation on how an OSM approach could address the challenges facing the industry as it tries to adapt to an uncertain future.

Construction: sustainable?

Sustainability is a nebulous term with many meanings. To some it is the tool for advancing a strong environmental agenda, for others an equality agenda, and for yet others it is another term for 'business longevity'. All agree that the term refers to planning for the future, whether or not there is agreement on what the challenges of the future may be. When addressing sustainability in construction and the built environment, it is useful to view the concept in the broadest sense that does not simply take an environmental slant. Although simplistic, 'triple bottom line' thinking offers a framework for analysing OSM with a view to assessing the extent to which it can be proposed as a feasible and sustainable solution for construction.

The three 'bottom lines', namely financial, environmental and social aspects, are used to represent some of the challenges of construction and some of the engineering solutions that OSM can offer. Each of these aspects cannot be adequately expressed as a single measure, as their scope is extensive. Financial aspects could range from macroeconomic productivity to work process efficiencies. Social aspects are likewise complex and broad, ranging from individual impacts such as work hours to broader aspects such as workforce dynamics and employment. Similarly, environmental impacts cover, among others, land use, waste, material choice, emissions and energy use, both within the process of creating the final product and in the operation of that product. In the following sections, only a limited number of indicators will be used to demonstrate the extent to which traditional on-site construction is in need of innovation which could be realized through OSM.

Financial aspects

Viewed critically, construction fails on all three 'bottom lines'. It is known as a relatively low-profit industry with a significant level of inherent risk. It is inefficient and fragmented, resulting in high levels of waste and low levels of productivity. Coupled with the cyclic nature of construction in most economies (Bon 1992; Raftery 1992; Hillebrandt 2000; ABS 2003), the long-term sustainability of construction firms is limited. The industry has an entrenched 'boom and bust' reputation that has come to expect significant influx and outflow of persons and organizations. The financial implications are obvious, but these too impact greatly on environmental sustainability and society more broadly.

The typical supply chains used in Australian construction are segregated. The building process tends to procure all necessary base materials, together with numerous subcontracted trades, to construct an independently acquired design, usually under the direction of the main contractor. Very little interaction takes place between the entities in the supply chain prior to works on site. This largely accounts for the adversarial and inefficient state of the industry. The traditional division of professions and trades, coupled with the complex contractual arrangements to designate risk between these parties, often results in sub-optimal products and fragmented processes. Further, the adversarial nature of construction serves to entrench these divisions. Construction needs a more stable environment, with long-term supply chain relationships, based on mutual gain rather than mistrust.

OSM in construction does, however, suggest a solution to these issues. Eastman and Sacks (2008: 517), for instance, state that the productivity of assembling common building components off site is higher than equivalent on-site construction:

> The off-site sectors, such as curtainwall, structural steel, and precast concrete fabrication, consistently show higher productivity growth than on-site sectors. Furthermore, the value-added content of the off-site sectors is increasing faster than that of the on-site sectors, indicating faster productivity growth.

Figure 19.1 illustrates their point.

OSM, as will be further illustrated in the following sections, purports to offer significant benefits to productivity/process efficiencies, waste minimization, profitability and, thereby, long-term financial sustainability.

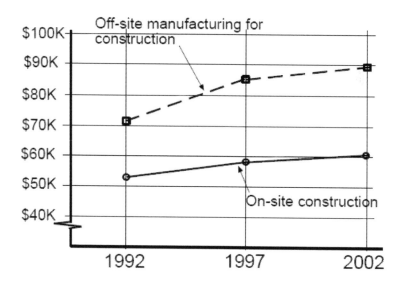

Figure 19.1 Aggregate value added for off-site construction manufacturing for construction and comparable on-site industry sectors (source: Eastman and Sacks (2008: 525)).

Social aspects

The second 'bottom line', that of social impact, is especially marked in construction. The industry creates the urban environment in which people live and work, while also generating a significant portion of GDP (Hillebrandt 2000). In Australia, for instance, construction employed 870,000 in 2005–06 and consistently contributes around 6 per cent of GDP (ABS 2008, 2004). Although having such a significant influence in the economic and social fabric of any society, it has some features which give it a generally negative social 'image'.

The 'boom and bust' cycles typical of the industry result in large flows of personnel into and out of the industry. Its low barriers of entry encourage many unskilled and semi-skilled persons into the industry during 'boom' times, with resultant compromises in safety and process efficiencies. Conversely, 'bust' times see many lose employment, with very few transferable skills. Further complications may arise with the increased global movement of both skilled and unskilled labour. This unprecedented movement of international skills has yet to be tested under strained economic conditions.

Of most significant social impact is the poor safety record of construction. The International Labour Organization estimates that there are at least 60,000 fatal accidents on construction sites around the world each year – one every ten minutes. Construction accounts for 17 per cent of all fatal workplace accidents (International Labour Organization 2005). Workers' compensation statistics (Australian Safety and Compensation Council 2008) show that the fatality rate in the Australian construction industry is 9.3 per 100,000 workers, compared to 3.1 for all industries, and since 1997–98 an average of 49 construction workers have been killed each year – nearly one per week (Fraser 2007). By contrast, the manufacturing industry fatality rate is approximately 4 per 100,000 workers, even though it includes high-risk sub-sectors such as petro-chemical (Australian Safety and Compensation Council 2008). Clearly, the construction industry's occupational health and safety performance is unacceptable.

Further proxy measures of social impact are provided by the unfavourable work-life balance and stress-related statistics for the industry. Lingard and Francis (2004) found that male employees in site-based roles reported significantly higher levels of work–family conflict and emotional exhaustion than male employees who worked in regional or head offices. This is compounded by a culture of long hours and weekend work, in which employees struggle to achieve a balance between their work and personal lives. Evidence also points to a comparatively higher level of 'burnout' among construction employees than those of other professions such as military, technology and management.

The inherent risks and long work hours, together with the cyclic and project nature of work, make construction an industry with high social impact. OSM can ameliorate some of these negative social features.

Environmental aspects

The third 'bottom line' covers environmental aspects. Two facets need to be differentiated when addressing the environmental performance of construction: the process of constructing a facility, and the environmental performance of the facility over its useful life. The impact of the latter is greater when considering material selection, design, energy use and emissions, whilst the former is more concerned with impacts on the industry as a whole and the construction site environment more specifically. The performance of construction, on both fronts, is considered poor. Unfortunately, the division of production and product that is typical of construction perpetuates the division of responsibility. By contrast, manufacturing designs products concurrently. It ensures that the process of production is efficient without compromise to the product. This philosophy has much to offer construction, by enabling environmentally sustainable solutions to be developed in a holistic manner.

For brevity, only two indicators are used to demonstrate some of the challenges that the construction industry faces. Although these are encountered in all industries, including manufacturing, the nature of construction dictates that the measurement and actions to address the problems are more difficult. It should be borne in mind that the examples below are merely indicators of broader issues. Statistics are particularly nebulous regarding environmental performance, and should always be read with caution.

The first indicator is physical waste. It is generally agreed that waste production in construction is significant. In 2002–03, Australia generated approximately 32.4 million tonnes of solid waste, of which approximately 42 per cent was from the construction and demolition sector (Productivity Commission 2006). Manufacturing principles concentrate on waste elimination through engineering design and process efficiency. The waste minimization aspect of construction can be significantly addressed within an OSM approach.

A second, although oblique, reference to construction's environmental challenges concerns direct and indirect greenhouse gas emissions (Australian Greenhouse Office 2005). Figures 19.2 and 19.3 indicate that the group to

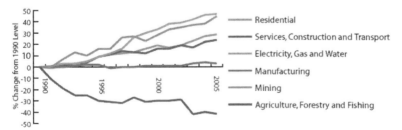

Figure 19.2 Percentage change in direct emissions by economic sector, 1990–2005 (source: Australian Greenhouse Office (2005), reproduced with permission).

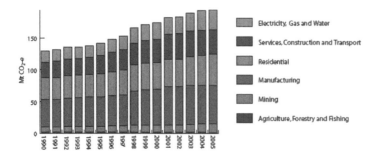

Figure 19.3 Indirect greenhouse gas emissions from generation of purchased electricity by economic sector, 1990–2005 (source: Australian Greenhouse Office (2005), reproduced with permission).

which construction belongs has an increasing emissions trend. This is in contrast to manufacturing's slowing emissions trend. Although it is difficult to draw conclusions from such figures, it suggests that the controllable environment of the 'factory' will allow better regulation of emissions.

As has been briefly demonstrated, construction faces challenges on all major fronts when it looks to its future sustainability. There is immense potential for efficiency improvement through both process and product design. It is held that the financial gains offered by these solutions would help address the environmental concerns of construction. Further, these changes should enable the social impacts of construction to be improved. Solutions for construction using many of the processes of manufacturing and engineering can basically address sustainability concerns that current approaches cannot. The inevitable call to reduce consumption and environmental emissions, without engineered technological solutions, is bound to be resisted by both developed and developing economies. OSM is proposed as a method for 'rethinking' construction – of modernizing and technologically revolutionizing it. Some indication of how this could be achieved and its associated challenges is provided below.

Off-site manufacture

The call for OSM to play a major role in transitioning construction into a modern and efficient industry is not new. The Australian construction industry recently identified OSM as a key vision for improving the industry over the next decade (Hampson and Brandon 2004). This echoes sentiments in other parts of the world, specifically the UK where government-commissioned reports have proposed OSM as an important contributor to progress in the construction industry (e.g., Egan 1998; Barker 2004).

Manufacturing principles offer great efficiency for delivering a high-quality product. Often these principles are misconstrued as simply producing large volumes of standard products. However, industrialization – the broader term that incorporates manufacture – encompasses many concepts and initiatives. The PATH project (2002) summarized some examples of industrialization concepts that have been successful in other industries and that should have application in construction. Briefly, these include (but are not limited to):

- just-in-time manufacturing that includes effective supply chain management;
- flexible, agile, lean production systems;
- concurrent engineering and design for manufacturers that use various techniques and processes to enhance the manufacturability of the product;
- manufacturing requirements planning, manufacturing resource planning and enterprise resource planning systems, which are processes that are enabled by information technology;
- concurrent design, where communication among designers and producers (construction foremen, site supervisors, trade contractors) can significantly improve the efficiency of production;
- time- and space-based scheduling that facilitates keeping track of who is where, doing what, and when. This is especially appropriate for construction activities, as crews move among sites.

These approaches broadly aim to give more control over value specification and demand, whilst designing the process to eliminate waste and optimize efficiency: empowering workers and seeking continuous improvement (Roy *et al.* 2003). These principles address the main aspects of the sustainability challenge.

OSM *in construction*

Given the high profile offered to OSM in the UK, activities to encourage its adoption are considerable, involving several research initiatives, communities of practice and government-sponsored forums (e.g., Accelerating Change). Approximately £5 million had been invested by the government in research projects that included construction OSM between 1997 and 2001, growing to £10 million when industry funding is taken into account (Gibb 2001). Large industry-level studies identified an abundant array of benefits and barriers to OSM, with the hope that these would spur activity. Despite these well-documented benefits, uptake is limited (Neale *et al.* 1993; Bottom *et al.* 1994; CIRIA 1999, 2000; BSRIA 1999; Housing Forum 2002; Gibb and Isack 2003; Goodier and Gibb 2007). Goodier and Gibb (2004) suggested that OSM accounted for approximately 2 per cent of the £106.8 billion UK construction sector in 2004.

A major reason advanced for the reluctance of clients and contractors to adopt OSM is that they have difficulty in ascertaining the benefits that this would add to a project (Pasquire and Gibb 2002). The use of OSM is poorly understood by many of those involved in the construction process, and is based on anecdotal information rather than data-supported intelligence (CIRIA 2000). Given the nature of the industry, this response is not surprising. Each link in the supply chain is concerned with the benefits accruable to that level, while ignoring any broader systemic benefits to the project.

Much research has been undertaken to enable OSM benefits to be better assessed and realized within projects. One report concludes that a deliberate and systematic use of OSM, commencing early in the process, would increase predictability and efficiency, and ultimately add value to the process (Gibb 2001). Others, such as Blismas *et al.* (2003), developed a tool enabling a comparison between traditional methods and OSM options, highlighting that a holistic evaluation would provide a more accurate and realistic assessment than is commonly used in the industry. Blismas *et al.* (2006) found that cost factors were often buried within the nebulous prelim-inaries figure, with little reference to the building approach taken. Further, issues such as health and safety, effects on management and process benefits were either implicit or disregarded within these comparison exercises. Their attempt to incorporate a broader view of value, including social and environmental value, is one move towards understanding the implications of adopting an OSM approach.

Blismas (2007) scoped the state of OSM in Australia for the CRC for Construction Innovation to highlight the major benefits and barriers in the construction industry. The major barriers were identified through a series of interviews and workshops (Figure 19.4). The barriers largely correlate with those of other countries, and essentially hinge on the industry's lack of production knowledge across the supply chain and its cultural inertia.

Scrutiny of barriers such as cost, program and process reveal that these mainly reflect the industry's traditional fragmented structure. For instance, the program constraints were identified as longer lead time, inability to change designs and high fragmentation. These barriers indicate that OSM is basically viewed as an alternative method of construction within the current industry structure and professions, and not as a distinctly different procure-ment route. Added to this, OSM has a stigma inherited from poor-quality buildings of past generations.

Notwithstanding the poor understanding of OSM concepts, the benefits appear to be recognized by the industry. However, a continued reductionist or substitutionary approach to benefit evaluation will invariably have difficulty assigning benefit and therefore fail to convince the industry of its usefulness. OSM cannot deliver any benefit unless the processes, procurement

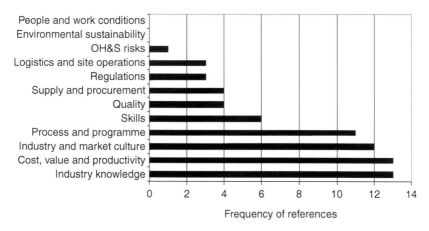

Figure 19.4 Relative significance of barriers to OSM identified for Australia (source: Blismas (2007), reproduced with permission).

routes and supply chains change in concert in order to realize the benefits. Such industry-level changes make implementation particularly challenging. The following section elaborates on the extent to which the Australian construction industry recognizes the potential of OSM, and how this could address sustainability challenges.

Sustainability through OSM

The potential benefits of OSM are clear in all three aspects of sustainability. Its widespread use is dependent on the drivers outweighing the constraints. Figure 19.5 illustrates the main drivers in the Australian construction industry as identified by Blismas (2007). The drivers deliver the benefits that will ultimately make it viable. The shortage of trade skills in construction is clearly the main driver. As costs for these increase, the viability of producing alternative systems that do not require specialized skills becomes more attractive. The drivers and their associated benefits are categorized according to the three key dimensions of sustainability. The black bars reflect financial aspects, the lined bars social, and the hatched bar environmental. Interestingly, all three feature prominently in the higher frequencies of the chart. This suggests that the industry is already aware of the 'triple' benefits that OSM can offer. Table 19.1 provides a summary of the drivers and their benefits highlighted (X) according to significance of reference to the sustainability categories.

Not surprisingly, the financial-related benefits of OSM dominate the three main dimensions of sustainability. Generally, it is seen to offer several benefits over current building approaches that can directly improve the financial sustainability of the industry.

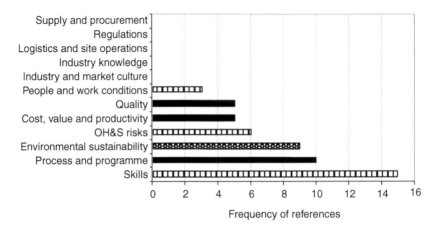

Figure 19.5 Relative significance of drivers of OSM identified for Australia (source: Blismas (2007)).

The process of construction is entirely transformed by OSM, which fundamentally reshapes the building process by ensuring that design includes both product and process design. The benefits listed in Table 19.1 can only be realized through processes that do not rely on simply improving the way current building systems are used, but also endeavour to rethink the process in such a way that all non-value-adding steps are eliminated, thereby improving efficiency. These efficiencies transfer throughout the construction process, from manufacture to logistics and on-site placement. The elimination of trades from site combined with interface reductions simplifies the process so that it is delivered faster and with a high degree of quality. The highly engineered designs can ensure that waste is eliminated and energy consumption minimized.

These improvements further translate into other aspects such as safety, whereby a meticulously engineered process ensures that the most efficient and safest methods are designed into the product and process systems developed. The control available within the workplace not only improves productivity but also ensures OH&S work conditions can be improved.

The future for construction

The vision for OSM in construction postulated in this chapter is ambitious, but not new. Much has been written about the lessons of manufacturing for construction. The current economic drivers, particularly concerning trade skill shortages, are making OSM increasingly attractive. However, adoption of OSM without appreciation of its underlying principles can be damaging. Anecdotal evidence from the United States housing industry showed that a misapprehension of OSM can lead to financial losses. The US housing

Table 19.1 Summary of benefits of OSM identified in a study scoping the state of OSM in Australia

Drivers and benefits	Financial	Social	Environmental
Skills and knowledge		X	
– Significant shortage of skilled trades in construction is a major driver. OSM offers employment to unskilled and semi-skilled workers, fulfilling roles undertaken by trades			
– Revitalization of traditional manufacturing regions with high unemployment can be realized with OSM plants			
Process and program	X		
– Reduces construction time on site and overall			
– Simplifies construction process with, for example:			
– fewer trades and interfaces to manage and coordinate on site			
– ability to transport large loads easily			
– fewer disruptions on site			
– reduced wet and dirty trades on site			
– removing difficult operations off site			
– site can continue to operate when using OSM			
– Guarantees delivery time and quality			
– Program is centrally driven rather than independently			
Environmental sustainability			X
– Waste is reduced both on and off site by being 'designed out'			
– Better housekeeping on site due to removal of trades			
– Sustainable solutions better incorporated through design, so that buildings can achieve better energy performance			
– Product adaptable/reusable into the future			
OH&S risks		X	
– Reduced on-site risks due to lower likelihood and exposure due to fewer people on site			
Cost, value and productivity	X		
– Lower costs can be achieved where work is under resource pressure			
– Lower costs of workforce in remote areas			
– Lower whole cost of construction with an integrated solution			
– Increased cost certainty through better risk control			
– Lower overheads, less on-site damage, less wastage			

Continued

Table 19.1 Continued

Drivers and benefits	Financial	Social	Environmental
Quality – Higher quality and better control is possible in the factory – High levels of consistency can be attained – Products are tried and tested in the factory	X		
People and work conditions – Improved working conditions for labour whose work environment is controlled and specifically and ergonomically designed		X	

Source: Blismas (2007), reproduced with permission.

market saw the widespread adoption of panelized systems by larger builders as being a more profitable option. However, the level of process change that is needed to make OSM viable made the approach unsuccessful with many builders. In order to emphasize the whole-of-life and whole-of-process integration needed to make OSM truly sustainable, an example of a panelized OSM system for construction is presented in Table 19.2, illustrating how an engineered solution might deliver the benefits identified above. The system is described using the typical project phases as a framework.

Conclusion

Construction faces numerous challenges in the twenty-first century. It is an industry upon which all of society is dependent; but its financial, social and environmental credentials are continually being brought into question. Some indicative evidence has been presented to demonstrate some of the issues faced by the industry. It is argued that the current approaches to addressing these are piecemeal and will only provide marginal gains. An integrated OSM approach is proposed as the framework for substantially advancing construction and addressing these issues.

The proposition asserts that OSM has the potential to address the 'triple bottom line':

- *financially*, by improving productivity through reducing fragmentation, improving profitability, creating new markets and releasing innovative capacity, thus making construction a modern industry;
- *socially*, by improving safety and work conditions, stabilizing workforces and improving economically impoverished regions, thus making construction a more attractive industry to work in;
- *environmentally*, by creating a cleaner, more efficient industry, delivering highly engineered, high-quality buildings with superior performance, and ensuring the building stock remains useful well into the future.

Table 19.2 Integrated OSM system with proposed sustainability benefits

Description	Design	Off-site production/manufacture	Logistics and delivery	On-site production/construction	Operation/maintenance	Adaptability
	Integrated flexible design facilitated by IT platforms allows all information of the building to be captured at design	Integrated platform supporting the system automatically communicates production information to suppliers with required specifications for delivery by a given date	Panels and modules are produced and completed in a predetermined order for delivery and placement	Placement from the truck should be direct and ensures that on-site works are quick with minimal on-site labour	Precise and quality-assured product ensures high levels of performance during the life of the product	Sophisticated OSM system is designed to be 'future proof' so that panels and modules can be moved and replaced over time
	Building, process and performance are all modelled and known; conflicts are eliminated	Information is also communicated directly to plant that can produce the components in the order and dimensions required	Tracking systems ensure every component is tracked at every stage, including placement	Precise factory-quality panels fit without need for adjustment on site	Engineered system eliminates reliance on trades' quality and ensures better performance	Grid-based universal connection system enhances the future adaptability of the products, thus ensuring a significant advance in product sustainability
	Integrated system determines the components required for manufacture, and all production information is automatically generated	Off-site work maximized and all trades taken off the site	Process modelling ensures that delivery is just-in-time and placement is direct from truck into the structure	Placement process is well sequenced without any system conflicts	Automated information systems permit real-time information on the building to be collected for performance monitoring and optimization	Fully integrated computer models, accurate at time of design, provide precise 'as-built' models to be used to design changes to existing buildings
		Panels made with high level of finishes	Real-time information on the project is accessible by all parties	All remaining trades and placements are recorded in real-time so that all parties can know the precise project stage	Energy and water consumption are examples of the more basic measures that would be recorded	
				PDAs and RFID systems provide actual component information at any time		

Continued

Table 19.2 Continued

	Design	Off-site production/ manufacture	Logistics and delivery	On-site production/ construction	Operation/ maintenance	Adaptability
Financial	Duplication eliminated; Design conflict minimized; Mistakes minimized; Waste eliminated in product and process	Process time reduced; Waste eliminated in product and process; Can utilize cheaper labour	Precise management reduces mistakes, rework and costs; Just-in-time delivery reduces costs; Delivery costs can be optimized	Process time reduced; Precise management reduces mistakes, rework and costs	More efficient product has lower cost to operate	Lower refurbishment and extension costs due to modularity
Social	–	OH&S risks reduced in factory; Better work conditions for staff; Can utilize unskilled and semi-skilled labour	OH&S risk exposure reduced with fewer on-site deliveries; Fewer deliveries, with less disruption to neighbourhood	OH&S risk exposure reduced with fewer on-site works; Disruption to site environment reduced	Safe to operate and maintain by users and occupants	–
Environmental	Building performance modelled and known; Performance optimized; Waste minimized in design optimization	Waste minimized; Environmental damage on site minimized	Highly controlled supply chain and process can ensure minimization of deliveries and emissions	Environmental impact on site reduced	Optimized environmental performance of the building due to quality of product and design	Future-proof buildings that can be reused extend life of the building and ensure all types of environmental impact are minimized

Sustainability solutions can be engineered through OSM. However, it remains to be seen whether the bold leadership and strong collaborative efforts required will be forthcoming.

Bibliography

ABS (2003) 'Understanding the building life cycle by its cyclical nature', *Australian Economic Indicators, December 2003*, Canberra: Australian Bureau of Statistics.

—— (2004) *Australian System of National Accounts, 2003–04*, Canberra: Australian Bureau of Statistics.

—— (2008) *Year Book Australia, 2008*, Canberra: Australian Bureau of Statistics.

Australian Greenhouse Office (2005) *Australia's National Greenhouse Accounts: National Inventory by Economic Sector, 2005*, Canberra: Department of Environment and Water Resources.

Australian Safety and Compensation Council (2008) *NOHSC Online Statistics Interactive – NOSI*, Canberra: National Occupational Health and Safety Commission. Online. Available at HTTP: <http://nosi.ascc.gov.au/> (accessed 29 May 2008).

Barker, K. (2004) *Review of Housing Supply*, London: Office of the Deputy Prime Minister.

Blismas, N.G. (2007) *Off-Site Manufacture in Australia: Current State and Future Directions*, Brisbane: CRC for Construction Innovation.

Blismas, N.G., Pasquire, C.L. and Gibb, A.G.F. (2006) 'Benefit evaluation for off-site production in construction', *Construction Management and Economics*, 24: 121–30.

Blismas, N.G., Pasquire, C.L., Gibb, A.G.F. and Aldridge, G.B. (2003) *IMM-PREST: Interactive Method for Measuring Pre-assembly and Standardisation Benefit in Construction* (CD format), Loughborough: Loughborough University Enterprises.

Bon, R. (1992) 'The future of international construction', *Habitat International*, 16 (3): 119–28.

Bottom, D., Gann, D., Groak, S. and Meikle, J. (1994) *Innovation in Japanese Prefabricated House-Building Industries*, London: Construction Industry Research and Information Association.

BSRIA (compiled by Wilson, D.G., Smith, M.H. and Deal, J.) (1999) *Prefabrication and Pre-assembly: Applying the Techniques to Building Engineering Services*, Briefing note ACT 2/99, Bracknell: Building Services Research and Information Association.

CIRIA (compiled by Gibb, A.G.F., Groak, S., Neale, R.H. and Sparksman, W.G.) (1999) *Adding Value to Construction Projects Through Standardisation and Pre-assembly in Construction*, Report R176, London: Construction Industry Research and Information Association.

—— (principal author Gibb, A.G.F.) (2000) *Client's Guide and Toolkit for Optimising Standardisation and Pre-assembly in Construction*, Report CP/75, London: Construction Industry Research and Information Association.

Cole, T.R.H. (2003) *Final Report of the Royal Commission into the Building and Construction Industry*, vol. 6 (*Reform: Occupational Health and Safety*), Canberra: Commonwealth of Australia.

Eastman, C.M. and Sacks, R. (2008) 'Relative productivity in the AEC industries in the United States for on-site and off-site activities', *Journal of Construction Engineering and Management*, 134 (7): 517–26.

Egan, J. (1998) *Rethinking Construction*, London: Department of Trade and Industry.

Fraser, L. (2007) 'Significant developments in occupational health and safety in Australia's construction industry', *International Journal of Occupational and Environmental Health*, 13: 12–20.

Gibb, A.G.F. (2001) 'Standardisation and pre-assembly: distinguishing myth from reality using case study research', *Construction Management and Economics*, 19: 307–15.

Gibb, A.G.F and Isack, F. (2003) 'Re-engineering through pre-assembly: client expectations and drivers', *Building Research and Information*, 31 (2): 146–60.

Goodier, C.I. and Gibb, A.G.F. (2004) *The Value of the UK Market for Offsite*, Buildoffsite.

—— (2007) 'Future opportunities for offsite in the UK', *Construction Management and Economics*, 25: 585–95.

Hampson, K. and Brandon, P. (2004) *Construction 2020: A Vision for Australia's Property and Construction Industry*, Brisbane: CRC for Construction Innovation.

Hillebrandt, P.M. (2000) *Economic Theory and the Construction Industry*, 3rd edn, London: Macmillan.

Housing Forum (2002) *Homing in on Excellence: A Commentary on the Use of Off-Site Fabrication Methods for the UK House Building Industry*, London: Housing Forum.

International Labour Organization (2005) *Prevention: A Global Strategy: Promoting Safety and Health at Work*, Geneva. Online. Available at HTTP: <www.ilo.org/public/english/protection/safework/worldday/products05/report05_en.pdf> (accessed 27 November 2006).

Latham, M. (1994) *Constructing the Team: Final Report of the Government/Industry Review of Procurement and Contractual Arrangements in the UK Construction Industry*, London: HMSO.

Lingard, H. and Francis, V. (2004) 'The work-life experiences of office and site-based employees in the Australian construction industry', *Construction Management and Economics*, 22: 991–1002.

Neale, R.H., Price, A.D.F. and Sher, W.D. (1993) *Prefabricated Modules in Construction: A Study of Current Practice in the United Kingdom*, Ascot: Chartered Institute of Building.

Pasquire, C.L. and Gibb, A.G.F. (2002) 'Considerations for assessing the benefits of standardisation and pre-assembly in construction', *Journal of Financial Management of Property and Construction*, 7 (3): 151–61.

PATH (2002) *Technology Roadmap: Whole House and Building Process Redesign: One Year Progress Report*, Washington, DC: Partnership for Advancing Technology in Housing, report prepared for Office of Policy Development and Research, US Department of Housing and Urban Development.

Productivity Commission (2006) *Waste Management*, Report no. 38, Canberra: Commonwealth of Australia.

Raftery, J. (1992) *Principles of Building Economics*, Oxford: BSP Professional Books.

Roy, R., Brown, J. and Gaze, G. (2003) 'Re-engineering the construction process in the speculative house-building sector', *Construction Management and Economics*, 21 (2): 137–46.

Part V

Facilities management and re-lifing

20 Towards sustainable facilities management

Lan Ding, Robin Drogemuller, Paul Akhurst, Richard Hough, Stuart Bull and Chris Linning

About 1 per cent of all Australian commercial building stock is new each year. This means that if new building works are relied on to provide significant change in the sustainability of the stock, then it will take many years before a significant impact is achieved. Fortunately, commercial buildings are refurbished at relatively frequent intervals, so their rate of improvement is much faster than the rate of new construction. However, there are other drivers which could lead to significant improvements in our building stock. These fall under the aegis of facilities management (FM).

In this chapter, FM is defined to cover the entire project life cycle after handover from the initial construction. This includes not only the day-to-day operations involved in running a facility, but also major refurbishments and upgrades.

Sustainable FM is a process that manages the efficient operation of facilities in support of service delivery, operational life cycle costs and the use of resources, and minimizes their impact on built environment.

Digitization, procurement and benchmarking have been found to be beneficial to continuous improvement of FM functions and operations in support of objectives and service delivery of the organization. The effective integration of these three elements will support the delivery of FM services in a more sustainable way, contributing to the productivity and efficiency of both the FM team and the enterprise as a whole, as well as improving return on investment for business.

This chapter uses the Sydney Opera House (SOH) as an example of the types of actions that a facilities manager can take to improve performance. While the initial reaction may be that this is an exceptional building and hence may not be an appropriate example, it still has to meet all the requirements of a building that is used intensively. The factors that make it an iconic building actually make it a good case study, as the FM processes also have to be of a high standard.

The SOH complex needs to be managed in a sustainable way from a number of perspectives. As with all other structures, cost is a significant motivator, as there is a budgeted amount of money available to operate the SOH over any year. Maintenance funding is provided as a 'ring-fenced' grant by the New South Wales state government, and is not available for other activities more closely aligned with its purpose. The SOH also plays a key social role through fulfilling its primary function as a performing arts centre and via its secondary functions as a social hub and tourist draw-card. The environmental impacts are also important, given its placement on Sydney Harbour.

This chapter describes the activities that are underway to improve the FM processes at the SOH. These have included improvements in access to information about the complex, changes in the relationships within the FM activities, and an assessment of how various factors impact on the FM processes. The use of 'towards' in the chapter title acknowledges that the existing processes are early steps along the path to a more sustainable facility.

This chapter reports on the results of a number of projects at the SOH. The longest-standing projects involve the gradual conversion of paper documents, together with on-site surveys, to eventually build accurate as-built data on the structure itself. Parallel to this work and dependent on its results, the Sydney Opera House FM Exemplar project aimed to achieve integration of procurement, benchmarking and digitization as a basis for promoting FM as a business enabler and developing sustainable practices.

The procurement research developed a performance-based procurement framework for improving service delivery. FM requirements are defined in terms of performance objectives and the use of multi-criteria decision-making strategies. The benchmarking research developed performance benchmarking that comprised methods, procedures and data collection, and delivered internal benchmarks or key performance indicators (KPIs) which enable the identification and improvement of critical success factors. The digitization research developed a digital FM model based on the building information model (BIM) to assist in the integration and automation of FM. The digital FM model provided an integrated information environment which allowed alignment of services, performance and information with organizational business objectives.

Context

Each year, over 1,500 performances and 1,000 other events are held at the SOH. This makes it one of the most heavily utilized performing arts centres in the world. Its smooth functioning and the maintenance of a satisfactory operating environment would be impossible without an appropriate FM system. As with almost all existing facilities, the FM systems do not contain all of the available information, and neither do they capture all of the potentially useful information from FM and business activities in an integrated reusable way.

The SOH consists of seven performance spaces, 37 plant rooms, 12 lifts and over 1,500 rooms, all underneath the pure geometric forms of its curved profile. The supporting structure, the equipment and the activities to maintain them are all complex.

The SOH FM team has developed an asset strategy and total asset management plan (TAM) which align the asset and facility service levels with the corporate business objectives. As with any successful business enterprise, the Sydney Opera House Trust has defined corporate business goals, as follows:

- to be Australia's pre-eminent showcase for performing arts and culture and an international leader in the presentation and development of artists and their work;
- to attract and engage a broad range of customers and provide compelling experiences that inspire them to return;
- to maintain and enhance Sydney Opera House as a cultural landmark, performing arts centre and architectural masterpiece;
- to create a customer-focused workplace where people are recognized for their contribution, realize their potential and are inspired to achieve outstanding results;
- to invest in the performing arts, cultural activities and audience development by maximizing business results for Sydney Opera House and leveraging its assets, resources and brand.

Recent initiatives have included:

- a program of consolidating the existing documentation on the building and surrounds and filling in gaps;
- a maintenance management software system used for both in-house and contracted FM activities;
- proposals for a series of major refurbishment projects, including the Opera Theatre renewal.

Two technical activities that support these initiatives are:

- definition of an 'as maintained' BIM – this activity is continuing;
- participation in the Sydney Opera House FM Exemplar project through the CRC for Construction Innovation (CRC CI).

The FM team has been working on building 'as constructed' and 'as maintained' data for the complex over a number of years. The aims and proposed extent of this work have changed over time as the capability of computer systems has improved. This work is underway, using the team of Arup (structural), Utzon Architects, JPW Architects and Steensen Varming (building services). Specialist consultants are providing services such as laser scanning to rapidly generate 'as constructed' surfaces of the facility. For such a large and complex facility, this is a major undertaking in itself.

The Sydney Opera House FM Exemplar project used some of the available information to provide a demonstration of what functions FM systems may provide in the near future. The project grew out of a discussion arising from the Construction 2020 (Hampson and Brandon 2004) series of seminars. In subsequent discussions between Paul Akhurst (then Director of Facilities, SOH) and CRC CI personnel, a project scope was developed. Additional funding support was obtained, with the assistance of the Facility Management Association of Australia, through the Australian government's Facilities Management Action Agenda.

Sydney Opera House FM exemplar: integrated solution for FM

Corporate objectives and business priorities determine service delivery for the organization and bring special requirements to facilities functions. As an iconic performing arts facility, the SOH has special needs and services requirements, resulting from its business characteristics. The FM strategy identified the linkages between services and assets by assessing the dependency of service delivery on asset performance.

In the TAM plan, results and services are clustered as performance arts service, visitor experience service and building and property service. Asset dependency to the service clusters is grouped into:

- theatre technical facilities (audio, lighting, staging and stage management);
- security systems (CCTV, cyberlocks and swipecards);
- staff and presenter facilities (offices, dressing rooms and toilets);
- visitor facilities (retail, catering, bars and public toilets);
- mechanical, electrical and hydraulic systems;
- fire services;
- information and communication technology;
- building and site (structural and building fabric).

The assets are managed across multidimensional FM streams to ensure asset functionality and capability to best support each of the dependent services.

Figure 20.1 presents an integrated FM solution model, developed through the Sydney Opera House FM Exemplar project, which integrates results and services, asset dependency and FM streams into an overall business framework to provide a generic model that can be applied by the FM industry.

Procurement

The choice of appropriate procurement methods for the required services and the selection of suppliers whose goals and methods complement

Sydney Opera House FM Exemplar

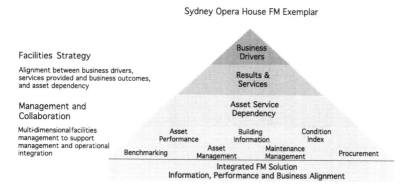

Facilities Strategy

Alignment between business drivers,
services provided and business outcomes,
and asset dependency

Management and
Collaboration

Multi-dimensional facilities
management to support
management and operational
integration

Figure 20.1 Development of an integrated FM solution model.

those of the SOH were seen as critically important to sustainable FM practices. Business objectives provided targets for the performance requirement at each level, including service procurement performance and asset performance. Figure 20.2 presents a performance alignment framework, which integrates business drivers, KPIs and benchmarking, to support service procurement.

In the service procurement process, business drivers provided demand requirements as performance assessment criteria. Provision options for required maintenance services and works were assessed against KPIs, and optimized procurement routes were identified. The performance assessment criteria considered the requirements defined in the strategic asset management plan in the context of the market and the service provision reliability and risk. These criteria were applied in the outsourcing process.

KPIs were identified as multidimensional, including non-financial criteria – for example, quality, time, risk-sharing, response to urgent work, partnership and understanding corporate business expectations. Selection of KPIs and the weightings to be assigned to them was important in determining the drivers and priorities, and to keep service providers focused on achieving excellent performances in key areas.

Benchmarking played an important role in support of procurement decision-making. It enabled the examination of service procurement and building assets and their management against the KPIs, through a comparison with similar services and assets. It provided feedback to facilities managers for refining the selection of procurement methods and for identifying asset management gaps. Whilst internal benchmarking examined historical data and defined internal performance standards, external benchmarking compared them with a group of peers involved with similar FM services and functions.

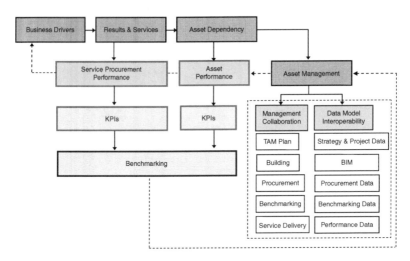

Figure 20.2 Aligning service procurement performance with business drivers supported by benchmarking.

Benchmarking

Benchmarking is important when assessing sustainable performance, as it provides a comparison with the performance of other facilities. It provides a useful tool for measuring the effectiveness of an organization in achieving performance targets, and also in making comparisons with other organizations with similar services and assets. Performance benchmarking can be conducted, based on a set of KPIs which are generally used to monitor strategically important activities in an organization. KPIs may be used to monitor performance trends and set targets that are central to an organization's success. Internal benchmarking identifies internal benchmarks over time, while external benchmarking is considered a comparative process between organizations that can be used to improve FM performances against best practice. Both internal benchmarking and external benchmarking are conducted at the SOH (Figure 20.3).

An example of internal benchmarking at the SOH is the use of building condition indices (BCIs) for establishing internal benchmarks of the overall building presentation. The BCIs are used to underpin the performance requirements in the building maintenance and cleaning contracts. They are defined to measure building fabric condition, cleanliness, tidiness and overall presentation. They use a percentage-based scoring system, with 100 per cent equalling 'as new', and defined standards below this in 10 per cent decrements.

The FM team agree performance targets and benchmarks with the service providers for cleaning and building fabric maintenance, using the BCI

Figure 20.3 Internal benchmarking and external benchmarking at the SOH.

method. The benchmarks are then linked to the flow of business to fully align with procurement operations and decision-making. They are subsequently reflected in the contract specification and evaluation criteria. For example, toilets are required to be at the stated cleaning standard when a foyer is first opened to patrons and at the start of each interval. The benchmarks are set for the period immediately after cleaning. The benchmarks and historical data are stored in a digital database at SOH and used for reference to define contracts.

External benchmarking provides a measurement of the organizational performance and compares it with similar measurements from other organizations with similar or competing services in the same market. External benchmarking enables the identification of best practice in FM and high-lighting of required areas of improvement.

The initial contact with national and international organizations sought an expression of interest in being part of a strategic partnership with the Sydney Opera House FM Exemplar project, an audit of the data that they were currently collecting, and its availability for benchmarking. Both existing and potential data within partner organizations were investigated. To assist partner organizations in developing better data collection, quality assurance of data, standardized vocabulary and comparable collection methods were recommended. It was suggested that the data collection be kept in a digital format for comparison and be able to integrate with the digital BIM.

Fifteen national and international organizations participated in benchmarking and presented their performance metrics for comparison. Four KPI areas were selected:

- service provider's performance – quality of service, safety, timeliness and compliance;
- facility condition assessment – building presentation and services;
- energy management – rate of consumption and management;
- accessibility – security provision and information for visitors.

It was also recommended that the FM benchmarking for iconic facilities and performing arts facilities incorporate electricity, water and gas consumption, and that consumption should be measured against business outputs such as number of performances and patrons and/or opening hours to assess FM management and operation sustainability.

Facilities management and building information modelling

Sustainable management of a facility requires as full a knowledge of the facility as is appropriate and feasible. This matches the goals of building information modelling. Since the SOH has a long expected life, there are projects to be managed under the general umbrella of 'FM' that are of a larger scale than the more routine aspects of inspection, maintenance and upgrade of building services systems and fixtures and fittings. Such larger projects have recently included the opening up and integration of the western foyers, and improvements to disabled access and to toilet facilities. One proposed significant project is the refurbishment of the Opera Theatre, including auditorium, stage and scenery-handling facilities.

To maximize the value and use of the BIM tool, it is important that it can be deployed for larger projects. In FM terms, the subsequent benefits include the following:

- consultant design teams can develop refurbishment proposals directly within the framework of the 'existing conditions' 3D CAD model;
- works contractors can then use the model for visualization during tendering, and potentially for quantities estimation, and then for 4D visualization and programming during the works;
- contractors' as-built documentation can be uploaded to update the evolving model;
- the FM team has easy access to current information.

The asset management systems at the SOH are not significantly different to those used in many other buildings. Figure 20.4 illustrates an integrated information alignment system that could be developed from the current system. It comprises strategic and operational FM information, building information and performance information in a hierarchy. Performance information is supported through the development of building condition index KPIs and service provider performance KPIs, which are compared against actual performance data and benchmarking data to identify gaps and opportunities. Performance information in turn provides the information necessary for FM operational requirements, including procurement information relating to service providers, building information and maintenance and service information. FM operational information in turn supports the FM planning information required for operational and capital expenditure maintenance tasks and projects, and associated budget information. FM planning

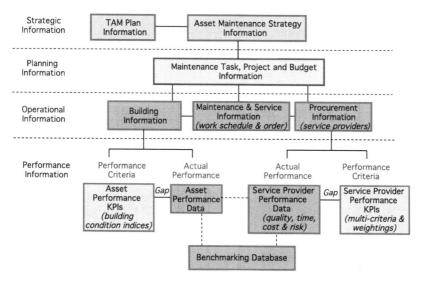

Figure 20.4 Integrated information alignment to support the FM process.

information is linked to the strategic asset maintenance plan that contributes to the total asset maintenance plan.

An example of the information required by the maintenance management and operational processes, which then supports service procurement, is as follows:

- visualization of the management of infrastructure, asset, space and furniture;
- historical data of maintenance services, cost and service providers;
- life cycle data of projects;
- generation of real-time reports of ongoing FM activities and tasks;
- work-order generation and work prioritization;
- tracking of scheduled and unscheduled maintenance activities;
- tracking of capital and labour costs;
- asset registers.

The life cycle information refers to data related to the life cycle management of a maintenance project, and includes the total life cycle cost of the maintenance project, and the aspects of operability and quality.

The integrated FM information alignment can be supported by development of BIM that is linked to the 3D building model.

Business case for BIM for facilities management

The management team of the SOH has found that a number of key issues affect FM activities through their response to normal business needs:

- the building structure is complex, and building service systems – already the major cost of ongoing maintenance – are undergoing technology change, with new computer-based services becoming increasingly important;
- the current 'documentation' of the facility comprises several independent systems, some of which are overlapping, and is not ideal to service current and future FM requirements;
- as with all performing arts centres, the condition and serviceability of key public areas and the functionality of performance spaces require periodic review;
- many business functions such as space or event management require up-to-date information about the facility that could be more efficiently delivered;
- major building upgrades are being considered that will put considerable strain on existing Facilities Portfolio services and the capacity to manage them effectively, and create a demand for accurate information on the current condition and as-built status of the building;
- increasing use of external service providers to deliver maintenance requires accurate information for contract specifications and a common (internal) database for maintenance records across all providers.

Building a BIM model

There were several drivers that led to the decision by the Sydney Opera House Trust to begin transferring historical records into a 3D CAD model, including the following.

- *Consolidation opportunity*. Hardcopy and electronic scanned drawings were located among a range of consultants and government departments.
- *Design management opportunity*. Funds had been committed by the New South Wales state government for commencement of upgrade projects to enhance facilities and operations, including refurbishment of the Utzon Room, construction of the western colonnade, improvements to disabled access and integration of the western foyers. Plans for substantial enhancements to the Opera Theatre auditorium, including orchestra pit enlargement and stage and scenery-handling upgrades, were also being developed in parallel. A 3D CAD model of existing conditions promised a valuable base for insertion and 'virtual testing' of refurbishment proposals, small or large (Figure 20.5).
- *Opportunity for management of buildings services maintenance and operations*. As noted previously, the scale and complexity of building services provide an excellent argument for conversion over time to a more automated management process with a 3D graphical interface and direct links to recording for operational and maintenance purposes.

Figure 20.5 BIM model of Opera Theatre (source: Copyright Sydney Opera House).

- *Contract management.* A coordinated 3D model with linked database promises improved coordination and consistency of documentation among the large number of contractors participating in modification and upgrade works on site over a given period.
- *Potential future opportunities.* A well-developed 3D CAD model offers a range of other opportunities, depending on the level of detail to be provided. These could include virtual tours of the facility, virtual trials by the performance companies of scenery installation and reconfiguration, even virtual rehearsals where lighting, acoustic and environmental (thermal, ventilation) parameters can be simulated, as experienced at a particular seating location.

The software used to initiate the project was Bentley's 3D suite of BIM tools, because this was in use by both of the project architects (Utzon Architects and JPW Architects) and the structural consultant (Arup) for the refurbishment projects noted above.

A substantial effort went into assembling the available drawings, both structural and architectural, and into discovering information considered to be the most current, given the large number of interventions over the years. This often led to site visits to resolve uncertainties, and on-site surveys.

The model varies in detail, based on the immediacy of proposed future work, in that the most information is provided in those areas proposed for refurbishment – so that the Opera Theatre, for instance, is treated in more detail than the Concert Hall (Figure 20.6).

Historical data sources were not abandoned in the process, but cross-referenced from elements in the 3D model to the relevant drawings or survey data on which they were based. Future users will be able to revisit sources and examine how conflicts among source data were resolved.

While record drawings existed for the auditorium ceiling, a multiple station laser-based site survey carried out by Hard and Forrester was used in preference to obtain a more accurate as-built surface. This in turn offered improved knowledge of the clearances between the ceiling structure and the soffit of the concrete shells above – one of the critical aspects of the building's geometry. The laser-based survey was used to a level of 1 m above finished floor level due to the seating that remained in situ and obstructed the laser. Traditional survey methods were then used to complete the lower wall and floor areas.

Other surfaces examined and documented by the architects for addition to the structural 'base model' were the precast concrete cladding panels, both floor- and wall-mounted, based again on record drawings and on-site inspections. With the building services systems, only the larger duct runs

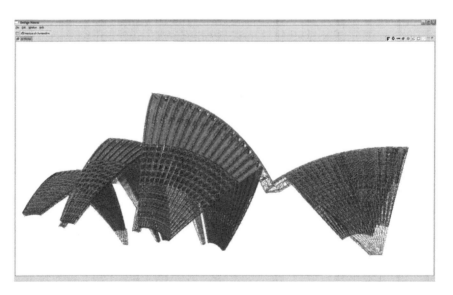

Figure 20.6 3D CAD record images of SOH shell primary roof elements (source: Copyright Sydney Opera House).

for supply and exhaust air and associated plant were captured in the initial stage, in areas of particular design interest, such as loft, flytower and beneath the balconies. Building services consultant Steensen Varming provided Bentley HVAC Building Services 3D modelling of existing conditions for integration into the growing model.

Modelling of the highly complex shell rib elements, primary arch support structure and long span concourse beams within Bentley Structural proved straightforward, due to the large amount of accurate historical documentation still available.

The shells are all created from spheres of the same radius but with the centre points at different elevations. This allowed a single typical tapering rib of the longest length, 191 feet long for the largest shell on the Concert Hall, to be created, arrayed around the sphere surface and trimmed to suit the height of each shell. A similar method was used in creating the continuously changing profile of the long span post tensioned concourse beams.

Testing the model

While it was convenient for the contributors to the model to share the same software during its creation, it was not intended that its future use should be restricted to that software. For example, to capture and integrate documentation of new works by design consultants and contractors, some flexibility would be needed, albeit within minimum compatibility requirements specified by the SOH as client.

To check the robustness of the model for interchange between CAD packages, an experiment was carried out to transfer the entire model from Bentley Structural to Graphisoft's ArchiCAD, then back to Bentley Structural, and to check if full functionality could be retained through the exchanges. The IFC 2×3 standard was used as the information exchange schema to effect the transfers, with Woods Bagot Architects, one of the research project participants, acting as the ArchiCAD host.

The full exchange was eventually completed with near perfect functionality from Bentley Structural into Graphisoft's ArchiCAD (Figure 20.7). Experiments have also been carried out with IFC exchanges to Autodesk's Revit Structure 4 and Revit Structure 2008, which proved fairly successful in terms of geometric representations but less successful in translating the information within the objects.

With the lifetime of most buildings likely to exceed that of many software companies, long-term use of BIM will be best served by the use of open standards and an upgrade and maintenance strategy both for the software systems used to access and manipulate data, and also for the data itself. The information (data) standards for FM will improve over time as commercially available software becomes more powerful and the information needs of facilities managers become better understood.

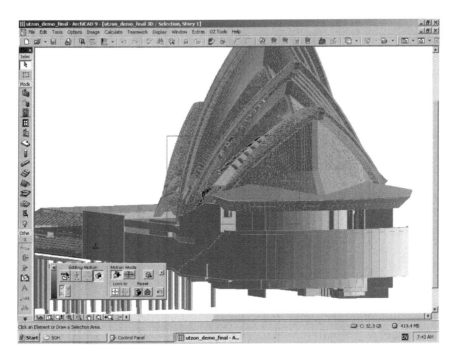

Figure 20.7 SOH model in ArchiCAD (source: Copyright Sydney Opera House).

Interoperability with analysis engines

Although current CAD and analysis tools are not yet well adapted to the frequent and radical reconfigurations that can occur in early design stages, this drawback is less pronounced in design for refurbishment, where at least the existing building can be safely modelled in full detail to provide the framework for the proposed intervention. The local portion of the larger BIM network relevant to such early design can be represented as in Figure 20.8.

The 'Design Manager' represents a file transfer program based on a chosen schema, and the plug-ins are written so that they can provide access to a wide range of analysis (and CAD) packages via application programming interfaces (APIs). The plug-ins need to be more or less elaborate depending on the size and sophistication of the Design Manager, which may not only provide the file transfer function, but also add value to the files being transferred by filtering them for simple optimization outcomes, thus minimizing the need for separate satellite optimization routines linked directly to the analysis engines themselves.

Arup has been researching the application of BIM to early stages of the design process, with RMIT University's Spatial Information and Architecture

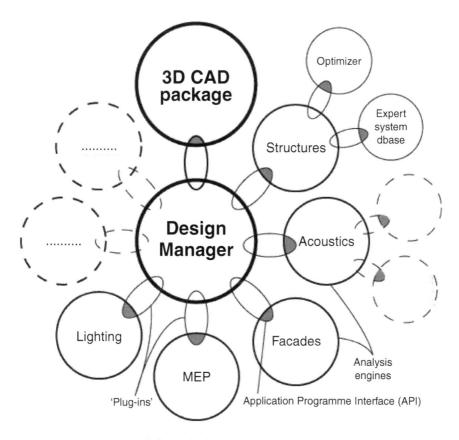

Figure 20.8 BIM network for early design.

Laboratory, using diagrams similar to Figure 20.8. An opportunity to apply the intent of the figure occurred during the scheme design stage for the refurbishment of the Opera Theatre, using the 3D CAD model of the building as the framework onto which a range of options for modifications to the auditorium could be overlayed. The options could then be checked against a wide range of architectural criteria and also against particular technical criteria, such as structural feasibility (extent and implications of internal demolition and reconstruction), acoustic outcomes (including volume and reverberation) and smoke management in fire mode (using computational fluids dynamics – CFD modelling).

Regarding structural analysis, automatic links had already been established via in-house Visual Basic programming between Bentley Structural models and structural analysis packages GSA and Strand 7, for the extensive structural design work carried out on the Water Cube, the Beijing National Aquatics Centre (Hough *et al.* 2007). Linking between the CAD model, the

Odeon package for acoustic analysis and the FDS package for CFD analysis was partially successful. Differing types of mesh are required by such analysis packages – for example, CFD modelling can require a solid block type of mesh, whereas acoustic models typically require a 'watertight' single surface type of construction. Exports of Bentley Structural and Architecture models provided a starting point for creation of these environmental models, but significant revisions were required outside of the BIM links for completion of these analysis models. Challenges included reducing the density of surface facets and triangulation, and ensuring joints between surface elements were closed 'watertight' for the acoustic and lighting analyses.

Links to FM databases

Beyond its use as a framework for modelling of major refurbishment projects, the detailed BIM of the existing building offered itself as the graphic interface for an automated FM tool, covering operations, maintenance and upgrading of building services systems and any other systems requiring inspection and maintenance, such as furniture, fixtures and fittings.

The development of links to other databases illustrates the versatility of the BIM concept and the benefits that will be realized from the introduction of interoperable standards for software. Currently, data require a secondary translation process before they can be moved from one system to another, which is time consuming and can introduce reconciliation errors. As an interim measure, the FM team has been developing its own 'language' so that there is a consistency between the various databases (for example, 'purchase order' is always abbreviated to PO). Similarly, the original room-numbering system is being re-established and reinforced so that each room has a unique identification, while allowing variable generic and colloquial names that will change over time.

There are, of course, countless ways of linking the CAD model to individual serviced components within the building. Some components will be too small or too well embedded within their sub-system to warrant their explicit graphical representation in the 3D model. Hand basins in washrooms, for example, can be easily referenced by a specifier plus room identification.

Given Bentley's basis of model organization by geometric objects rather than by intelligent associative objects, such as in ArchiCAD or by contained volumes (rooms), door openings were chosen as the key descriptor of a space. They offered the unique advantage over rooms that each door has been uniquely numbered since original construction, per standard SOH practice. The door location and number remain (virtually) unchanged, regardless of the physical rearrangement of doors, wall openings and rooms, which has of course been considerable during the life of the facility so far. Door numbers are only ever 'retired', never reused. When a new door opening is created, it is given a new number.

Trial links were created between the CAD model and existing facilities records to demonstrate the potential of a fully developed interface. So far, these links comprise the door opening in walls being invisibly tagged and linked to cells within the master door schedule, which indicates fire compartmentation, designation of space and functional use. This linking mechanism makes it possible to locate a door number in the external FM database and consequently access the database, which holds all the information about the wall containing that door, as well as compartmentation of the space and all relevant asset and FM information.

A single database holds room data with bi-directional links, either working from the BIM model to the FM database or from the FM database to the BIM model. Future links can provide a direct connection to FM/asset management software packages.

The benefit of this integrated process is that the client can use the BIM model for documentation as well as building management, thereby creating a powerful 'interactive window' between the physical model and the building information data sets.

Other examples of the BIM providing a graphical organizing function to assist FM processes include its use in the inspection, updating and recording of schedules of fire-rated walls, and its proposed use in the inspection and recording of sealant condition between tiles and between tile lids on the clad surface of the roof shells.

Providing FM information from BIM

The software deliverables from the Exemplar project used the Design View (CRC CI 2007) software platform that was developed in previous CRC CI projects. Only a few factors of interest to the FM team were explicitly modelled, as the deliverables were intended as proof of concept rather than 'production' systems. One significant measure for the SOH is the building presentation index (BPI). This is a measure of the visual presentation of key public areas, as part of a priority asset management initiative. The basic user interface, together with the results of a query against the BPI, are shown in Figure 20.9. The spaces shown in the plan are below the podium level. The plan-based display of the results gives a good overview of the results in the context of the surrounding spaces, without expecting the user to know all of the room (space) names or numbers or the adjacency relationships between them.

Another key parameter for SOH is energy usage. Proof of concept modelling demonstrated the possibilities created by combining spatial and energy data. The data are stored in the existing SOH databases and can be accessed for display. The results for a particular space are shown in Figure 20.10, together with the results from other queries. Queries can also search on multiple parameters – for example, all spaces that are used for rehearsals with an energy usage greater than 7,000 kWh could be identified.

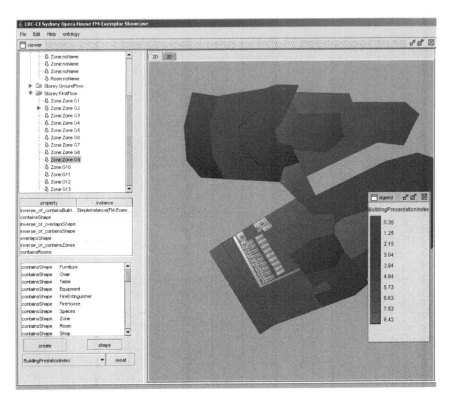

Figure 20.9 Results of building product index query (source: Mitchell and Schevers (2006), reproduced with permission).

Conclusion and future work

The SOH is an exceptional building in many ways. However, its FM requirements are not significantly different to many other buildings. The fact that it is proving cost-effective for the Sydney Opera House Trust to integrate changes within its processes to handle revised procurement methods, integrated with the identification of KPIs and a BIM, indicates that this would be something that other facilities with complex requirements or significant refurbishment programs should consider. Ideally a BIM would be produced during the design of a building, but this is only now becoming possible with the new generation of CAD and analysis software. Creation of a BIM for an existing facility will be a retrospective exercise, and the cost-benefit will need to be assessed for each facility.

The implementation of the software deliverables across a range of FM parameters has demonstrated that BIM data can be integrated with 'traditional' FM databases. The ability to show performance against KPIs also indicates that BIM and KPIs can be used jointly to analyse and display

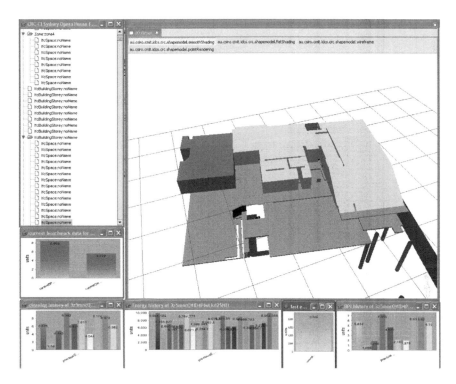

Figure 20.10 Results of queries against a range of implemented factors (source: Mitchell and Schevers (2006), reproduced with permission).

performance on performance-based service contracts, and consequently are useful across a range of contractual arrangements.

Further work is necessary to take the set of KPIs defined across the participants in the benchmarking survey and weave these into a sustainability assessment framework.

Acknowledgements

A large team of people have been involved in the projects reported in this chapter. Arup, Utzon Architects, JPW Architects and Steensen Varming have all been involved in the work funded by the Sydney Opera House Trust on building the BIM models of the SOH.

The Sydney Opera House Trust, the Australian government and the CRC for Construction Innovation funded the Sydney Opera House FM Exemplar project. The participating organizations were the CSIRO, Rider Hunt, the Sydney Opera House Trust, Transfield Services, Woods Bagot and the Facility Management Association of Australia.

Bibliography

CRC CI (2007) *Adopting BIM for Facilities Management: Solutions for Managing the Sydney Opera House*, Brisbane: CRC for Construction Innovation.

Hampson, K. and Brandon, P. (2004) *Construction 2020: A Vision for Australia's Property and Construction Industry*, Brisbane: CRC for Construction Innovation.

Hough, R., Downing, S. and Plume, J. (2007) 'Digital architecture and its implications for structural engineering', in *Proceedings of the 4th International Structural Engineering and Construction Conference*, Melbourne, 26–28 September.

Mitchell, J. and Schevers, H. (eds) (2006) *Sydney Opera House: FM Exemplar Project Report Number 2005–001-C-4: Building Information Modelling for FM at Sydney Opera House*, Brisbane: CRC for Construction Innovation.

21 Life cycle modelling and design knowledge development in virtual environments

Rabee Reffat and John Gero

As the construction industry adapts to new computer technologies, computerized design information, as well as construction and maintenance data, are all becoming increasingly available. Indeed, the growth of such data has begun to far outpace an individual's ability to interpret and digest the embodied information. Such volumes of data clearly overwhelm the traditional methods of data analysis, such as spreadsheets and ad hoc queries. For instance, current information technology applied to facility maintenance utilizes databases to keep track of information and notification of maintenance schedules. However, these databases are not well linked with interactive 3D models of buildings, and are mostly presented in tabular formats. Applying techniques from a new area, called data-mining and knowledge discovery, to the records of existing facilities has the potential to improve the management and maintenance of existing facilities and the design of new facilities. This will lead to more efficient and effective facilities management and maintenance through better planning, based on models developed from available maintenance data, resulting in more eco-efficient buildings. Furthermore, designers and facility maintenance managers will be better equipped to achieve higher performance by utilizing appropriate information technology techniques at their workplace.

Data-mining and knowledge discovery in databases are tools that allow identification of valid, useful and previously unknown patterns within existing databases (Witten and Frank 2000; Christiansson 1998; Frawley *et al.* 1992). These technologies combine techniques from machine learning, artificial intelligence, pattern recognition, statistics and visualization to automatically extract concepts, interrelationships and patterns of interest from large databases. Data-mining is capable of finding patterns in data that can assist in planning. Patterns and correlations identified from data-mining existing records of maintenance and other facilities management activities provide feedback, and can improve future maintenance operation decision-making and inform strategic planning as well as the design of new facilities. Most available computer tools for the building industry offer little more than productivity improvement in the transmission of graphical drawings and textual specifications, without addressing more fundamental changes in building life cycle modelling and management. Virtual environments (VEs) can provide designers and facility managers with a foundation to work distributedly. Designers, building owners, facility managers and technicians can visualize and navigate the virtual building modelled in distributed VEs without being co-located. Information can be shared in different ways, depending on the way in which and the extent to which the information must be coupled.

A virtual mining environment has been developed to integrate datamining with agent-based technology, database management systems, object-based CAD systems and 3D VEs. This chapter presents a system prototype of a virtual mining environment to provide dynamic decision support for improving building life cycle modelling and management.

Life cycle modelling (LCM)

The life cycle cost concept is addressed in the British Standards as 'Terotechnology', which is defined as a combination of management, financial, engineering, building and other practices applied to physical assets in pursuit of economic life cycle costs. Life cycle modelling (LCM) contributes to competitiveness of the company by providing enhanced information for decision-making. It helps facility managers evaluate alternative equipment and process selection based on total costs rather than the initial purchase price. The multidimensional information that LCM presents is merged from hybrid project domains, such as management, engineering and finance. LCM may be applied in a wide range of critical functions including:

a   evaluation and comparison of alternative designs;
b   assessment of economic viability of projects and products;
c   identification of cost drivers and cost-effective improvements;
d   evaluation and comparison of alternative strategies for product use, operation, test, inspection and maintenance;
e   evaluation and comparison of different approaches for replacement, rehabilitation/life extension or disposal of ageing facilities;
f   optimal allocation of available funds to activities in a process for product development;
g   assessment of product assurance criteria through verification tests and their trade-offs; and
h   long-term financial planning.

With the increased use of information technology in the 1990s, the increase of information availability has become a dilemma due to inefficiencies in processing information for decision-making. This problem becomes critical in the building industry when the high degree of complexity of work flows involved and the accompanying uncertainty for decision-making in the lifetime of a building are considered. Thus, efficiently dealing with information from different stages of a building's life cycle to improve profitability and productivity as well as strategic resource planning has become a business driver for life cycle modelling. However, the design of new buildings and facilities tends to focus on short-term cost and the immediate needs of the owner for a building that meets various business and functional requirements. Current technologies such as computer-aided design (CAD) have focused on the needs of designers to develop designs that meet design briefs that do not include significant life cycle design requirements. Very little attention has been given to modelling of the life cycle costs of buildings at the design and management stages. There are several life cycle models available for buildings as a whole and for their component systems and, although there is no one model that has been accepted as a standard, there are some areas of commonality.

Life cycle cost models form predictions based on several parameters, some of which include a degree of uncertainty, such as the reliability of a part. These inputs can range from the cost of installation to the cost associated with carrying spare parts in inventory (Siewiorek and Swarz 1982). By accurately predicting failure rates and repair costs, it is also possible to compute the optimal schedule of preventative maintenance for each asset. What can be predicted and the accuracy of those predictions depends, of course, on the availability and accuracy of the maintenance data. Furthermore, current life cycle modelling systems fail to provide a seamless integration of hybrid information that provides users access to previously inaccessible knowledge. It has been shown in the AEC (Architecture, Engineering and Construction) industry that major factors contributing to construction quality problems include inadequate information and poor communication (Burait *et al.* 1992; Arditi and Gunaydin 1998). The detection of previously undiscovered patterns in Building Maintenance System (BMS) data can be used to determine factors such as the cost-effectiveness and expected failure rate of assorted building materials or equipment in varying environments and circumstances. These factors are important throughout the life cycle of a building, and such information could be used in its design, construction, refurbishment and maintenance, representing a potential decrease in cost and increase in reliability. Such knowledge is significant for saving resources in construction projects.

LCM in virtual environments

Virtual environments (VE) are computer-generated synthetic environments in which users are provided with multi-modal, highly natural forms of computer interaction. VE research is concerned with creating artificial worlds in which users have the impression of being in that world, and with the ability to navigate through the world and manipulate objects in the world. Four existing life cycle modelling prototypes based on buildings with an object-oriented database in VE platforms were surveyed: Future Home (Murray *et al.* 2000), Virtual Building Life cycle (VBLC) (Linnert 1999), LICHEE (Life Cycle House Energy Evaluation) (CSIRO 2003) and Product Model and Fourth Dimension (PM4D) (Fischer and Kam 2002). The various benefits derived from these projects include the following:

- use of VEs as platforms to provide an interactive interface to improve communications between all team members as well as a simulation tool in enhancing predictable strategic planning within the whole life cycle of selected buildings;
- application of the IFC-compliant (Industry Foundation Class) object-oriented database in standardizing the data exchange and facilitating the manipulation and reusability of project information;
- linkage of maintenance data to the 3D CAD model providing the potential for future development of intelligent life cycle analysis and control capability.

However, these prototypes do not provide the capability of data-mining of the hybrid data gathered from different stages of a building's life cycle, and do not provide performance-gaining life cycle analysis.

A data-mining approach to LCM

Data-mining has been defined as the 'nontrivial extraction of implicit, previously unknown, and potentially useful information from data' (Frawley *et al.* 1992). It enables people to understand how systems that were once thought to be completely chaotic have predictable patterns (Peitgen *et al.* 1992). Patterns and causal relationships can be found behind apparently random data in AEC projects. By applying data-mining to identify significant novel patterns, project managers will be able to build knowledge models that may be used for recurrent activities of ongoing construction projects, as well as for future project activities, and possibly avoid unanticipated negative consequences (Soibelman and Kim 2002).

Data-mining also presents the potential to address the problem of transforming knowledge implicit in data into explicit knowledge for decision-making. In contrast to traditional methods of statistical data analysis, it is an automated process that discovers trends and patterns without the need for human intervention. It takes, as input, variables whose relevance may not be obvious to a designer but which becomes evident as a result of this process. In addition, data-mining makes no prior assumption about the probability distribution of the input variables (Gaussian, Poisson, etc.), as is required in statistics, and is therefore more robust and universal in application. However, like other methods, the process of transforming the data into a format suitable for knowledge discovery is not automated and has a large impact on the results obtained. Thus, the approach to be taken here is based on a comprehensive view of the building management problem. It views the process of building design, maintenance and replacement as a process generating an enormous amount of information. While current practice addresses parts of this information generation and management, our approach attempts to account for the entire life cycle flow of this information.

The costs of designing and building structures are much smaller than the costs of operating a building or other structure over the course of its lifespan. The knowledge that becomes available through data-mining enables a building owner/developer to make important decisions about life cycle costs in advance, thereby significantly affecting and improving design decisions. The rich set of building data that is created during the design and documentation phase remains relevant even after the building is constructed, and only becomes richer as maintenance data are added. Architects, interior designers and engineers, as well as contractors, marketing and sales personnel, building managers and owners can extract information from the databases for the building's renovation, maintenance and operation. Figure 21.1 outlines the proposed model of the flow of information in building design and maintenance. The bold arrows depict the functionality provided by our proposed approach, while the dashed arrows describe the scope of present approaches to building information management.

Data-mining techniques can be used effectively on data stored in a building maintenance system by creating knowledge that can be used in future management and design decision-making. Knowledge that implicitly resides in building maintenance system databases includes information about:

1 components that frequently need maintenance and therefore need to be inspected carefully;

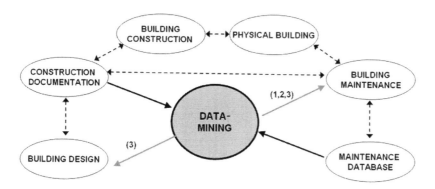

Figure 21.1 Integrating data-mining within the life cycle of building information management.

2   historical consequences of maintenance decisions that may inform future decisions;

3   components of buildings that significantly determine maintenance cost and therefore may inform future building designs, as well as refurbishment of the building in question.

This information can be extracted using data-mining techniques and used to improve all phases of the building life cycle, for both current and future buildings, as indicated by the numbers (1, 2, 3) and (3) on the arrows in Figure 21.1, which correspond to those above. The processing and transformation of data into a format suitable for data-mining is an important step in a successful application of data-mining, and the stages for extracting knowledge from data using data-mining techniques are shown in Figure 21.2.

The data-mining approach for life cycle modelling in this work views the process of building design, maintenance and replacement as a process that generates high volumes of information. While current practices address only parts of this information generation and management, this approach attempts to account for the life cycle flow of this information. The integration of data-mining within the process of information flow provides the opportunity to increase the value of building data, and to feed back and improve the processes of building design and maintenance.

Incorporating data-mining in virtual environments for LCM

A significant need exists for the application of new techniques and tools to automatically assist in the analysis of increasing volumes of data for useful knowledge building. The increasing use of databases to store information about facilities, their use and their maintenance provides the background and platform for the use of data-mining techniques for future projections. The current technology for facility maintenance uses databases to keep track of information, and for notification of maintenance schedules. These

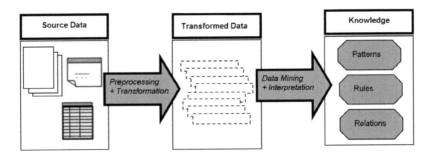

Figure 21.2 Stages for extracting knowledge from data using data-mining techniques.

are not well linked with an interactive 3D model of the building, and are generally presented in tabular form.

A virtual mining environment for providing dynamic decision support

The virtual mining environment promotes real-time support and multiple team participation and involvement. It can be remotely accessed synchronously by different users, who will be aware of the presence of others and communicate with them. Users might mine the same or different building elements based on their focus and interest. Each might be looking at different building assets and using different mining and discovery techniques than others within the same VE. Within the virtual mining environment, data-mining techniques are utilized to discover rules and patterns of useful knowledge from the maintenance records of a building to help improve the maintenance and management of existing and future buildings. Although it may at first sound appealing to have an autonomous data-mining system, in practice such a system would uncover an overwhelmingly large set of patterns, most of which would be irrelevant to the user. Therefore, it is important to provide a user-focused approach to mine the maintenance data of buildings.

The virtual mining environment prototype system described in this chapter includes an object-oriented 3D CAD model of a building modelled by ArchiCAD software and a maintenance database in a standard SQL (standard query language) format. The architecture of this virtual mining environment has been developed to include three agents – interface, maintenance and filter agents – as illustrated in Figure 21.3. Their roles include the following:

- appropriate mapping between the building assets of the building model in the VE is maintained by the maintenance agent that connects data contained within the maintenance database with data contained within the Express Data Manager (EDM) database via the VE (Active Worlds);
- linking data-mining techniques to building models in Active Worlds is achieved via the maintenance agent that accesses the maintenance database and applies its mining algorithms to it;
- linking knowledge development with the building model in VEs is carried out by the filter agent which assists in improving maintenance management by providing life cycle implications as feedback, whenever building assets (mechanical and electrical elements) are selected in the building model in the VE.

Feedback of useful knowledge can be discovered by the maintenance agent in the application of the four data-mining techniques and algorithms

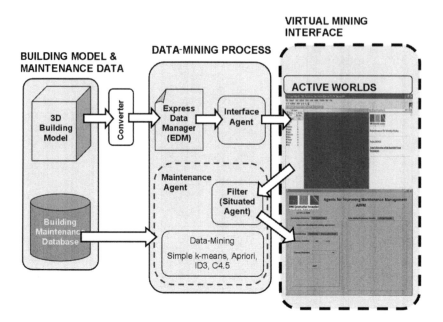

Figure 21.3 Architecture of the virtual mining environment.

(simple Kmeans, *apriori*, C4.5 and ID3) (Witten and Frank 2000). The data-mining algorithms and the link between its knowledge development and the building model in a 3D VE has been fully implemented. The use of the virtual mining environment requires the utilization of the following four phases progressively:

• *Phase 1* involves the manual pre-processing of data, which removes noisy, erroneous and incomplete data to derive important attributes from original raw data – for example, the raw text description of time of work orders '1/12/2001' which need to be converted to a meaningful attribute such as 'month'. Moreover, various 'testing' algorithms are run through the maintenance data to find out the suitable data-mining approaches. As a result of Phase 1, the quality of the data is improved.

• *Phase 2* adopts the EDM interface agent developed by Maher and Gero (2002) in converting IFCs (Industry Foundation Classes) object model into a RenderWare (RWX) format, so as to be available in the VE. The VE provides a collaborative multi-user interface and, more importantly, a means for the user to walk-through 3D object models in real time. The user is able to navigate and select a building asset type to explore useful knowledge.

- *Phase 3* instantiates the maintenance interface agent and the maintenance agent. Once the user decides to select a certain building asset, the maintenance interface agent is invoked to load related data from the database. The maintenance agent performs data-mining on the selected asset type. The four mining algorithms that have been implemented in the system prototype are: clustering using simple Kmeans, associative rules learning in *apriori*, and classification using C4.5 and ID3.
- *Phase 4* is where a filter agent is activated. This is a software agent that performs post-processing of the mined results. The filter agent filters out irrelevant patterns based on heuristic rules.

The virtual mining environment interface and user scenario

The user is able to navigate a 3D model within a real-time VE, as shown in the top left of Figure 21.4. The user is able to instantiate the virtual mining environment prototype system by clicking on the desired building asset or component, invoking the main maintenance interface as shown in the bottom right of Figure 21.4.

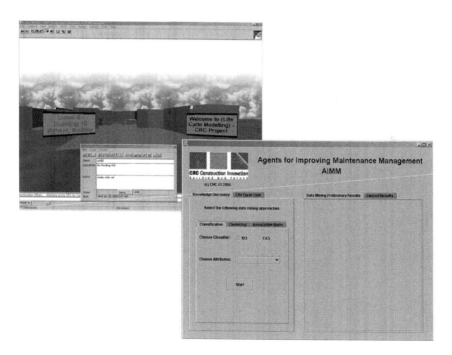

Figure 21.4 Selecting an asset type in Active Worlds instantiates the maintenance interface agent.

Once the interface agent is activated, it pops up the knowledge discovery panel, which has three stacked sub-panels: classification, clustering and associative rules. These provide a range of ways for using each different algorithm. This gives users greater flexibility and scope, since they may test a variety of data-mining approaches for each type of algorithm. On the right-hand side are another two stacked panels that are dedicated to reporting results. Results are reported in two ways. The panel named 'Data Mining Preliminary Results' displays the results of the chosen algorithm in their 'raw' form. The panel named 'Filtered Results' displays the results in their interpreted form using domain-derived heuristics. The overall data-mining interface is illustrated in Figure 21.5(a), and illustrates the hierarchy of stacked panels for the different data-mining scenarios.

In this scenario, an Air Handling Unit (AHU) is selected as the building component that a user wishes to apply data-mining to. The following sequence of actions is then followed:

1   The user navigates the building in a real-time and online 3D VE, as shown in Figure 21.5(a).
2   Once the user selects a building asset type (such as the Air Handling Unit) the object property window pops out, describing general information of the selected object, as shown in Figure 21.5(b).
3   The interface agent is then invoked and the main window pops up to allow selection of algorithms, as illustrated in Figure 21.6(a).
4   The user explores a variety of data-mining algorithms, chooses the desired algorithm and runs it, as shown in Figure 21.6(b).
5   The maintenance agent running the algorithm is invoked and results are reported first in the Data Mining Preliminary Results panel, as illustrated in Figure 21.6(c).

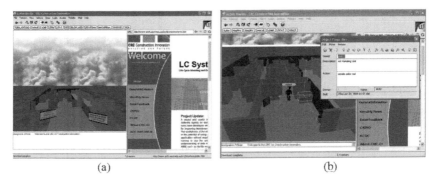

(a)   (b)

Figure 21.5 (a) The primary interface of software agents prototype system (AIMM) in an interactive network multi-user environment; and (b) the user selects a building asset type (the Air Handling Unit) and an object property window pops out describing general information of the selected object.

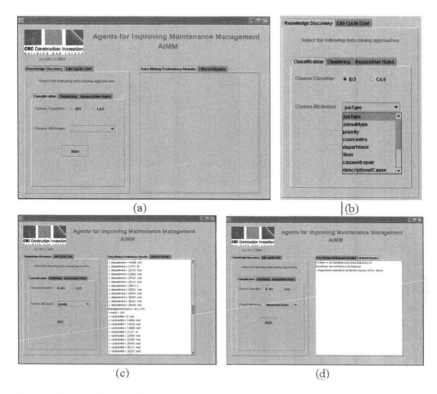

Figure 21.6 (a) The AIMM prototype system is instantiated once a building asset type
has been selected; (b) data-mining techniques and different attributes for
the user to choose from based on focus and interest; (c) preliminary
results of applying the ID3 with the 'Priority' attribute on the maintenance data of Air Handling Unit; and (d) an example of the filtered
knowledge presented to the user from the preliminary results of applying
the ID3 with the 'Work Order Status' attribute on the maintenance data
of Air Handling Unit.

6 The user selects Filtered Results in order to access post-processed knowledge; an example of filtered knowledge is shown in Figure 21.6(d).

Data-mining techniques assisted in identifying critical cost issues. For
instance, discovering that corrective maintenance accounts for approximately
55 per cent of all work orders implies a high level of unplanned maintenance
that contributes to increasing the operational cost. The maintenance services required for the air conditioning system were related to thermal sensa-
tion complaints (too_hot 32 per cent, too_cold 28 per cent, not working
7.5 per cent; total 67.5 per cent). Hence, applying data-mining techniques
assists facility and building managers to identify the crucial maintenance

issues, and directs the improvement of strategic planning to add value to the life cycle of buildings and their occupancy.

Other benefits include constructing predictive plans based on correlations obtained from applying data-mining techniques on the maintenance data sets of buildings – for instance, considering the role of potential correlations between seasons and malfunction rates in guiding the allocation of maintenance resources; also, investigating any abnormal phenomenon discovered from the maintenance data set, such as 'all outstanding works took place in December', to study the reasons for the cause of increase in outstanding maintenance jobs taking place in December (e.g., Christmas holiday or any other causes). Appropriately addressing such problems leads to better processes for improving the maintenance management of existing facilities, and will also better guide the design of future facilities (Reffat *et al.* 2004a, 2004b; Reffat and Gero 2005).

Discussion

The development of data-mining agents of facilities and building maintenance data in a 3D VE provides a new approach for improving the management and maintenance of existing building facilities as well as guiding future design decisions. VE building models offer the opportunity for practitioners to navigate through the model, to manipulate and to interact with its objects. The integration of facilities databases with interactive 3D VEs containing building models and data-mining techniques provides a visual modelling tool for the simulation and projection of the financial and operational impact of maintenance, refurbishment and major replacement and extension of a building and its components over its life cycle. The virtual mining environment has the potential to change the conventional approaches of applying data-mining and to provide a cutting-edge technology to transform the way decision support for building maintenance is carried out. It facilitates real-time assistance for decision-making, and provides an interactive platform for designers, building owners, facility managers and technicians to communicate and interact collaboratively and virtually within a 3D real-time and multi-user VE. The building model presented in the real-time multi-user VE is composed of sets of 3D building elements, including walls, doors, floors, windows, roof and mechanical and electrical equipment. Each element is linked to a knowledge base (which includes the maintenance records of that element) and to a data-mining agent capable of being triggered based on a user's request to mine the knowledge base and provide useful knowledge that helps to enhance the decision-making process for building facility management.

Data-mining of building maintenance information can help discover: procedures that reduce future failures; repairs or maintenance operations that are being executed improperly; ways to improve repairs that reduce subsequent downtime; undocumented methods being used by experienced personnel that result in reduced downtime; and advance notice of likely failures before they occur. Such discoveries can be used to modify building maintenance and repair procedures, thereby reducing downtime, increasing uptime, and significantly reducing the costs of maintenance and repair. Furthermore, the knowledge base of building elements is not static, since all daily maintenance actions that take place are updated in the knowledge base. Hence, the knowledge acquired by data-mining agents is active and dynamic. Therefore, the virtual mining environment facilitates a live, active and dynamic decision support for building maintenance and facility management over the building life cycle.

Acknowledgements

The research presented in this paper is funded by CRC-Construction Innovation, Project No. 2001–002-B 'Life Cycle Modelling and Design Knowledge Development in Virtual Environment'. The industrial maintenance data are provided by Central Sydney Area Health Service, Royal Prince Alfred Hospital, Sydney, Australia. The assistance of Lan Ding, Julie Jupp, PakSan Liew and Wei Peng in the development of this project is gratefully acknowledged.

Bibliography

Arditi, D. and Gunaydin, M. (1998) 'Factors that affect process quality in the life cycle of building projects', *Journal of Construction Engineering and Management*, 124 (3): 194–203.

Burait, J., Farrington, J. and Bedbetter, W. (1992) 'Causes of quality deviations in design and construction', *Journal of Construction Engineering and Management*, 118 (1): 34–49.

Christiansson, P. (1998) 'Using knowledge nodes for knowledge discovery and collaboration', in I. Smith (ed.) *Artificial Intelligence in Structural Engineering: Information Technology for Design, Collaboration, Maintenance, and Monitoring*, Berlin: Springer-Verlag.

CSIRO (2003) *Annual Report 2002–2003*, Canberra: Commonwealth Scientific and Industrial Research Organization. Online. Available at HTTP: <www.csiro.au/files/files/p2kv.pdf> (accessed 20 February 2008).

Fischer, M. and Kam, C. (2002) *PM4D Final Report*, Technical Report no. 143, Stanford, CA: Center for Integrated Facility Engineering, Stanford University.

Frawley, W., Piatetsky-Shapiro, G. and Matheus, C. (1992) 'Knowledge discovery in databases: an overview', *AI Magazine*, 13: 57–70.

Linnert, C. (1999) *Virtual Building Life Cycle: Visualization of Life Cycle Data on a Virtual Model of a Building in a 3D-CAD Environment*, Diploma thesis, Karlsruhe: Institute for Industrial Building Production, University of Karlsruhe.

Maher, M.L. and Gero, J.S. (2002) 'Agent models of virtual worlds', *ACADIA 2002: Thresholds*, Pomona, CA: California State Polytechnic University.

Murray, N., Fernando, T. and Aouad, G. (2000) 'A virtual environment for building construction', *Proceedings of the International Symposium on Automation and Robotics in Construction*, Taipei: National Taiwan University.

Peitgen, H.O., Jurgens, H. and Saupe, D. (1992) *Chaos and Fractals: New Frontiers of Science*, New York, NY: Springer-Verlag.

Reffat, R. and Gero, J. (2005) 'A virtual mining environment for providing dynamic decision support for building maintenance', *Proceedings of the 23rd Conference on Education in Computer Aided Architectural Design in Europe*, Lisbon: Technical University of Lisbon.

Reffat, R., Gero, J. and Peng, W. (2004a), 'Improving the management of building life cycles: a data mining approach', paper presented at CRC Research Conference, Brisbane.

—— (2004b) 'Using data mining on building maintenance during the building life cycle', *Proceedings of the 38th Australian & New Zealand Architectural Science Association Conference*, Hobart: School of Architecture, University of Tasmania.

Siewiorek, D.P. and Swarz, R.S. (1982) *Theory and Practice of Reliable System Design*, Bedford, MA: Digital Press.

Soibelman, L. and Kim, H. (2002) 'Data preparation process for construction knowledge generation through knowledge discovery in databases', *Journal of Computing in Civil Engineering*, 16 (1): 39–48.

Witten, I. and Frank, E. (2000) *Data Mining: Practical Machine Learning Tools and Techniques with Java Implementations*, San Francisco, CA: Morgan Kaufmann.

22 Right-sizing HVAC

Steve Moller and P.C. Thomas

Many heating, ventilating and air conditioning (HVAC) systems have more capacity than is ever required to keep its occupants comfortable. Such 'over-sized' systems can have negative effects on the environment and on occupant comfort, as well as on the economic outcomes for the building. This chapter describes the factors that lead to over-sizing and suggests measures to ensure that HVAC systems have sufficient but not excessive capacity – that is, are 'right-sized'. This work is based on a research project funded by the CRC for Construction Innovation (Thomas and Moller 2007).

What do we mean by right-sized HVAC?

Over-sized HVAC systems

An over-sized HVAC system has significantly more capacity than is required to meet the design brief. For comfort cooling applications, standard design practice uses the concept of a 'design day' – conditions that are exceeded, on average, on ten days per year. Thus the HVAC system in a building will be fully loaded – that is, will run at full cooling capacity – on these 10 days, providing the building is fully occupied. Cooling systems can also operate fully loaded when removing heat built up after a hot weekend. If an HVAC system is under-sized there will be more hours per year when the plant is running fully loaded, and the system will not be able to hold indoor design conditions even on a design day – that is, temperatures in these air conditioned spaces will rise. If an HVAC system is over-sized, it will never run fully loaded; a log of chiller loading will reveal that utilized capacity will almost always be less than 80 or 90 per cent of installed capacity. Other symptoms that over-sized HVAC systems may exhibit are:

- frequent short cycling of individual chiller machines;
- difficulty in maintaining design conditions during periods of high humidity;
- excessive re-heat energy consumption.

Some of the conditions that must be fulfilled for the label of 'over-sized HVAC' system to be warranted are as follows:

- System components and controls must have been correctly designed and commissioned. If chilled water system flow rates are lower than designed, the chiller set will never be fully loaded. Cooling loads can also be reduced if outside air rates are low or if space temperatures are not controlled within designed limits.
- The building must be fully occupied.
- False loading of heating and cooling systems must not be occurring – that is, if heating is fighting cooling, due to dampers or valves leaking, the chiller plant may be artificially loaded, giving the impression of correct sizing.
- Spare capacity, intentionally included in the design, must be excluded from the analysis. This spare capacity provides redundancy to cover breakdowns or an allowance for future increased tenant loads.

Evidence of over-sizing

HVAC over-sizing is common in the UK. Knight and Dunn (2004) reported on a study of over 30 air conditioning systems in office buildings in Wales: 'Analysis of the part-load chiller energy consumption profiles … revealed that virtually all the systems were over-sized for the loads they actually encountered in practice.' Crozier (2000) reported on a survey of 50 HVAC systems in the UK, which showed that 80 per cent of the heating plant, 88 per cent of the ventilation plant and 100 per cent of the chiller plant incorporated capacity above that needed to meet design requirements. Approximately one-third had more than twice the required amount of capacity. Deng (2002) presented a case study from Hong Kong where the original design included four 2,000-kW chillers (total 8,000 kW), but operating records showed that the highest cooling load recorded was 3,516 kW.

Right-sized HVAC systems

A right-sized HVAC system has adequate capacity to meet the loads defined in the design brief. The actual total internal loads experienced on peak summer days are likely to be less than specified in the brief; therefore, a right-sized HVAC system will be able to maintain comfort even when the mercury rises above the design temperature. A key feature is consideration for future high-load areas. Rather than providing for these throughout the building, a right-sized approach will include provision of tenant condenser or chilled water systems.

Causes of over-sizing

Occupancy loads

Knight and Dunn (2004) conclude that high design guide values for estimating internal heat gains (occupancy, lighting and equipment) are one important reason why commercial HVAC systems are being over-sized.

These values are based on high occupancy levels (persons/m^2) which are rarely reached.

Traditionally, design loads for occupancy in office buildings are set at 10 m^2/person. A report by Australian state governments (Government Real Estate Groups 2000) reveals that the internal targets for net lettable area (NLA) per full-time equivalent employee ranged from 15 m^2 to 18 m^2. Actual numbers were even higher, ranging from 17.6 m^2 to 21.2 m^2. From this report it seems that government offices may be designed for occupancies that are 50 to 100 per cent larger than required.

There may be some automatic application of the default figure of 10 m^2/person, required by AS1668 for calculating fresh air requirements when occupancy is unknown, to calculation of heat gain from occupants. There is no reason why the same figure should be used for both.

Internal loads

The Property Council of Australia has developed a grading matrix for office buildings that is the benchmark for space quality in the real estate industry. This matrix quotes internal load capability for a particular grade of building to be *more than* a designated W/m^2 value. This type of grading encourages 'chest beating' exercises within the industry, with each property manager promoting higher and higher internal load capabilities. Anecdotal reports of managers advertising capabilities of 40 W/m^2 are known to the authors.

Anecdotal reports also seem to suggest that activities in the top end of the Australian real estate market are diverging. For open-plan office space in general, there seems to be evidence of the internal load reducing. This is due to uptake of much more efficient LCD monitors that also offer other advantages through glare and contrast performance. Many employers are providing employees with laptops, and requiring them to take these home. Both these actions would tend to reduce internal loads, because the equipment is much more efficient, and portable computers are generally not left on overnight. Komor (1997) reported measured loads from office equipment in 44 buildings in the USA. The simple average was 8.9 W/m^2 and the highest value was 12 W/m^2.

However, there are higher energy density requirements for areas dedicated to IT which house computer servers and other high-power equipment in a confined space, due to security reasons. These areas require 24-hour cooling, and are usually conditioned by additional supplementary HVAC systems supplied by a dedicated tenant condenser water system. These specialized internal loads should not be included when calculating the size of the base building chilled-water system.

Temperature setpoints

Close control of internal temperatures (e.g., 22.5°C ± 1°C) is difficult to maintain in practice, and can lead to excess capacity and higher energy use. The energy efficiency provisions introduced in Section J of the Building Code Australia (2005) require HVAC systems to be designed to maintain a temperature range between 20°C and 24°C for 98 per cent of operating time. Such wider thermostat settings can improve stability of operation due to a larger 'deadband' provision, and also result in a smaller system capacity requirement.

Discrete design process

Concept designs can be carried out independently by project team players. For example, an HVAC designer may use overly conservative glazing characteristics very early in the project and develop high cooling load estimates. If these are not later revised, due to paucity of time or budget, for example, there is a good chance that the installed HVAC system will be over-sized.

Overshadowing

Not considering the impact of surrounding buildings when doing the cooling load calculations can have a significant impact on peak demand. This is sometimes done because the client or engineer takes the view that buildings around the project may be demolished at some later date. Such a situation could lead to a significant increase in design cooling load on the project, for considering a scenario that might never eventuate.

Unknown tenants

Most buildings in Australia are speculative. Tenant requirements are unknown until late in the project, when a property agent may be successful in finding an anchor tenant. The Property Council of Australia's grading matrix recommendation provides minimum internal load capability, and encourages larger, rather than smaller, internal load assumptions for design calculation.

Contractual obligations

A correctly designed system will not maintain temperatures in the worst hours of the year, when conditions go beyond the design day and other internal loads are at high levels. Engineers are conscious that building use changes frequently, and design their HVAC systems to be able to cope with this by over-sizing. Design fees for engineers are based on competitive tender, and do not generally allow for iterative or integrated design solutions. Ultimately, they feel their reputations would suffer should a building HVAC system fail to maintain temperatures even on days when the ambient temperature increases beyond the design conditions.

There is also the 'split incentive' impact, identified as one of three major reasons for the government to propose a mandatory regulation of energy efficiency via a new section of the Building Code of Australia. This split incentive exists because the developer does not generally reap the economic benefits of an energy-efficient, lower greenhouse impact design, since the developer does not normally own and operate the building. Given two alternative design solutions, the developer will pick the least-cost solution.

'Design and construct' contracts generally discourage right-sizing approaches. The contractor tenders on a rough design load that is to be confirmed before detailed design and construction. The competitive tender situation under which these jobs are won means there is little incentive to review and optimize design calculations.

This type of risk-averse, aggressive, commercial environment discourages right-sizing by using a 'worst case' approach as a convenient, no-hassle solution to these issues. Unfortunately, it is the environment, society and tenants that are penalized by such wasteful building practice.

Over-sizing in smaller buildings

Hourahan (2004) lists the reasons for over-sizing of unitary air conditioning systems, which are typically found in residential and small commercial buildings.

1 Peak load estimate was not made:

  - prior experience was used to 'guess' the load;
  - simple replacement of 'like for like', assuming the original installation was correct; this ignores whether building functions have changed or the building has been upgraded (lighting, insulation, etc.);
  - use of obsolete and inadequate 'rules of thumb'.

2 Incorrect observance of procedures:

  - mistakes in the load calculation;
  - overly conservative assumptions of building attributes;
  - use of safety factors;
  - designers compensate for air distribution problems by over-sizing;
  - not allowing for load diversity between multiple zones.

3 Safeguarding call-backs:

  - designers seek to minimize occupant complaints on days that exceed design conditions.

Impacts of over-sizing

Component over-sizing

Over-sizing of cooling plant can mean that at least some of the following items are over-specified due to the flow-on effect across the whole HVAC system:

- chillers and cooling towers;
- pumps, pipes and valves;
- plant rooms, risers and ductwork;
- fans, motors, cables, switchboards and sub-station;
- variable air volume (VAV) boxes, grilles and registers.

Reduction in the size of the main ducts could potentially reduce riser size requirements, with a possible increase in NLA. However, in the authors' experience, many buildings have insufficient riser areas, forcing the mechanical engineer to design with high air velocities (and high static pressure) in associated ductwork. The result is significantly increased fan motor sizing and energy use for every single hour of fan operation. Reducing riser sizing is false economy.

Cable sizes and switchboards would be affected in a similar way to pipes and ductwork. The potential for a commercially viable reduction in cable diameter specification for larger components is more likely, as compared to cable size reductions for on-floor electrical circuits.

Energy consumption

Over-sized plant will operate inefficiently on a number of levels, leading to increased energy consumption. Since the entire system is over-sized to handle even the peak design day conditions, almost all installed equipment and machinery will operate at low part-load factors for large percentages of their total operational time. Chiller logs provide proof that a significant fraction (20 per cent or more) of installed chiller capacity is never used. Figure 22.1 presents results from a building energy simulation model that show estimated run time at various part-load capacities for a single chiller plant room. The graphs display a right-sized chiller (Chiller 1) and an over-sized chiller (Chiller 2). The over-sized chiller runs most of the time at below 30 per cent of rated capacity.

Another symptom of over-sizing is intermittent plant operation. This occurs when the diversified loads, during off-peak conditions, are so small that the cooling plant is unable to turn down to low-load levels that allow the plant components to operate in a stable manner. This results in frequent stop-and-start operation, which could lead to shorter mean time between failures and higher maintenance costs. Over-sized fans on variable-speed drives (VSD) will turn down to the lowest speed possible and work as a virtual constant speed system. There is no control of flow in such a situation.

Figure 22.2 shows simulated results depicting annual operation of a VSD controlled, over-sized AHU (air handling unit) fan. The fan turns down to the lowest allowed part-load (30 per cent in this case) for approximately 60 per cent of operational hours. Air volume at this lowest speed setting may still be too high, leading to excessive re-heat energy or loss of temperature control.

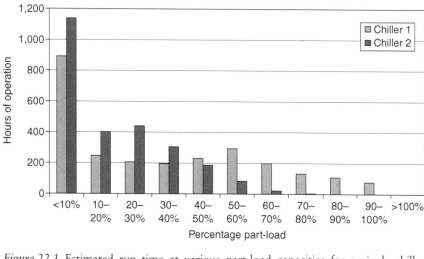

Figure 22.1 Estimated run time at various part-load capacities for a single chiller plant room design: Chiller 1 (right-sized chiller) and Chiller 2 (over-sized chiller).

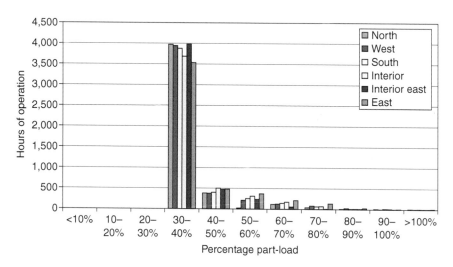

Figure 22.2 Simulated results depicting annual operation of a VSD controlled, over-sized AHU fan.

Younes and Carter (2006) presented a sensitivity analysis that showed that larger systems can result in air conditioning energy use up to 20 per cent higher than if the same system were right-sized. Significantly, the lower the mechanical efficiency, the more sensitive the plant energy use will be to the degree of over-design.

414 S. Moller and P.C. Thomas

Maximum electrical demand

Over-sized equipment can also impose a penalty in terms of peak demand charges for electricity, as the larger motors would draw higher currents when loaded. The impacts can flow through the energy supply chain with thicker cables, larger switchboards and sub-stations.

There will be occasions when a building with over-sized HVAC plant will run at full capacity – for example, when the plant starts after a hot weekend it may ramp up to full load while bringing the space temperature back into the comfort zone. These periods of non-steady state operation only need to happen once in a month to have a direct impact on maximum electrical demand charges.

Capital cost

Right-sizing would lead to energy-efficient operation, resulting in annual energy and energy cost saving. Additionally, savings in maintenance from lower equipment breakdown rates would result as the systems would operate for a greater number of hours nearer their design point.

Thermal comfort

Over-sized plant can result in increased discomfort in the air conditioned space. Space temperatures will be difficult to control when minimum flow rates result in significantly more cooling than is required. This would result in the space temperatures being rapidly driven below the thermostat setting. When the spaces become too cold, electric or gas energy would be used to re-heat the air in an effort to maintain comfort conditions. The result is a very significant impost on energy use, quite easily doubling it in such spaces.

Since flow rates will be at minimum flow as explained above, over-sizing of diffusers will lead to poor air distribution in the space for most operational hours. This could result in cold air dumping just below the diffusers, and stagnant areas within the air conditioned spaces.

Air conditioning systems for office building applications are generally designed for 'comfort cooling' and are not required to maintain 'critical' processes. No explicit humidity control component (dehumidifier, humidifier) is generally specified in these types of applications. In some system configurations, space temperature at low load is controlled by allowing the chilled water temperature at the cooling coil to increase. In over-sized systems this would happen most of the time, since the system will always be over-cooling. Higher chilled water temperature severely reduces the dehumidification potential of the cooling coil, and the relative humidity in the space increases when ambient humidity is high. Spaces served by such systems will feel 'clammy' or 'muggy' when climate conditions are warm (not hot) and humid.

Hourahan (2004) lists the effects of over-sizing of unitary air conditioning systems, which are typically found in residential and small commercial buildings:

- marginal part-load temperature control;
- large temperature differences between rooms;
- degraded humidity control;
- draughts and noise;
- occupant discomfort and dissatisfaction;
- larger ducts installed;
- increased electrical circuit sizing;
- excessive low-load operation;
- frequent cycling (loading/unloading);
- shorter equipment life;
- nuisance service calls;
- higher installed costs;
- increased operating expense;
- increased installed load on the public utility system;
- increased potential for mould growth;
- potential to contribute to asthma and other respiratory conditions.

Proctor *et al.* (1996) explain why air conditioners that cycle ON/OFF to control temperature are less efficient when they are over-sized. Air conditioners are very inefficient when they first start, only reaching peak efficiency after about ten minutes of operation. An over-sized air conditioner will cool the space quicker, and may often operate in this inefficient 'zone'. Proctor *et al.* (1996) estimate that if an air conditioner were double the required size, the energy consumption would increase by 10 per cent.

Case study

This case study examines the extent of over-sizing in a commercial office building, and the impacts of the resulting equipment size on capital and operating costs. The building and services were surveyed, staff interviewed and documentation obtained. Where possible, operating data were obtained to provide a baseline for computer modelling.

Building description

The building is a prestige office building in Melbourne with more than 50 levels of office space, a large basement car park, and retail tenancies at street level. The facade comprises full-height double glazing and insulated spandrel panels.

Air handling plant

The building is divided into four rises, each of which is served from central variable volume AHUs to meet internal loads, with skin system AHUs to meet external loads. AHUs utilize outside air economy operation to reduce cooling loads during mild weather. Fan-assisted variable volume boxes provide zone temperature control and ensure minimum outside air and air change rates. Humidifiers are used to maintain minimum relative humidity levels.

Refrigeration plant

A central chilled water system provides refrigeration to all the AHUs. Chillers are water-cooled reciprocating and centrifugal. The chilled water circulation system is primary/secondary, with two-way valves and variable speed secondary chilled water pumps. Condenser water to the system is supplied from induced draught cross-flow cooling towers.

Heating plant

Heating to the AHUs is supplied from high efficiency gas-fired boilers with variable speed heating water pumps.

Other air conditioning systems

The building has the following additional systems which have significant energy consumption:

- tenant condenser water systems, with closed circuit cooling towers to meet the needs of tenants with high electronic equipment loads;
- car park and loading dock exhaust systems;
- stand-by diesel generators;
- lift motor room air conditioning systems;
- toilet exhaust systems;
- tenant tearoom and kitchen exhaust systems.

Extent of over-sizing

A typical floor plate was modelled, using a whole-building energy simulation program. Estimates of peak demands with detailed glazing and geometry were entered into the model, and the results used to recalculate plant capacities. On-floor loads were also calculated and AHU sizes were recalculated.

Impacts

Table 22.1 gives an indication of the extent of over-sizing.

Table 22.1 The extent of over-sizing

Item	Extent of over-sizing
Chillers	34%
Ductwork	25%

Right-sizing would lead to energy-efficient operation, resulting in annual energy and energy cost saving. There would also be savings in maintenance from lower equipment breakdown rates, as the systems would operate for a greater number of hours nearer their design point.

Table 22.2 shows estimated cost savings attributable to right-sizing. The estimated savings on main plant are seen to be substantial: almost 29 per cent for cooling towers and 21 per cent for chillers. Savings on air handlers are estimated at 24 per cent. In this project, the predicted absolute savings are well in excess of $0.5 million.

The cost savings for the estimated energy savings are smaller than the fraction of energy savings itself, because most of the saving would occur at shoulder and off-peak rates.

The above costs were factored into a 15-year life cycle cost analysis, with zero escalation, and the results indicate significant savings, as shown in Table 22.3. The order of savings due to system components is more than double the capital cost saving estimate, at more than $1.3 million; however, this saving is the same order of magnitude as the energy saving. The total saving over 15 years is approximately $2.5 million for this case-study example.

Right-sizing recommendations

Challenge 'rules of thumb'

Challenge 'rule of thumb' load calculations and/or brief requirements which may be out of date or copied from a previous project specification. For example, obtain current equipment load data that matches the intended use. Younes and Carter (2006) recommend allowing $15\,W/m^2$ for small power heat loads, with an additional $10\,W/m^2$ designed into the risers and plant where specialist and intensive uses are anticipated. Use the 'return brief' process to challenge requirements that may be inappropriate, highlighting the implications for the design.

Accurate load estimation

Conduct an accurate load estimate, using established design data (e.g., from AIRAH or ASHRAE) specific to project location, and then resist the temptation to apply 'safety factors'. Do not use W/m^2 loading or other approximate methods for sizing equipment.

Table 22.2 Estimated capital and energy cost savings

System component	As-installed	Right-sized	Saving $	Saving within component, %
Cooling towers	$401,600	$285,600	$116,000	28.9
Chillers	$1,112,000	$879,000	$233,000	21.0
Condenser water pumps	$49,500	$41,900	$7,600	15.4
Chilled water pumps	$41,000	$33,400	$7,600	18.5
Air handling units	$1,164,900	$885,600	$279,300	24.0
Total equipment	$2,769,000	$2,125,500	$643,500	23.2
Energy – electricity	$1,889,546	$1,811,405	$78,141	4.1
Energy – gas	$50,647	$45,583	$5,065	10.0
Total energy	$1,940,193	$1,856,988	$83,206	4.3

Table 22.3 Estimated life cycle cost savings

System component	Saving
Cooling towers	$272,020
Chillers	$535,900
Condenser water pumps	$17,480
Chilled water pumps	$17,480
Air handling units	$505,334
Total equipment	$1,348,214
Energy – electricity	$1,172,114
energy – gas	$75,971
Total energy	$1,248,085
Total savings	$2,596,299

Dynamic calculation methods

Use of computer-based load estimation programs that account for thermal storage and diversification of peak loads for each zone and air handling system should be encouraged. Static methods cannot properly account for the daily diversity (e.g., between east and west zone cooling loads), and the final block load estimate will be over-sized. The new methods allow for the impact of innovative shading schemes that are difficult to quantify using static methods. The solutions generated by such calculations can be significantly smaller than numbers considered acceptable in traditional practice.

Systems approach

Use a systems approach, rather than a component-based approach, to designing the HVAC system. The designer should consider the overall system configuration, and explicitly design the low-load operation strategy. Select chiller sizes and staged- or variable-speed pumps and fans to ensure good part-load performance. Most buildings spend the bulk of their operating hours running at less than 50 per cent load. Individual chillers usually become less efficient once the load falls below 50 per cent. This problem is exacerbated if the chiller is over-sized. Pumping systems and cooling tower configurations should also be designed to operate efficiently at low loads.

Modern chilled water plants have many configuration and control options that are designed to improve low part-load operation. These include the use of chilled water and/or condenser water temperature resets, primary/secondary pumping loops with variable speed control, chillers that can handle variable flow or have in-built variable speed control, and variable speed control of cooling tower fans. An experienced engineer with the ability to critically use dynamic simulation analysis will be able to select and optimize an appropriate mix of equipment and control strategies to arrive at a cost-effective energy-efficient solution.

While heating systems have not been the focus of this study, the authors note that similar concerns exist in this area. Heating systems (particularly for office buildings that run mainly during the day) are generally sized on extremely conservative design assumptions. These may include unrealistic night-time design temperatures, with zero occupancy, equipment and lighting loads. Such processes result in selection of boilers running at extremely low part-load factors for the majority of operating hours. Most boilers, even with modulating burners, run extremely inefficiently below 25 to 30 per cent capacity.

Design for flexibility

Allow for unknown future tenancies by designing flexibility into the system, and not by over-sizing the entire system. For example, generous sizing of distribution pipework and main ductwork will allow available capacity to be redistributed in the future.

Separate high-load areas

A significantly greater amount of internal heat is generated in specialized areas such as IT server rooms, bank trading floors and call centres, as compared to general open plan office areas. Since these specialized areas are usually a small percentage of the total lettable area of a building, one solution is to add auxiliary stand-alone systems plumbed into a dedicated tenant condenser water loop. A systems approach needs to be applied in designing these supplementary HVAC systems, with the system able to efficiently handle a wide diversity in operating capacity.

Integrated design process

The authors strongly recommend the use of an integrated load and energy-use simulation analysis. Validated whole-building simulation programs provide the opportunity to model and test the proposed design in an integrated manner, and offer sophisticated insights not available by any other method. In the hands of an expert, these models can represent the building and its interactions in far greater detail than other methods, and allow description and testing of different operational scenarios.

Whole-building energy simulation offers an interactive approach to design where the review responsibility for all major energy sub-systems under a single entity (e.g., an ESD consultant with relevant expertise in facade, electric lighting and building services) would lead to right-sized outcomes in all these disciplines. In 2006, the Building Code of Australia introduced mandatory energy efficiency provisions for non-residential buildings (ABCB 2006). Assessment methods include 'deemed to satisfy' solutions and alternative verification methods using whole-building simulation. It has provided an opportunity to change design and procurement methods for buildings from a linear to an iterative process with right-sized HVAC outcomes.

Recognize the value of design

The case-study building provides a clear indication of the lost opportunity for capital cost saving and longer-term financial benefit. This could have been avoided or significantly reduced with application of an integrated, iterative design process. This necessarily means a longer design lead time, and higher design fees for architects and engineers, but the savings far outweigh these increases as a fraction of the total project price. The environmental benefits related to a reduction in demand for energy are also significant, and will inject increased benefit financially with the entry of carbon pricing.

Commissioning and maintenance

Current commissioning procedures (using either CIBSE or ASHRAE methods) concentrate on confirming system design parameters at simulated design conditions. However, energy-efficient operation depends on the ability of the system to operate in a stable manner at low-load conditions. Therefore, commissioning tests should calibrate and test the tracking of fan speeds (supply and return fans) at lower speeds, stable part-load operation of pumping systems, correct functioning of chiller scheduling algorithms, and correct implementation of chilled water and/or condenser water temperature reset strategy. All these functions, when implemented correctly, improve the energy efficiency of the HVAC system.

As systems drift from their calibrated value over time, regular maintenance should include calibration of controls and system parameters to ensure long-term part-load operation.

Life cycle cost analysis

Consider the use of a comprehensive life cycle cost (LCC) analysis for selection of the optimal design solutions. Use an integrated design process to test the impact of a change in one system across the whole building. For example, it is possible to justify the use of high-efficiency lighting systems by reviewing the drop in AHU coil and fan sizes, and potential reductions in frame sizes for chillers and cooling towers. A detailed simulation model will allow these interactions to be tested reasonably quickly, and the associated costing to be reviewed. A design process that uses LCC will lead to a change in the design and procurement process. As seen in the case-study example, there can be significant benefits by going through the process. However, the authors include a note of caution here: using a LCC analysis methodology that does not provide a truly integrated approach can potentially lead to inefficient outcomes for the project.

Conclusion

Right-sizing is an important component of cost-effective and energy-efficient air conditioning system design. This chapter has illustrated the impacts of over-sizing and recommended a series of measures to ensure that new HVAC in all types of buildings are correctly sized.

Bibliography

ABCB (2006) *BCA, Vol. 1: Class 5–9 Buildings*, Canberra: Australian Building Codes Board.

Crozier, B. (2000) *Enhancing the Performance of Oversized Plant*, Application Guide AG1/2000, Bracknell: Building Services Research and Information Association.

Deng, S. (2002) 'Sizing replacement chiller plants', *ASHRAE Journal*, 44 (6): 47–9.

Government Real Estate Groups (2000) *National Office Accommodation Benchmarking*.

Hourahan, G.C. (2004) 'How to properly size unitary equipment', *ASHRAE Journal*, 46 (2): 15–17.

Knight, I. and Dunn, G. (2004) 'Size does matter', *Building Services Journal*, 08/04: 34–6.

Komor, P. (1997) 'Space cooling demands from office plug loads', *ASHRAE Journal*, 39 (12): 41–4.

Proctor, J., Katsnelson, Z. and Wilson, B. (1996) 'Bigger is not better: sizing air conditioners properly', *Refrigeration Service and Contracting*, 64 (4).

Thomas, P.C. and Moller, S. (2007) *HVAC System Size: Getting It Right*, Brisbane: CRC for Construction Innovation.

Younes, A. and Carter, G. (2006) 'Internal heat load allowances: is more actually better?', *Ecolibrium*, Oct.: 26–34.

23 Evaluating the impact of sustainability on investment property performance

Terry Boyd

Sustainability is currently one of the most popular research fields for academics. It has become the catchword for most new concepts and it has strong emotive and ethical connotations. There is clear evidence that human behaviour has severely damaged the Earth's ecosystems and that, looking forward, there is a need to ensure that future generations are not compromised by our actions today.

The impact of sustainability on real property is a major concern at present, but it is unfortunate that sustainability has, in many cases, been trivialized by overemphasis and generalization. Real property takes its worth from its utility, and the impact of humans on property directly affects its sustainability. Understandably, there is growing awareness that property development and management should be evaluated against criteria that embody sustainability measures. This means that realistic sustainability evaluations of property assets will balance economic and social performance measures with environmental protection. This is the triple bottom line (TBL) evaluation approach.

The TBL approach has, historically, had little impact on the property marketplace, possibly because there has been no substantial change driver to broaden the existing financial emphasis in the assessment process. While incremental regulatory changes may force the marketplace to make some improvements, existing regulations have not caused investors and developers to change their fundamental approach of focusing on the financial bottom line.

The author believes it is the space occupier – the tenant or lessee – who will cause the shift to more healthy and productive buildings. This will not simply mean an improvement in the indoor air or lighting quality, but rather a much broader demand for comfortable, user-friendly and adaptable space. In other words, improvements in environmental factors are not enough; there must also be major improvement in the social and cultural aspects of the built environment. In Australia, the states (especially New South Wales and Victoria) are leading the way by specifying that new space occupied by their departments must have a high sustainability rating.

This is a strong encouragement for developers to develop 4- and 5-star (Green Star – GBCA) rated buildings.

While the trend towards accommodating more user-focused demand has started, it is difficult to evaluate its impact on the marketplace as there is very little historical evidence. Despite the difficulties, we must attempt to evaluate the emerging impact of these sustainability factors. These will influence the decision of space occupiers whether to lease or not to lease, and also to set the acceptable level of rental.

It raises the question of whether we can, realistically, quantify the impact of environmental and social characteristics on investment property. If so, what is the impact on investment property performance?

This chapter is focused on this challenge. Initially it considers the impact of sustainability on valuation methodology – the assessment of the TBL – and whether traditional valuation methodology can assess the TBL. This requires the identification of measurable indicators for environmental and social characteristics. Thereafter, the impact of the various indicators on the financial model is discussed with reference to current studies. Finally, a pilot case study is presented that is based on market data and a survey. The results of the study quantify the probable impact of enhanced environmental and social factors on the expected return from commercial buildings. Simulation is used to estimate the risk/return profiles of the different scenarios. The specific outputs from the cash-flow exercises should be treated with caution because of the limited available data and the ongoing need to refine and expand this research field.

Evaluating sustainability in property performance

Concepts of sustainable development have been highlighted over the past 25 years. Stern (1997) refers to the World Conservation Strategy put forward by the International Union for the Conservation of Nature in 1980. The impact of buildings on the environment has been researched over this period, with findings from the USA that non-residential buildings:

- consume 30 to 40 per cent of all the nation's energy;
- add 30 to 40 per cent to atmospheric emissions;
- use 60 per cent of all electricity;
- use 25 per cent of all water;
- take up 35 to 40 per cent of the municipal solid waste stream (von Paumgartten 2003: 26).

However, there is limited empirical research into the impact of environmental and social factors on the economic performance of property assets. The reasons for this will be further discussed below.

The impact and assessment of sustainability on building design and construction is a fertile area of research, and many interesting findings are being published in journals such as *Building Research and Information* (Routledge). In a paper on building environmental assessment methods, Cole (2005: 460) describes the evaluation of building environmental assessment methods over the past 15 years, but rightly comments: 'Given the uncertainties of climatic change and associated social, economic and political consequences, there will be no single or easy path to a sustainable future.'

Lui (2006) deals with the key conflicting elements in the formulation of sustainable development policies. She identifies the key elements as the economic, the socio-environmental, the socio-economic and the legal system, and discusses how these elements represent competing values and that there is a need to develop harmonious strategies to resolve the conflicts.

Many other disciplines allied to the property industry are also debating the impact of sustainability, with literature in the fields of planning, economics and facilities management: refer to Griffiths (2003), Stern (1997), von Paumgartten (2003), Bakens *et al.* (2005) and Bullen (2007). Most of the authors emphasize the complexity of the evaluation process and the need to consider the interests of the key stakeholders.

The question is often raised in the literature as to whether it is feasible, at present, to evaluate the impact of sustainability on investment property assets. Without doubt, corporations are including a focus on sustainability in their strategies and there is growing interest from the public in socially responsible entities. Cassidy (2004: 5) states that, by 2010, 'Ninety percent of Fortune 500 companies will adopt the so-called "triple bottom line reporting". The growth of interest in green building among major corporations is one expression of this phenomenon.'

Several property valuation specialists are examining the ability to evaluate TBL techniques for investment property. In the UK, the Royal Institution of Chartered Surveyors has recently prepared a report on *Surveying Sustainability – A Short Guide for the Property Professional* (RICS *et al.* 2007) and sponsored research work in the field by, *inter alia*, the Upstream Group (2003), Kingston University (Sayce and Ellison 2003) and Liverpool John Moores University (Boughey 2000). Sayce and Ellison use the traditional cash-flow approach to assess the market value of property when different sustainability criteria are incorporated.

More recently, Lutzkendorf and Lorenz (2005) examined the valuation approach for sustainable buildings. They consider that traditional valuation methods may not be suitable for valuing a building's performance taking account of environmental and social requirements: 'Relying on historical valuation methods will lead to an unbalanced approach for determining a property's exchange price or market value' (Lutzkendorf and Lorenz 2005: 228).

They continue that the 'worth' from the viewpoint of the user should be examined, and recommend advanced valuation methods such as 'artificial neural networks, hedonic pricing methods, special analysis, fuzzy logic, autoregressive integrated moving averaging and rough set theory'. They place emphasis on the hedonic pricing techniques, but note the difficulties of describing and assessing the different building features and the lack of large and appropriate data sets. They also consider that 'many valuers might not have the facilities, required skills and, probably, motivation to use those techniques. They are therefore likely relying on traditional valuation approaches in the foreseeable future' (Lutzkendorf and Lorenz 2005: 228–9).

The challenge by Lutzkendorf and Lorenz means that the appropriateness of the traditional investment approach for a TBL application should be critically examined. The traditional approach, for major investment properties, is the discounted cash-flow approach that identifies the expected cash flow from the property over a future period of time and then assesses the current value or the total return over the selected period. This approach can accommodate the different sustainability scenarios by varying the key inputs for differing environmental or social benchmark levels. However, life cycle costings and assessments are limited to the period of the cash-flow study, often five to 10 years. The advanced valuation methods described by Lutzkendorf and Lorenz may be capable of quantifying the impact of specific environmental or social factors, provided adequate market data can be identified and analysed.

The author believes that the traditional investment valuation method is capable of assessing the impact of environmental and social factors on the financial performance, and that the advanced methods, while capable of assessing aspects of worth, should not be seen as a replacement for the traditional approach. It is probable that advanced methods derived from environmental economics may supplement the traditional method, but they are unlikely to replace it.

The balance of this chapter will demonstrate the use of the traditional investment valuation method to assess the performance measures of investment property if the TBL approach is included or excluded. However, prior to discussing the valuation exercise it is necessary to identify the environmental and social factors and their performance indicators.

The environmental and social benchmarks

Valuable work identifying appropriate environmental indicators for built assets has already been undertaken internationally. Most developed countries have some form of Green Building Council that has identified local environmental benchmarks for buildings. For example, the Green Building Council of Australia (GBCA) has developed a Green Star rating system, and there are other similar assessment models. The USA has, *inter alia*, the Leadership in Energy and Environmental Design (LEED) Green Building Rating System, and the UK has the Building Research Establishment

Environmental Assessment Method (BREEAM). The main tools used in the UK and Europe are listed in RICS *et al.* (2007).

However, many of the building measures deal with design and new construction, and lesser consideration has been given to existing building stock as operating entities. Cole (2005: 456) describes building rating measures from countries such as Japan, Hong Kong and South Africa, and comments that: 'There is little doubt that building environmental assessment methods have contributed enormously to furthering the promotion of higher environmental expectations and are directly and indirectly influencing the performance of buildings.'

As this study is undertaken in Australia, it is logical to examine the Green Star rating of GBCA and the emerging combined environmental rating measures being researched by the Australian CRC in Construction Innovation. The Green Star rating system is described by the GBCA as:

> Australia's first comprehensive rating system for evaluating the environmental design and performance of Australian buildings based on a number of criteria, including energy and water efficiency, indoor environmental quality and resource conservation.
>
> The Green Star rating system has eight environment impact categories which are: Energy, emissions, transport, materials, water, land use and ecology, indoor environment quality, and management.

The GBCA's office rating tool formed the basis for the environmental factors selected in the CRC for Construction Innovation's research project (2001–11-C) on 'The Evaluation of the Functional Performance of Commercial Buildings' (2004), which is earlier research undertaken by the author and industry partners. The project also considered the work undertaken by Environment Australia (2003) and by the Upstream Group (2003), Sayce and Ellison (2003) and others in the UK.

While environmental benchmarking is well advanced, the benchmarks for social factors are not yet established. Many existing benchmarks include some social characteristics within the environmental benchmarks, such as Sayce and Ellison (2003) and Lutzkendorf and Lorenz (2005). The author considers that it is necessary to have a distinct set of benchmarks for social factors in order to examine the meaningful TBL assessment of built assets.

The environmental and social benchmarks developed and tested in the CRC Construction Innovation Project (2004) are used in this study, and are set out in Tables 23.1 and 23.2.

It is accepted that the specification of environmental and social indicators can take many forms. These tables represent one attempt to identify the major characteristics of an operational nature with particular reference to the utility of the building. The selection of benchmark indicators should be evaluated against the market's perception of value of the individual

Table 23.1 Recommended environmental benchmarks: existing buildings

Field/topics	Measures
Resource consumption	
Energy	• Net fossil fuel energy use (assessed on an intra-building and market comparison basis) • Effective action to reduce greenhouse gas emissions (particularly from energy use) • Office lighting power density and peak energy demand reduction strategies • Evidence of alternative energy supplies from renewable sources or from co-generation
Air conditioning	• Condition of air-conditioning plant • Use of ODP or GWP refrigerants
Water	• Water consumption (potable, hygiene and cooling towers), • Recycling and water capture measures • Wastewater reduction • Hazardous and non-hazardous waste and effluents recycling or removal strategies
Design and use	
Transport	• Public transport availability and standard of service • Strategies to discourage single-occupancy vehicle journeys, including cyclist facilities
Building fabric	• Age of building (obsolescence or depreciation of materials) • Reuse or upgrade history or potential • Suitability of original materials for refurbishment and facade retention • Ecological impacts of materials used
Interior	• Indoor quality measured by ventilation, natural lighting, individual thermal control, noise abatement • Absence of indoor air pollutants
Environment	• Quality of overall built environment and site use in relation to aesthetics, visual blending and connection contribution of its street frontage and wider precinct
Governance	
Awareness	• Maximization by management of the potential of the environmental design features through awareness programs
Disclosure	• Disclosure and transparency of environmental data, regulation compliance, awards and environmental expenditure of any type

Table 23.2 Proposed social benchmarks: existing buildings

Topic	Measures
Health and Safety	• Compliance with H & S regulations and appropriate signage • Adequate public liability and service provider insurance • Awareness and training of emergency evacuation and accident first aid procedures for all floor wardens • First aid station accessible to all building users
Stakeholder relations	• Monitoring of stakeholder concerns, views and provisions • Transparency and disclosure of landlord/tenant contracts and marketing agreements • Supportive use and occupation guidelines for tenants • Appropriate training for security and public relations personnel
Community engagement	• Encouragement of employment of local residents within the building • Provision of accessible public facilities • Promotion of and linkage to local service providers • Accessible communication channels with building stakeholders
Accessibility	• Connections to designated green spaces • Proximity to urban spaces (town centres, malls, etc.) • Wheelchair access • Proximity to childminding facilities
Occupier satisfaction and productivity	• Quality of communal service areas • Complementary usage of building (compatible tenants) • Occupant productivity in terms of satisfaction and physical well-being
Cultural issues	• Recognition of indigenous people through cultural space and communication of site history • Consideration of gender equity and minority group requirements • Preservation of heritage values • Value of artwork as percentage of the fit-out
Local impacts	• Aesthetic implications (compliance with precinct theme, building scale, etc.) • Practical implications (traffic generation, off-street emergency parking and pedestrian management) • Nature of tenant businesses and naming rights • Community linkages and sponsorship of local neighbourhood activities

measures. Once the appropriate indicators and their component character-istics have been selected, the next challenge is to determine a grading or weighting for the indicators.

The environmental benchmarks, described in Table 23.1, were used in the case study (described later), but the author considers that future research should utilize one of the acceptable industry rating systems, such as the GBCA's Green Star rating system for 'Offices as built V2 tool', <www.gbca.org.au/green-star/rating-tools/green-star-office-as-built-v2/ 1533.htm>.

When considering the social indicators, as shown in Table 23.2, the ranking of the measurement categories was examined by surveying several building managers and users. Each measure was assessed using a five-point Likert scale, and the scores were summed to identify possible ranking and weighting. However, the lack of uniformity meant that no consistent pattern was achieved and consequently, at this developmental stage, it was decided that the social measures could, at best, only be applied on a broad scale, with buildings placed in three categories:

- not socially responsible;
- acceptable socially responsible standard – for a private corporation or individual;
- acceptable socially responsible standard – for a public body.

However, when testing the TBL effect in the case study, it was decided to limit the grading of each social measure to a two-way variable, being either 'enhanced' or 'not enhanced'. This simple alternative was used because of the difficulty of proving the impact of varying levels of social enhancement for each individual measure.

Before considering this case study, the international research in this field is described.

Triple bottom line evaluations: current literature

The impact of sustainability measures on the return from investment prop-erty is, clearly, an important issue for investors. They need to know whether the application of advancements in environmental and/or social factors will result in improved returns from the property assets. In short, the current literature is inconclusive.

Brenchley (quoted in Ryder 2004) predicts that, with an increasing number of ethical investment funds emerging, it is inevitable that investors will begin to look more seriously at property over the next five to ten years, and that this increased demand for environmentally and socially sus-tainable buildings is likely to result in premium values. Even today, a good energy rating on a building may give it a market edge. There is some evid-ence that, for public sector tenants at least, a fall in the rating during tenancy can actually trigger a diminution in rent. An example is a New South Wales Police Services lease with Multiplex that states that the rent will be reduced if the building's 4.5-star rating falls (Dorfling 2004). This suggests that a premium rent may be achieved based on an expectation of lower occupancy costs or a better working environment.

Research into the impact of TBL considerations has been undertaken in both the USA and the UK. A report to California's Sustainable Building Task Force (Kats 2003) quantifies the costs and financial benefits of green buildings. It places a strong focus on productivity and health, and quotes from the BOMA/ULI (1999) Office Tenant Survey Report that respondents attributed the highest importance to tenant comfort features, including comfortable air temperature and indoor air quality.

Kats (2003) found that the financial benefits of building green include savings from reduced energy, water and waste; lower operation and maintenance costs; and enhanced occupant productivity and health. The findings are summarized in Table 23.3.

While this study refers to a 10-fold benefit on the initial estimate of the cost premium, the figures indicate that the majority of the benefits will come from 'productivity and health value', being 70 per cent of the total benefit for certified and silver class, and 87 per cent for gold and platinum class. The report refers to the difficulty of assessing the productivity and health value components within this exercise, and this means that the results should be viewed with caution. The author considers that the most appropriate way to assess the productivity and health value is to find rental evidence from the marketplace, and that the large productivity and health benefits require further examination.

Rawstron and Francis (2007) surveyed UK institutions and investors on attitudes towards sustainable property investment, asking them to rate the importance of eight sustainability factors. The weighting of these key attributes is shown in Table 23.4.

Table 23.3 Financial benefits of green buildings: summary of findings

Category	20-year NPV
Energy value	$5.79/sq ft
Emissions value	$1.18/sq ft
Water value	$0.51/sq ft
Waste value (construction only) – 1 year	$0.03/sq ft
Commissioning O & M value	$8.47/sq ft
Productivity and health value (certified and silver)	$36.89/sq ft
Productivity and health value (gold and platinum)	$55.33/sq ft
Less Green Cost Premium	($4.00)/sq ft
Total 20-year NPV (certified and silver)	*$48.87/sq ft*
Total 20-year NPV (gold and platinum)	*$67.31/sq ft*

Source: Kats (2003: 84).

Table 23.4 Key attributes for a sustainable property

Attribute	Weighting (out of 5)
High energy efficiency	4.0
Low levels of pollution	3.7
Good access to public transport	3.6
Efficient monitoring/management of the building systems	3.6
Use of sustainable materials in construction	3.5
Working environment promotes staff health and well-being	3.3
Construction brownfield/eco-friendly site	3.2
High water efficiency	3.1

Source: Rawstron and Francis (2007: 10, 11).

They conclude by stating:

> We believe increasing emphasis on sustainability will be a long-term trend, driven by a combination of increased legislation and regulation, a greater demand by occupiers, increased market knowledge with benchmarking of sustainability criteria, and more emphasis on Corporate Social Responsibility issues.
>
> (Rawstron and Francis 2007: 12)

Another recent study looks at the impact of sustainability standards when assessing the market value of property assets. This research by Sayce *et al.* (2004) is a continuation of their Sustainable Property Appraisal Project, and describes the findings of their pilot studies on four different properties. They adjust four key variables (rental growth, depreciation, risk premium and cash flow) for various sustainability criteria. The sustainable criteria weighting in this exercise is shown in Table 23.5.

They assess the value of each property using standard valuation factors and thereafter incorporate the weighted sustainability criteria in addition to the standard valuation factors. Their finding is that the incorporation of the weighted sustainability criteria reduces the value of any property that fails to meet sustainability criteria. They conclude that 'the new net present values produced in these examples suggest the standard appraisal process is over-valuing, if sustainability is taken into account' (Sayce *et al.* 2004: 233).

The results of the study by Sayce and colleagues are interesting, and progress the research in this field. The author's concern with the alternative valuation approach (that adjusts for sustainability criteria) is that the rent currently being paid relates to the existing level of sustainability of the building. If the building had a higher level of sustainability, the rental level may be higher. Therefore, it could arguably be suggested that the 'less sustainable' building is correctly valued by the market and that a 'more sustainable' building would have a higher value.

Table 23.5 Weighting of sustainable criteria

Sustainable criteria	Weighting
Assessibility	1.0
Building quality	0.9
Adaptability	0.8
Occupier satisfaction	0.7
Pollutants	0.6
Contextual fit	0.5
Energy efficiency	0.4
Water and waste	0.3
Occupier impact	0.2

Source: Sayce *et al.* (2004: 227).

Myers *et al.* (2007: 7) also examined the recent literature and concluded that:

> Current research suggests that there is still an increased cost for sustainable features in a building and possibly more longer-term maintenance and management costs which further reduce the potential profit or return on investment for stakeholders, making it inherently difficult to convince stakeholders of the economic benefits of sustainable buildings.

The current literature on the impact of sustainability on property values is inconclusive, and indicates the exploratory nature of this research. This chapter will now demonstrate a process to evaluate the impact of sustainability measures.

Evaluating the impact of environmental and social measures

There is little doubt that environmental and social measures will have an impact on investment property performance. An illustration of the probable impact of enhanced environmental characteristics on investment-type buildings is shown in Figure 23.1.

Figure 23.1 indicates that there are four expected results from greater environmental efficiency, and that three of these should have a positive effect on the capital value of the building. However, the degree and timing of the impact is complicated, and will differ according to the type of environmental improvement. It is too simplistic to conclude that the change will always, or even frequently, have a positive impact on the capital value. What is important is that the impact of enhanced environmental factors on space users is explicitly examined.

The author, probably because of the limited number of these buildings, has found limited literature on post-occupancy evaluations on enhanced

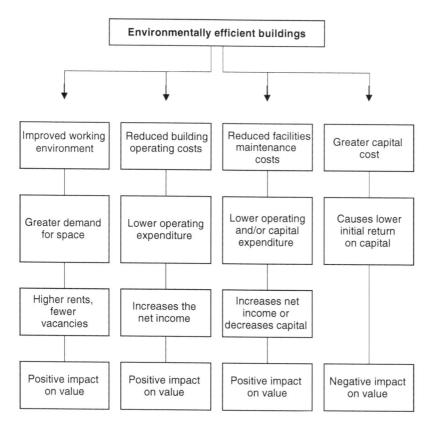

Figure 23.1 Value impact of environmentally efficient buildings.

buildings. A good commentary on the difficulties of occupier comfort indices is given in Humphreys (2005).

While acknowledging that it is difficult to find market-based evidence of the impact of environmental and social factors on the return from an investment property, the original CRC-CI Project (2004) examined market data that were available in Queensland. Fortunately, this project had the assistance of the Queensland Department of Public Works, Rider Hunt and Arup in analysing this information.

After completing the CRC-CI project, the author had further discussions with a group of property managers, valuers and office tenants to gauge their willingness to pay for environmental and social improvements to buildings. This pilot study followed a contingent valuation approach, and the results are used in the case study below. The author also discussed the feasibility and cost of environmental and social criteria with building owners in order to estimate the impact of these factors on the performance of commercial buildings. It is acknowledged that there are very limited market data on the performance of environmentally and/or socially enhanced buildings to support the opinions of the property experts. Consequently, the testing of the impact of the indicators has been restricted to a pilot study. This examines whether a cash-flow valuation model, as used in the market to assess the financial performance of investment property, is capable of determining the TBL return. In addition, the study will assess the risk parameters of the different scenarios based on probability profiles of the key variables.

Case study

The case-study building is an existing prime office CBD building in Brisbane, Australia, which was worth approximately $112 million in July 2005. The building was ten years old at the time, and was expected to show a total return of approximately 9.5 per cent per annum over the next seven years.

The first step of the exercise was to identify the key input variables that would be affected by environmental or social enhancement of the building. Five key variables were selected and identified: the construction cost, the initial (current) rent level, the rental growth rate, the operating expenses and the capital expenses. Thereafter the author undertook a contingent valuation approach, with an industry focus group of six property professionals (managers, valuers and lessees), in order to assess the impact of enhanced environmental and social features on this office building. In this study, enhanced features were classified as equivalent to a 4-star Green Star rating (or above) for environmental features and an acceptable or not acceptable (average) 'socially responsible standard (private owner)' for the social features described in Table 23.2.

Because of the hypothetical nature of the questions on the probable impact of enhanced environmental and social features, there was a high degree of subjectivity in the responses. It was necessary to average the range of responses from the professionals, and the results are shown in Table 23.6.

The ranges of the variable changes shown in Table 23.6 refer to the percentage change of that particular variable within a cash-flow study of the case-study property under the different scenarios. It should be noted that these variables relate to the specific property used in the case study, and could change according to the type and characteristics of the building.

Having undertaken the difficult part of the study, the inputs were applied to cash-flow studies to assess the total return from the case-study property under four different scenarios:

1   The existing building condition without environmental or social enhancement.
2   The building with enhanced environmental (but not socially enhanced) features.
3   The building with enhanced social (but not environmentally enhanced) features.

Table 23.6 Breakdown of the impact of specific variables in cash-flow study (percentage variable change)

Input variable	Enhanced environmental features	Enhanced social features	Combined: environmental and social features
Construction cost	+5% to +7%	+3% to +5%	+8% to +11%
Initial rent (first year)	+3% to +5%	0% to +3%	+3% to +8%
Rental growth rate	+4% to +10%	+1% to +3%	+5% to +13%
Operating expenses	−4% to −10%	+1% to +3%	−3% to −7%
Capital expenses	−2% to −10%	0% to +1%	−2% to −9%

4   The building with both enhanced environmental and social features, the TBL approach.

A simulation exercise, using @Risk, was undertaken for each scenario, with probability profiles based on the opinions of the focus group members. The initial cash-flow study of the case-study property (in its existing condition), over a seven-year period using monthly intervals, showed a total annual return from the property of approximately 9.5 per cent. The results from the alternative exercises, incorporating the risk simulation results, are shown in Table 23.7.

Table 23.7 shows the change in expected return and ranges of the returns under the different scenarios. It will be seen that the case study, with enhanced environmental features, is likely to have an increased return, from 9.53 to 9.70 per cent p.a., and that the possible range of returns will be between 9.27 and 10.21 per cent p.a. The other two scenarios, the enhanced social features and the combined TBL features, show the return to be lower than the initial financial return, but there is not a significant difference between them. The case-study exercise demonstrates that the mean return with enhanced environmental and social features (TBL) is 9.32 per cent p.a., and that the possible range of returns is 8.90 to 9.69 per cent p.a. The variability of each exercise is shown by the standard deviation measure, and the greatest variation is in the environmental enhancement scenario because of the larger range of inputs, particularly rental growth and capital expenditure rates.

The measurement of performance change due to environmental and social enhancement demonstrates that there are several balancing factors in each exercise (more cost and greater net income), and that the resultant returns show only minor differences. It is possible that future studies may show greater input variable differences, and consequently greater differences between the results of the four scenarios.

The author's reaction to the case-study findings is positive in that they support the proposition that sustainability features may not have a major negative impact on property performance. There is also the inference that

Table 23.7 Changes in building performance (IRR) for enhanced environmental and social features

Brisbane case study office building	Resultant IRR	Standard deviation	Range minimum	Range maximum
In existing condition and net income	9.53%	–	–	–
With enhanced environmental features	9.70%	0.15%	9.27%	10.21%
With enhanced social features	9.26%	0.12%	8.87%	9.65%
With both enhanced environmental and social features	9.32%	0.13%	8.90%	9.69%

the impact is likely to be positive in the longer term. Whilst not supporting some international findings that suggest strong benefits from enhanced sustainability features, this study uses a conservative approach, as it is based solely on the measurable impact within a financial study over a seven-year period.

The case-study exercise has demonstrated that the existing cash-flow investment valuation model can be used to assess the TBL value/return. The basic cash-flow model is not the main concern with the TBL approach; it is the specification of the inputs (for the specific scenarios) that is difficult to determine from market evidence, and more research is required in this field.

Conclusions and recommendations

The objective of this chapter was to examine whether the impact of sustainability features on investment property can be quantified and, if so, what would be the likely effect. There is currently limited international literature that demonstrates the actual impact of sustainability on property asset performance. In addition, there is some concern in the literature that traditional valuation methods are inadequate to undertake this exercise.

This study concludes that the environmental and social indicators of sustainability can be evaluated using investment cash-flow studies and the appropriateness of this methodology is demonstrated in a practical case study. However, there are two difficult steps in the data collection and analysis process that require careful consideration:

- the identification and quantification of the key performance indicators (KPIs) for the environmental and social characteristics;
- the measurement of the impact of these environmental and social KPIs on the input variables of the investment cash-flow study.

The first step is receiving attention internationally, with the environmental standards being reasonably well established, but more work is required on the social standards. The second step is a field that requires further research. The environmental and social features will impact on building costs, operating and capital expenses (including depreciation) and rental income. The impact on building cost is relatively easy to assess, and the future operating and capital expenses can be estimated through life cycle costing exercises. However, the future rental income is the most difficult variable to assess because of the lack of market evidence.

Further research is required into the willingness of tenants to pay higher rents for space that provides improved health and productivity conditions. Both willingness-to-pay and post-occupancy evaluation surveys are needed because there is, as yet, very little evidence of market rent differentials. With time this situation will change, and we are likely to see the benefits of sustainable features translated into increased rents.

Having determined the inputs for the cash-flow performance measurement model, it is not difficult to run the model to assess either a change in value or rate of return. However, a further risk assessment step should be incorporated to profile the uncertainty in the exercise.

The results of the case-study exercise demonstrate that the TBL assessment is achievable and that, based on current market evidence, the difference between the returns achievable on an existing prime-grade office building, and a similar building which is environmentally and socially enhanced, is minimal. However, indicators of future demand suggest that sustainability enhancement will, in the future, provide a better return from investment property.

Bibliography

Bakens, W., Foliente, C. and Jasuja, M. (2005) 'Engaging stakeholders in performance-based buildings: lessons from the Performance Based Building Network', *Building Research and Information*, 33 (2): 149–58.

Boughey, J. (2000) 'Environment valuation, real property and sustainability', *Proceedings of Cutting Edge 2000*, London: RICS Research Foundation.

Boyd, T.P. (2003) 'Model consistency and data specification in property DCF studies', *Australian Property Journal*, 37: 553–9.

Bullen, P.A. (2007) 'Adaptive reuse and sustainability of commercial buildings', *Facilities*, 25 (1/2): 20–8.

Cassidy, R. (2004) 'Corporate real estate circa 2010', *Building Design and Construction*, 45 (6): 5.

Cole, R.J. (2005) 'Building environmental assessment methods: redefining intentions and roles', *Building Research and Information*, 35 (5): 455–67.

CRC Construction Innovation Project 2001–11-C (2004) *The Evaluation of the Functional Performance of Commercial Buildings*, Brisbane: CRC for Construction Innovation.

Dorfling, M. (2004) 'Buildings put to the greenhouse test', *Australian*, 6 May: 40.

Environment Australia (2003) *Triple Bottom Line Reporting in Australia: A Guide to Reporting Against Environmental Indicators*, Canberra: Department of the Environment and Heritage.

Green Building Council of Australia, *Green Star Rating Tools*, Sydney: Green Building Council of Australia. Online. Available at HTTP: <www.gbca.org.au/green-star/rating-tools/green-star-rating-tools/953.htm> (accessed 16 July 2006).

Griffiths, J. (2003) 'Sustainability appraisals – should be an intrinsic part of plan-making', *Planning*, 1540: 28.

Humphreys, M.A. (2005) 'Quantifying occupant comfort: are combined indices of the indoor environment practicable?', *Building Research and Information*, 33 (4): 317–25.

Kats, G. (2003) *The Costs and Financial Benefits of Green Buildings: A Report to California's Sustainable Building Task Force*, Washington, DC: Capital E.

Lui, A.M.M. (2006) 'The framework underpinning conflicting keys in sustainability: harmony-in-transit', *Property Management*, 24 (3): 219–28.

Lutzkendorf, T. and Lorenz, D. (2005) 'Sustainable building investment: valuing sustainable buildings through performance assessment', *Building Research and Information*, 33 (3): 212–34.

Myers, G., Reed, R. and Robinson, J. (2007) 'The relationship between sustainability and the value of office buildings', paper presented at the Pacific Rim Real Estate Conference, Perth, 21–24 January.

Rawstron, M. and Francis, D. (2007) 'Green Investments', *RICS Commercial Property Journal*, Nov./Dec.: 10–12.

RICS, Gleeds and Forum for the Future (2007) *Surveying Sustainability: A Short Guide for the Property Professional*, London: Royal Institution of Chartered Surveyors.

Ryder, T. (2004) 'Facing up to the future', *Property Australia*, 18: 4.

Sayce, S. and Ellison, L. (2003) 'Integrating sustainability into the appraisal of property worth: identifying appropriate indicators of sustainability', paper presented at the American Real Estate and Urban Economics Association Conference, Skye, Scotland, 21 August.

Sayce, S., Ellison, L. and Smith, J. (2004) 'Incorporating sustainability in commercial property appraisal: evidence from the UK', *Australian Property Journal*, 38 (3): 226–33.

Stern, D.I. (1997) 'The capital theory approach to sustainability: a critical appraisal', *Journal of Economic Issues*, 31 (1): 145–74.

Upstream Group (2003) *Sustainability and the Built Environment: An Agenda for Action*, London: RICS Foundation.

von Paumgartten, P. (2003) 'The business case for high-performance green buildings: sustainability and its financial impact', *Journal of Facilities Management*, 2 (1): 26–32.

24 Estimating residual service life of commercial buildings

Sujeeva Setunge and Arun Kumar

Commercial buildings are an extremely important part of the built infrastructure assets of a region. Buildings deteriorate with age and more importantly may not serve the purpose of original intent. Developing cost effective management strategies for the long term, ensuring sustainable use of physical resources and providing a defined level of service require an understanding of the residual service life of the building and/or its components. Residual life may serve as a benchmark for planning refurbishment strategies. In an advanced infrastructure management model, residual life of components and elements of the building would be estimated based on data collected over a period of time. In a basic or core infrastructure management model, a facility for at least an overall estimation of the remaining life of the total building stock would be required.

This chapter covers a review of generic tools available for deterioration modelling of infrastructure, identifies methods previously adopted for management of commercial building assets and provides details on specific methods developed for estimation of residual life of buildings. An example of adaptation and application of these specific methods to a seven-storey case-study building in Melbourne as part of a Re-Life project funded by the CRC for Construction Innovation is presented. Suggestions for an improved method for estimating the residual life of buildings are also discussed.

Residual service life and deterioration modelling

ISO 15686 (2000) defines 'service life' as the period of time after installation during which a building or its parts meets or exceeds the 'performance requirements'. Residual service life (RSL) has been defined as the remaining service life at a certain point of consideration. RSL prediction of a building is an estimation of the remaining period of time during which a building or its parts meet or exceed the performance requirements at any given time. In simple terms, in order to estimate the RSL of a building, knowledge of its existing condition, its past/future deterioration trend and minimum acceptable performance level for each of the components would be required. Identifying performance requirements therefore forms an important part of estimation of RSL. Performance can be divided into two areas: physical condition of building assets, and functionality provided by the assets. Therefore service life may be defined as the lesser of the two limits: (a) based on end of physical life, (b) based on end of functional performance.

Predicting the RSL of buildings requires selection of a suitable deterioration modelling technique. Techniques used for deterioration modelling can be broadly categorized as follows.

*   *Deterministic models.* Deterministic models are often useful when the relationships between different components are certain. These are the most basic form of deterioration models. Often, problems associated with the influence of various external parameters on the relationship is taken into account by grouping systems with similar attributes such as size, material and service type (Madanat *et al.* 1995).
*   *Statistical models.* Statistical models are useful when the random variation in input parameters can affect the behaviour of an infrastructure system significantly. Some common methods include Markov process (Madanat 1993), ordinal regression models (Johnson and Albert 1999) and linear discriminant models (Huberty 1994).
*   *Artificial intelligence based models.* These are models where the model structure is completely determined by data. No assumptions are made regarding the model structure. These are designed to mimic the human brain which is learning all the time. They have the advantage of being able to improve when the data collection matures. Some examples are fuzzy logic approach (Setunge *et al.* 2007) and neural networks (Hassoun 1995).

Specific models reported in literature for building deterioration fall into the first two categories. The application of any of these models requires a well-developed data collection regime. In the absence of a large quantity of data, models of the third category become almost impossible to implement.

Assessment of condition of building assets

In Europe some of the decision-making software tools developed include TOBUS (Flourentzou *et al.* 2002), MEDIC (Flourentzou *et al.* 2000a), EPIQR (Flourentzou *et al.* 2000b) and INVESTIMMO (2001). EPIQR (for apartment buildings) and TOBUS (for office buildings) have been developed for the assessment of retrofitting needs of buildings. Their use can facilitate a quick and accurate diagnosis of the condition of an existing building in terms of its major areas, including construction, energy performance, indoor environmental quality and functional obsolescence. The main advantages of using these tools are the ability to evaluate various refurbishments and retrofit scenarios, and to estimate cost of induced works, in the preliminary stages of a project (Rey 2004; Chan *et al.* 2001).

In EPIQR and TOBUS, deterioration of building materials and components is described by the use of a classification system with four classes. The prediction of the 'period of passing' into the next deterioration state is of interest, as this is directly connected to higher refurbishment costs. The prediction of qualitative deterioration states is important and corresponds to key moments in the element's life where some refurbishment action has to be taken (Rey 2004; Chan *et al.* 2001).

European countries have used MEDIC to predict the future condition of a building or an element as one of four or five condition ratings allocated. It is intended for use with EPIQR, and is based on subdividing a building into 50 elements. MEDIC calculates the remaining lifespan of an element, not as a deterministic unique value but as a probability distribution. It can assist the building owner in deciding the most judicious moment to undertake refurbishment to achieve short- and long-term financial objectives (Steelcase 2005; Alkhrdaji and Thomas 2002). Based on EPIQR and TOBUS, a decision-making tool for long-term efficient investment strategies in housing maintenance and refurbishment – INVESTIMMO – has been developed. Its primary objective is to evaluate housing maintenance and refurbishment options that cover expectations of tenants, housing market, quality of building upgrading and environmental impacts in addition to the factors identified in TOBUS (Chan *et al.* 2001). All these methods are based on a condition monitoring method which ranks elements of buildings into discrete stages.

The widely used approach of condition assessment is to categorize building elements into four condition states:

1   Condition A – Element in 'good' condition. Element is as new, fully functional.
2   Condition B – Element in 'fair' condition. Element is not as new; however, there are no major signs of distress or defects, it remains functional with or without minor problems.
3   Condition C – Element in 'serious' condition. Element is not new, has major problems in the beginning to advanced stage, functional use is restricted or needs significant support.
4   Condition D – Element in 'needs replacement' condition.

The above classification has been used in most infrastructure rating schemes in many countries. A disadvantage is the subjectiveness of the descriptions. They do not consider the consequence of an element being rated in a particular condition. For example, a phone or a computer in a cubicle may be in 'needs replacement' condition, but the failure to replace will not have any dire consequence. However, this classification is easily applicable to any condition assessment procedure. An improvement to this methodology was proposed by Sharabah *et al.* (2008) which presents a consequence rating to be used in conjunction with the condition rating at the decision-making stage. A typical consequence grading method can be depicted as follows:

- Rating 1: No disruption to the business; no harm to humans and/or no harm to the environment.
- Rating 2: Some disruption to the business; minor injuries to humans and/or minimal damage to the environment.
- Rating 3: Moderate disruption to the business; recoverable injuries to humans and/or significant damage to the environment.
- Rating 4: Business disruption; permanent disability or fatalities; huge (and possibly irreversible) impact on the environment.

A basic building hierarchy that can be used for data collection is shown in Figure 24.1.

Specific methods covering estimation of residual service life

ISO factorial (ISO 2001)

The most widely accepted and also the most widely criticized RSL method is the ISO factorial method. This is a deterministic method which can be customized for different scenarios using a number of factors. It is based on the formula below:

$$ESL = RSL \times f_A \times f_B \times f_C \times f_D \times f_E \times f_F \times f_G \qquad \text{(Equation 24.1)}$$

where

ESL = Estimated service life
RSL = Reference service life (this shall be denoted as RFSL for clarity)
f_A = Quality of component
f_B = Design level
f_C = Work execution
f_D = Indoor environment
f_E = Outdoor environment
f_F = In-use conditions
f_G = Maintenance.

It is expected that any one (or combination) of these factors can affect service life. Therefore, suitable factors can be assumed (or derived) to estimate the ESL. Hovde and Moser (2004) have shown that the ISO methods can be used to incorporate a probability distribution for these factors and thus specify a distribution for ESL rather than a deterministic

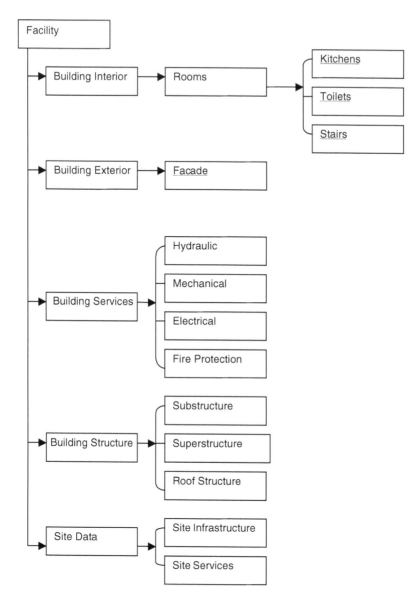

Figure 24.1 A basic building hierarchy for data collection.

estimate. Even under the conditions of rigorous analysis, it has not been possible to verify the accuracy of these predictions. The shortcomings in the ISO approach have prompted other researchers to develop new methods or models.

MEDIC method

The MEDIC method is based on a typical classification of a given element into four condition ratings that quantify the past and future degradation behaviour using a statistical approach. Thus the predictions are based on the combination of pre-defined probability distribution curves. Note that developing these curves requires considerable level of expertise and judgement. A typical set of condition curves for one element of a building is given in Figure 24.2. Figure 24.3 gives the reverse cumulative probabilities, which can be used to determine the condition at a given age. For example, at age 20, the element being considered has a 45 per cent probability of being in condition A and a 55 per cent probability of being in condition B.

Bamforth method

For Bamforth (2003), service life was defined by the time to achieve a maximum acceptable probability of the serviceability of a limit state being

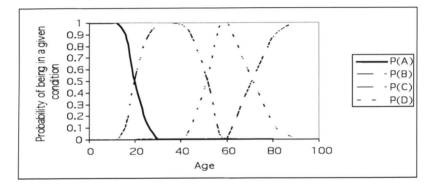

Figure 24.2 Probability of conditions A–D.

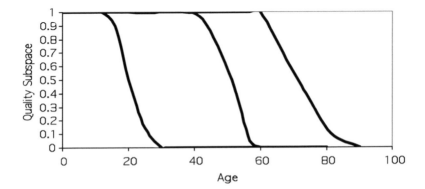

Figure 24.3 Reverse cumulative probability of condition data.

reached. This means that the margin against safety is no longer achievable. This method, although conceptually sound, does not specify the time at which the serviceability criteria would be reached.

Discrete-time Markov chain

Sharabah and Setunge (2007) present the application of a discrete-time Markov chain as a method for deterioration prediction of building assets: that, given the present state is known, the future probabilistic behaviour of the process depends only on the present state regardless of the past. If an element is in state 'i', there is a fixed probability, Pij, of it going into state j after the next time step. Pij is called a 'transition probability'. The matrix P whose ijth entry is Pij is called the transition matrix. Transition matrices consist of a set of states S (1,1,3 ... n) and a probability Pij to pass from state i to state j in one time-step, t.

Figures 24.4 and 24.5 show a typical transition matrix. The probability of an element being in a given state at a given point in time can then be depicted by the set of curves shown in Figure 24.4.

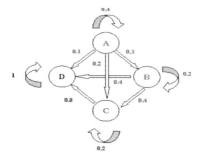

Figure 24.4 Transition from A to D.

	State A	State B	State C	State D
State A	0.4	0.3	0.2	0.1
State B	0	0.2	0.4	0.4
State C	0	0	0.2	0.8
State D	0	0	0	1

Figure 24.5 Transition matrix.

Application of existing methods to a case-study building

Description of the case-study building

The case-study building is a 40-year-old office building of seven occupied floors built over an older four-level car park. It is believed that a majority of services within the building are at or beyond their economic life time, and they have been in subsistence maintenance for quite some time. The building in its present stage is believed to be below standard in terms of acceptable indoor air quality levels, lighting and energy consumption, in addition to other issues such as user comfort and flexibility of changing the floor space to suit current operational requirements. Consequently, the options to redevelop or to demolish and rebuild are to be considered by management.

Structural aspects of the building in its present condition appear to be sound. However, signs of cracking, efflorescence, water stains, corrosion of reinforcement, spalling in concrete and minor deflections in slabs are noticeable. The facades have exhibited pronounced problems over the years, and subsequent maintenance actions have been undertaken. The facade of the building has been constructed from precast panels with a washed sand finish. The windows are inset into the precast as picture windows approximately two metres high. The authors collected the inspection reports, maintenance reports, drawings and all relevant data on the condition monitoring of facades that date back to the late 1980s. Figure 24.6 presents typical elevations of the building.

Figure 24.6 Typical elevations of case-study building and facades.

RSL of case-study building facade and north walls using ISO factorial method

Estimation of reference service life

The reference service life (RFSL) forms one of the key inputs in estimating the RSL. The information available on structural drawings, previous rehabilitation histories and user information were collected and processed to identify a RFSL for building facade and walls. The methodology and application was published in Venkatesan *et al.* (2006) and is summarized below.

Past reports were referenced and a report described the major defects observed in the building facade some ten years ago. Initially, visual inspections using binoculars were undertaken, followed by close-up inspections and tests using approved abseiling techniques. Electronic cover meter tests were then conducted over selected regions to determine the depth of cover to the reinforcement. Concrete samples had been extracted at different levels for laboratory examination. Results and observations were then collated and analysed.

During the visual inspection, at several places, disintegrated loose spalls had been observed. Other defects, such as cracks due to concrete shrinkage, exposed reinforcements and honeycombing, had been observed. Test results from cover meter and carbonation tests indicated that the average depth of carbonation was greater than the average cover at various locations. The tests considered the relationship between the depth of carbonation and the thickness of cover to the reinforcement as an important indicator of durability and cause of corrosion. Additional tests such as chloride ion concentrations were undertaken to identify the most probable cause of distress in the concrete facade visually observed in the form of mapped cracks. Samples from nearby locations of the same building were also extracted and tested. It was concluded that carbonation was the single most dominant factor leading to the development of loose spalls. The municipal authorities considered the threat of disintegrated loose spalls at significant heights from the ground level as a public safety issue. Based on this criterion of public safety, major repairs were undertaken some ten years ago. Since then, no further condition inspections have been conducted.

From the above discussion, it is apparent that decisions regarding repairing facades have been based on the issue of public safety. It is appropriate to state this as a limiting condition. That is, the facade has reached the limiting condition of public safety. This observation is not an opinion, but an analysis of what has happened in the case-study building. Thus, the Bamforth model for this case-study facade can be developed (as shown in Figure 24.7).

Preliminary estimates of RSL based on ISO factorial method

Results from the previous section indicate that a RFSL of 30 years can be used as a basis for estimating the remaining service life. This is the age at

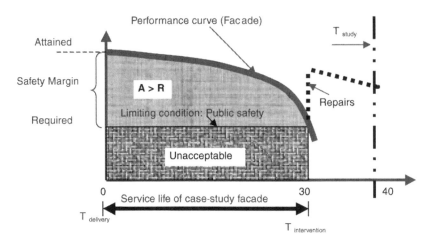

Figure 24.7 Bamforth's service life model for case-study building facade.

which the building facade required refurbishment. In particular, note that the repairs undertaken ten years ago improved the condition of the facade and walls, which have experienced deterioration since then. This is denoted by the dotted line (Repairs) in Figure 24.7. Therefore the issue was to investigate the time at which these elements would reach similar limit states in future. For this purpose, it is important to establish the factors that would affect the RSL of these elements. It was considered that the 'Outdoor environment' (factor f_E) and 'In-use conditions' (factor f_F) are the two most dominant factors that would influence the RSL (given that the carbonation has been identified as the single most dominant factor that caused the disintegration of facade elements). The rationale behind the analysis was further subdivided for each of the wall faces and factors arrived at, based on discussion and consensus opinions. This is summarized in Table 24.1.

Plain application of the ISO factor method results in the following values for the Estimated Service Life (ESL):

ESL (south) $= 25 \times 1 \times 1 \times 1 \times 1 \times 0.7 \times 0.8 = 14$ years
ESL (north) $= 25 \times 1 \times 1 \times 1 \times 1 \times 0.9 \times 1 = 22.5$ years
ESL (east) $= 25 \times 1 \times 1 \times 1 \times 1 \times 0.95 \times 1 = 23.75$ years
ESL (west) $= 25 \times 1 \times 1 \times 1 \times 1 \times 0.95 \times 1 = 23.75$ years   (Equation 24.2)

Note that a RFSL of 25 years has been used instead of 30 years. This is considered as a conservative option. Therefore, RSL based on Equation 24.2 can be estimated as follows:

RSL (south) $= 14 - 10 = 4$ years
RSL (north) $= 22.5 - 10 = 12.5$ years

Table 24.1 Evaluating factors for ISO method

Factor	Face	Relevant condition	Value
f_A: Quality of component	All	Generally good	1
f_B: Design level	All	Generally good	1
f_C: Work execution	All	Generally good	1
f_D: Indoor environment	All	Generally good	1
f_E: Outdoor environment	North	Not at risk due to rain leaks	0.9
	South	At a higher risk due to rain leaks (more windows)	0.7
	East	Heats up in summer	0.95
	West	Cools down in winter	0.95
f_F: In-use conditions	S	Frequent use/repair	0.8
	N,E,W	Consistent	1
f_G: Maintenance	All	Consistent	1

$$\text{RSL (east)} = 23.75 - 10 = 13.75 \text{ years}$$
$$\text{RSL (west)} = 23.75 - 10 = 13.75 \text{ years} \qquad \text{(Equation 24.3)}$$

The above results suggest that the given elements may have an estimated RSL of between four years (on the lower side of the estimates) and 13.75 years (on the higher side of the estimates), with an average of 12.5 years.

Rigorous estimates of RSL based on probability distributions of factors in the ISO method

The above discussions were based on simple multiplication of some notional factors derived on a judgemental basis. However, the factors influencing RSL are highly variable and therefore these factors should encompass a probability distribution as suggested by some researchers (Hovde and Moser 2004). Some of the factors obtained from the above reference were used in Table 24.2 to arrive at these results.

The RSL values estimated for the north, east and west side of walls appear less meaningful. Therefore, the RSL value of 29 years is chosen to be representative of the existing situation. RSL values estimated by the ISO factor methods need to be verified, and for this purpose the RSL estimate based on the MEDIC method was chosen for comparative analysis.

RSL estimates of case-study building facade and walls based on the MEDIC method

As noted earlier, the MEDIC method requires a predefined probability distribution curve for a given element based on experience and judgement. The RFSL of 25 years adopted in the previous sections was chosen as the basis of defining the four degradation schemes of the facade element. At the present stage, the facade and walls are in a 'Fair' condition. This can be confirmed by revisiting Figure 24.6. Since the time of study reported

Table 24.2 Factors for ISO method based on probabilistic distributions

Factor	Type of distribution	Face			
		South	North	East	West
ERSL	Deterministic	25	25	25	25
f_A	*Normal	1/0.1	1/0.1	1/0.1	1/0.1
f_B	*Deterministic	1	1	1	1
f_C	*Lognormal	1.05/0.1	1.05/0.1	1.05/0.1	1.05/0.1
f_D	*Lognormal	1.05/0.1	1.05/0.1	1.05/0.1	1.05/0.1
f_E	*Gumbel	1.25/0.2	1.25/0.2	1.25/0.2	1.25/0.2
f_F	*Lognormal	0.8/0.2	1.05/0.1	1.05/.1	1.05/.1
f_G	*Normal	1.05/.1	1.05/.1	1.05/.1	1.05/.1
RSL (years)		28.9	37.8	37.8	37.8

Note
* mean/standard deviation.

here is about 10 years since major repairs were undertaken, the element can be hypothetically assigned into the four possible schemes: Element in

- 'good' condition (A) 0–10 years;
- 'fair' condition (B) 5–17 years;
- 'minor deterioration' condition (C) 10–23 years;
- 'needing replacement or serious deterioration' condition (D) somewhere between 16 and 25 years.

Note that there is a significant overlap of these conditions, which is of practical significance. Building elements may not be characterized by exact transition from one condition to another; rather, the transition happens over time. Conditional probability curves and the quality subspace are shown in Figures 24.8 and 24.9.

If it is assumed that 50 per cent probability of the element being in stage D is the limiting condition, a reference line can be drawn as shown in Figure 24.9. The managing authority can decide on the definition of the limiting condition. The life expectancy of the element can be estimated as 20.5 years, with residual life being at 10.5 years.

RSL estimates using the discrete Markov chain

The transition matrix for a typical precast concrete facade has been developed, based on some initial approximations until the outcome matches the previous deterioration regime.

Probabilities given in Figure 24.10 can be used to identify a suitable threshold for maintenance action. If we assume 0.5 probability of being in Stage D as the threshold, the average remaining life would be about eight years. The advantage of the Markov chain is that once the transition

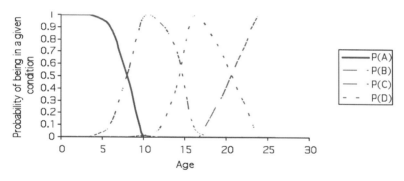

Figure 24.8 Typical conditional probability curves for building facade based on a reference service life of 25 years.

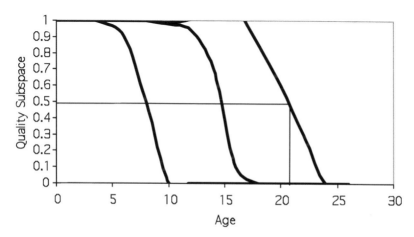

Figure 24.9 Quality subspace for the facade.

matrix is established, the user has a very powerful tool for prediction not only of the remaining life but also of the effect of maintenance activities and economic analysis of management decisions on property expenditure.

Comparison of residual service life estimates with expert opinions

Key results from the previous section can be summarized as follows:

1   RSL estimated by the ISO factorial method using less rigorous techniques resulted in:

   * four years on the lower side of the estimate;
   * 13.75 years on the higher side of the estimate;
   * 12.5 years as an average.

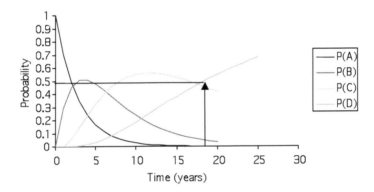

	State A	State B	State C	State D
State A	0.7	0.3	0	0
State B	0	0.8	0.2	0
State C	0	0	0.9	0.1
State D	0	0	0	1

Figure 24.10 Probability of change of facade condition.

2   RSL estimated by the ISO factorial method using rigorous techniques resulted in:

   •   19 years as the best possible estimate.

3   RSL estimated by the MEDIC method resulted in:

   •   11 years as the best possible estimate.

4   RSL estimated based on Markov chain:

   •   eight years as the threshold estimate.

The above results can be summarized as follows, regarding RSL estimated for the case-study facade and wall elements:

   •   five years on the lower side of the estimate;
   •   20 years on the higher side of the estimate;
   •   ten to 15 years on an average.

The above estimates of RSL were evaluated against expert opinion. For this purpose, persons who had worked in the case-study building and been involved in the maintenance regime were invited to provide their opinion on the case-study facade. These experts on a collective basis believed that the facade and walls would require major repairs in between ten and 15 years. Thus the methodology adopted in estimating the RSL suggests that the models are capable of providing meaningful estimates which can then be compared with expert opinions to validate the methods.

Conclusion: proposed method for estimating residual service life

It is clear that any of the methods considered can be used to estimate the RSL of a building element with some variability. The user should understand the variability of predictions and calibrate the selected method to the available condition data. Data required can be significant. The following process is proposed for estimating the RSL of buildings:

- utilize the range of condition assessment procedures covered in this chapter;
- define a condition rating for all the major elements of a building;
- prescribe threshold values for all major elements;
- develop the parameters required for undertaking the predictions, using any of the methods discussed in this chapter. The authors believe that the Markov process or a combination of Markov and MEDIC methods can be used if there is a significant uncertainty in data;
- develop the condition curves;
- estimate the difference between the predicted threshold period for action and the present as the RSL.

Bibliography

Alkhrdaji, T. and Thomas, J. (2002) *Upgrading Parking Structures: Techniques and Design Considerations*. Online. Available at HTTP: <www.structural.net/News/Media_coverage/media_csi.html> (accessed February 2005).

Bamforth, P. (2003) 'Probabilistic approach for predicting life cycle costs and performance of buildings and civil infrastructures', *Proceedings of the 2nd International Symposium on Integrated Life-time Engineering of Buildings and Infrastructures*, Kuopio, Finland.

Chan, A.P.C., Yung, E.H.K., Lam, P.T.I., Tam, C.M. and Cheung, S.O. (2001) 'Application of Delphi method in selection of procurement systems for construction projects', *Construction Management and Economics*, 19 (7): 699–718.

Flourentzou, F., Brandt, E. and Wetzel, C. (2000a) 'MEDIC: a method for predicting residual service life and refurbishment investment budgets', *Energy and Buildings*, 31: 167–70.

Flourentzou, F., Droutsa, K. and Wittchen, K.B. (2000b) 'EPIQR software', *Energy and Buildings*, 31: 129–36.

Flourentzou, F., Genre, J.L. and Roulet, C.A. (2002) 'TOBUS software: an interactive decision aid tool for building retrofit studies', *Energy and Buildings*, 34: 193–202.

Hovde, P.J. and Moser, K. (2004) *Performance Based Methods for Service Life Prediction*, Publication no. 294, Rotterdam: CIB.

Hassoun, M.H. (1995) *Fundamentals of Artificial Neural Networks*, Cambridge, MA: MIT Press.

Huberty, C.J. (1994) *Applied Discriminant Analysis*, New York, NY: John Wiley.

INVESTIMMO (2001) 'A decision-making tool for long-term efficient investment strategies in housing maintenance and refurbishment', Newsletter no. 1.

ISO 15686 (2000) *Building and Constructed Assets – Service Life Planning – Part 1: General Principles*, Geneva: International Organization for Standardization.

—— (2001) *Building and Construction Assets – Service Life Planning – Part 2: Service Life Prediction Procedures*, Geneva: International Organization for Standardization.

Johnson, V.E. and Albert, J.H. (1999) *Ordinal Data Modeling*, Amsterdam: Springer.

Madanat, S. (1993) 'Optimal infrastructure management decisions under uncertainty', *Transportation Research Part C*, 1C (1): 77–88.

Madanat, S., Mishalani, R. and Ibrahim, W.H.W (1995) 'Estimation of infrastructure transition probabilities from condition rating data', *Journal of Infrastructure Systems*, 1 (2): 120–5.

Rey, E. (2004) 'Office building retrofitting strategies: multi criteria approach of an architectural and technical issue', *Energy and Buildings*, 36: 367–72.

Setunge, S., Venkatesan, S. and Fenwick, J. (2007) 'Management of reinforced concrete bridges exposed to aggressive environments', in *Proceedings of the Biennial Conference of the Concrete Institute of Australia*, Adelaide.

Sharabah, A. and Setunge, S. (2007) 'A reliability based approach for service life modelling of council owned infrastructure assets', paper presented at the ICOMS Conference, Melbourne, 22–25 May.

Sharabah, A., Setunge, S., Karagiannis, C. and Godau, R. (2008) 'Effective condition monitoring and assessment for more sophisticated asset management systems', *Proceedings of the Clients Driving Innovation International Conference*, Gold Coast, Australia.

Steelcase (2005) *Reclaiming Buildings: Strategies for Change*, Grand Rapids, Mich. Online. Available at HTTP: <www.steelcase.com/Files/ de9c9b911ac84b3 bbdc25f352373316a.pdf> (accessed February 2005).

Venkatesan, S., Kumar, A. and Setunge, S. (2006) 'Assessment and integration of residual service life models', in *Proceedings of the Recent Advancements in Engineering Mechanics Conference*, Fullerton, CA: California State University.

25 Indoor environment quality and occupant productivity in office buildings

Phillip Paevere

Indoor environment quality (IEQ) is a generic term used to describe the physical and perceptual attributes of indoor spaces. These include the indoor air quality and the thermal, acoustic and visual properties of the environment, as well as various characteristics of the furnishings, facilities and fitout.

Good IEQ in a commercial building can deliver wide-reaching real and potential benefits to occupants, employers and building owners. These can be commercial in nature as well as intangible, and include:

- improved individual and organizational productivity;
- reduced illness and absenteeism;
- worker retention and attraction;
- reduced operational and maintenance costs;
- lower insurance costs, compliance costs and legal risks;
- improved organizational image and marketing potential.

Given that people spend much of their time indoors (see Figure 25.1), the characteristics of a building's interior spaces have the potential to impact on occupants' health, well-being and comfort, which in turn may impact on their productivity. Features of building interiors that enhance productivity include those that reduce discomforts and distractions, as well as those that enable more choice and control over the environment. Occupant productivity is a key driver in the business case for providing high-quality interior environments, since salary costs make up such a large proportion of overall business costs (see Figure 25.2), and so even small improvements in productivity can result in significant bottom-line business benefits.

One of the key challenges in assessing the link between enhanced productivity, IEQ and building and fitout design is being able to define and measure both 'good' IEQ and productivity in meaningful terms. This can be complex, because productivity indicators are highly specific to an organization's context and business goals, and the definition of 'good' IEQ is highly dependent on qualitative factors such as occupant satisfaction.

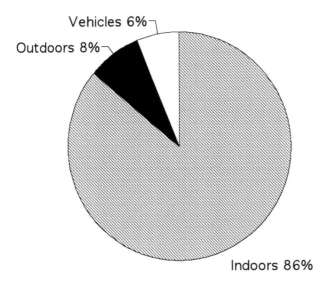

Figure 25.1 Breakdown of human activity by location (source: derived from Klepeis *et al.* (2001)).

This chapter presents a broad overview of IEQ in office buildings and the potential impact on occupant health, well-being and productivity, and provides some practical guidance for improving the indoor environment for enhanced occupant productivity.

Buildings, IEQ and productivity

Productivity assessment

One of the key challenges in assessing the link between enhanced productivity and improvements in IEQ is being able to define and measure both productivity and IEQ in meaningful terms. Defining productivity is not clear-cut because indicators are highly specific to an organization's context and business goals and, for an increasing proportion of creative and knowledge-based jobs, individual productivity may not necessarily be as important for business success as is the productivity of larger units or teams. There are no absolute measures or indicators of productivity that are valid across and between organizations. However, in many studies which examine IEQ impact on productivity, self-assessed or perceived productivity is used as a practical 'relative' indicator that can facilitate useful comparisons across building environments and work contexts.

Occupant questionnaires conducted by the UK consultancy Building Use Studies, with self-assessed ratings of productivity, show that the overall

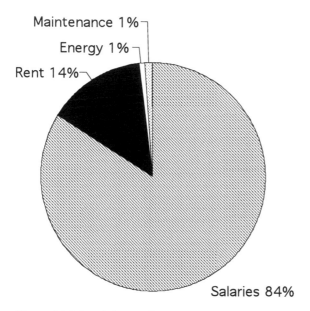

Maintenance 1%

Energy 1%

Rent 14%

Salaries 84%

Figure 25.2 Breakdown of typical business costs (source: derived from Browning and Romm (1994)).

effect of buildings on occupant productivity can range from approximately a 20 per cent gain to a 15 per cent loss. Most buildings fall into the +5 to −5 per cent range, with about two-thirds having negative perceived productivity scores. It should be noted that this effect relates not only to IEQ variables but also to a range of other issues, such as workspace design and facilities management response times, and that it is difficult to separate out the productivity impacts of IEQ and non-IEQ related issues (Leaman and Bordass 2006).

Productivity assessments are often conducted through 'before and after' case studies where a building has been renovated or occupants have moved premises and assessments are made on either side of a change. In these analyses, it is important to check that there have not been any major shifts in non-building related influences on productivity, such as management change, staff morale and quality of information systems. If it is assumed that these non-building related factors can significantly influence productivity, it is important to obtain some 'before and after' indicators of these, to be sure that the perceived productivity improvement is not swamped by any significant contextual shifts. As shown conceptually in Figure 25.3, it is quite feasible to obtain misleading results (i.e., 'false positive' or 'false negative') if these factors are ignored. More information on issues related to productivity assessment in buildings is given in Purdey (2005).

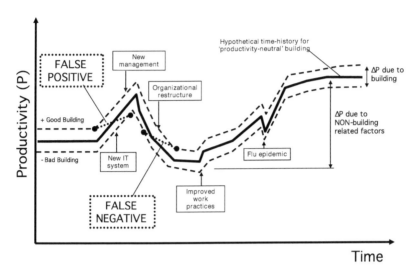

Figure 25.3 Conceptual diagram showing possible misleading effect of non-building factors on 'before and after' productivity assessments (source: Paevere and Brown (2008a), reproduced with permission).

Productivity enhancers in buildings

Productivity enhancers include features that reduce discomforts and distractions, encourage communication, and enable choice and control over the physical characteristics of the environment. Key productivity enhancers in buildings include:

- good IEQ;
- access to windows, daylight and sunlight where possible;
- personal control over temperature, ventilation, lighting and noise where possible;
- comfortable and adjustable furniture and equipment;
- flexible workspace layout and design;
- positive psychosocial features, such as connection with nature and pleasing aesthetics;
- opportunities for formal and informal social interaction.

Good IEQ

Although it is difficult to separate out the impact of individual elements of IEQ on productivity, provision of comfortable and appropriate thermal, acoustic and luminous conditions, adequate ventilation, and access to fresh air with low levels of pollutants, particles, toxins and odours are essential prerequisites for a productive and healthy working environment. Specific

IEQ impacts on productivity are discussed in more detail later in this chapter.

User control

As a general rule, occupants want their perceived needs to be met quickly and with as little intervention by themselves as possible. They normally respond well to IEQ features that enable more choice and control over their environment. Users tend to be more tolerant if they understand how things are supposed to work, and if they have a degree of control over them. Controls should clearly communicate to the user what they are for and how they are supposed to operate. They should provide feedback to the user that they have operated successfully after being used and, crucially, give some indication that something has happened as a result.

Furniture and equipment

Workstations and other furniture and equipment that provide a high level of user control will enhance occupant comfort and performance. Furnishings should support good posture, body mechanics and work techniques for the tasks to be accomplished – for example, ergonomically designed chairs and keyboards. Users should be able to adjust workstation configurations such as seating, computer equipment placement, light levels, work surface heights and local workspace layout. Use of opaque materials in workstations to provide access to daylight and views can also be beneficial.

Flexible workspace layout

Flexible spaces that provide for multiple spatial configurations and allow for rapid and easy change to meet the needs of changing workplaces are essential in facilitating organizational and individual productivity. Flexibility features that enhance productivity include:

- maximum flexibility for delivering power, voice and data;
- mobile and wireless technologies that support new work styles and work practices – this can enable workers to move effortlessly among spaces as their needs change;
- interior design that supports multiple spatial configurations, and a range of space types and densities to allow workers to move freely from solitary work to group action as required;
- informal workspaces in cafeterias and other social or public spaces. If open informal spaces are used, they should be separated from individual quiet spaces.

Flexible space design can also result in churn savings of 67 to 92 per cent due to the ease of relocation or addition of floor outlets, diffusers and power/data/voice outlets, without additional materials or external contractors (Loftness *et al.* 2002)

Positive psychosocial features

Workplace productivity can be potentially enhanced by incorporation of positive psychosocial features of workspace and interior design (Heerwagen 2000). For example, provision of opportunities to engage in social interaction, learning and information-sharing can result in improved communication and morale, which can in turn potentially enhance workgroup and organizational productivity. This can be achieved in practice through provision of multiple places to meet and greet, and a centrally located social space near well-travelled pathways within the building to encourage use and interaction. Provision of spaces for relaxation and psychological restoration, as well as individual concentration, and opportunities for regular exercise away from the desk, such as stair usage, can also have a positive effect on occupant well-being. Other positive features which can be incorporated into interior designs include connection to the natural environment, through internal gardens or indoor plants, and the provision of an interesting visual environment with aesthetic integrity.

Access to daylight and views of indoor and outdoor nature

A number of studies have shown that the negative effects resulting from sustained intense concentration levels can be partially overcome by attentional shifts, especially when the shifts are under the control of the person and when attention is shifted to a positive component, such as a view of nature (Tennessen and Cimprich 1995). Views of nature have also been found to reduce psychological stress and to enhance moods (Kaplan 1992; Heerwagen and Hase 2001). In a field experiment of the effects of daylight and views on cognitive performance, Heschong (2006) found that workers with visually interesting views, especially of natural vegetation, scored better on cognitive tasks than workers with no views or with less interesting views. There was a 6 to 7 per cent faster call-handling time for employees with seated access to views through larger windows with vegetation content from their cubicles, as compared to those with no view of the outdoors.

Aesthetics

Interior spaces should be designed with reference to basic human needs of shelter, comfort and communication, and connections to the patterns of nature. These and other psychosocial features of interior design can have a positive effect on productivity. A visually appealing environment can be provided through a well-balanced and appropriate use of scale, colours, textures, patterns, artwork and plants. Too much visual uniformity should be avoided, as should too much visual chaos.

Productivity inhibitors in buildings

Productivity inhibitors for buildings in which high-level cognitive work is undertaken are mainly related to characteristics that cause or contribute to discomfort, distractions and interruptions. These include:

- noise distractions and interruptions;
- visual distractions;
- high workstation density;
- poor IEQ, particularly thermal comfort levels and air quality.

Noise distractions and interruptions

The use of an open-plan layout to increase communication and flexibility can lead to a high level of complaints of distractions resulting from interruptions and people talking (Evans and Johnson 2000). It should be noted that distractions are less related to the actual noise level than to the degree of self-control over the noise, its content and its predictability (Kjellberg *et al.* 1996).

Interruptions are more likely to occur in open-plan offices where people can readily be seen and are thus considered 'available' for interactions. This presents a conflict for designers, because any productivity benefits from increased communication and interaction and from more efficient space usage must be traded off against the potential for increased noise levels and associated distractions and interruptions. A common solution is to provide small, enclosed 'concentration spaces' for individual concentrated work. However, in an open environment, people may have difficulty moving between their personal workspace and the concentration areas because they often need things that are at their desks, or they want to be able to use their computers, phones and paper documents simultaneously, and this can be difficult to negotiate.

Visual distractions

Visual distractions include both people and artefacts. Early research on the landscaped office showed that visual distractions associated with continual movement of people created high levels of dissatisfaction. This led to the widespread use of partitioned workstations. Although these have reduced visual distractions, they have not adequately reduced noise distractions. Research also shows that people are similarly distracted by artefacts on their desk, such as document piles, messages and 'to do' lists. The result in most offices is cognitive overload and inability to decide what to do and how to maintain focus (Lahlou 1999). People frequently switch attention from one thing to another as they are 'reminded' of the need to do something other than what they are currently working on, simply by looking at the note or pile of work left unfinished, or by looking up when someone interrupts them. Visual distractions are more likely to be prominent when workspaces do not have sufficient, readily accessible storage space.

High workstation density

Higher workstation density may lead to lower productivity due to reduced comfort levels and increased visual and acoustic distractions and interruptions (Fried *et al.* 2001). Things which assist productivity, such as availability of desk space and storage, become more scarce in higher density spaces, and cooling, ventilation and lighting systems need to work harder to achieve appropriate comfort levels.

Poor IEQ

There is a growing body of evidence that poor IEQ can negatively impact on individual and organizational productivity. Such effects are discussed in more detail in the following sections.

Indoor air quality

Indoor air quality (IAQ) refers to the totality of attributes of indoor air that affect a person's health, well-being and comfort. It is characterized by:

- physical factors, such as ambient temperature, humidity and ventilation rates;
- air pollutant factors, such as pollutant levels and exposure times;
- human factors, such as occupant health status, individual sensitivity and personal control.

Impact of IAQ on productivity

There is a mounting body of evidence that there is a clear financial motivation for ensuring good IAQ in office buildings. Wyon (2004) reviewed recent research findings and concluded that it was now beyond reasonable doubt that poor IAQ decreased worker productivity and caused visitors to express dissatisfaction. The effect on most aspects of office work performance was estimated to be as high as 6 to 9 per cent. A study of 39 schools in Sweden (Smedje and Norback 2000) showed a 69 per cent reduction in the two-year incidence of asthma among students in schools that received a new displacement ventilation system with increased fresh air supply rates, compared to those in schools that did not receive a new system.

Fisk (2002) estimates that improved heating, ventilation and cooling (HVAC) systems, which could limit the spread of contaminants and pathogens, could potentially reduce respiratory illnesses by 9 to 20 per cent. He estimates that in the US, the corresponding productivity increases from reduced absenteeism and illness could be as high as $6 to $14 billion from reduced respiratory disease, $1 to $4 billion from reduced asthma and allergies, and $10 to $30 billion from a reduction in symptoms associated with sick building syndrome.

Opportunities for improving air quality

Major contributors to poor IAQ include emissions from new building materials and furniture, emissions from office equipment such as photocopiers and printers, poor HVAC system performance and maintenance, and poor outside air quality. Other factors may include poor cleaning practices, poor moisture control which can lead to mould (e.g., water leaks or persistent damp surfaces), human occupancy effects (e.g., odours), poorly designed enclosed garages and poor overall building maintenance.

Improved IAQ can be best achieved by reducing or eliminating toxins and odours at their source, providing adequate ventilation rates, isolating office equipment into well-ventilated spaces, controlling moisture to reduce microbial growth and regularly maintaining the HVAC system.

Thermal environment

Thermal comfort refers to 'a condition of mind which expresses satisfaction with the thermal environment' (ISO 1994). It is usually described simply in terms of whether a person is feeling too hot or too cold. Thermal comfort can be difficult to define parametrically because a range of environmental and human factors need to be considered in order to determine what will make people feel comfortable. These include air and operative temperature, humidity, air velocity, level of personal control, and occupant factors such as clothing type and level of activity.

In practice, a high level of thermal comfort is considered to occur when a high proportion (e.g., 80 per cent or more) of building occupants are predicted to be satisfied with the thermal conditions, based on the above factors. A significant influence on thermal comfort is whether a space is mechanically conditioned or naturally conditioned; these are known to require different physical conditions for thermal comfort, since occupant expectations in the latter are shifted due to different thermal experiences and availability of individual control.

Impact of thermal environment on productivity

Some studies have shown a positive relationship between thermal comfort and occupant productivity. For example, in a controlled field experiment in Japan, Imanari *et al.* (1999) identified a significant improvement (up to 24 per cent) in measured work efficiency and accuracy among occupants working in an environment with superior thermal comfort conditions. Other studies have shown the link between user control of thermal conditions and productivity. Eight studies summarized in Kats (2003) show that provision of individual temperature control can increase individual productivity by 0.2 to 3 per cent.

As well as the direct effect on individual health and productivity, poor thermal conditions can potentially increase building maintenance outlays due to costs associated with occupant complaints. A study by Federspiel (2001), based on 575 buildings in the US, showed that nearly one-fifth of complaints to facilities managers were related to indoor environment issues, and most of these were related to thermal comfort.

Opportunities for improving thermal comfort

One of the most important things a designer can do to ensure high levels of thermal comfort is to establish appropriate criteria, based on standards and guidelines such as those developed by ASHRAE (2004) and ISO (1994). Where possible, provision of some level of personal control over the thermal environment can help improve occupant satisfaction. Examples include operable windows, personal ventilation controls or a personal fan or heater.

Once thermal comfort criteria have been established, these need to be monitored and maintained over time by building managers. Given that thermal comfort is largely defined by occupant satisfaction levels, monitoring requires the incorporation of occupant feedback to maintain comfort levels under different conditions and contexts. This should be actively sought by building managers, listened to, and acted on quickly where necessary.

Acoustic environment

Acoustic environment quality refers to the totality of the acoustic characteristics of a building interior that impact on occupants' aural perceptions. These can be affected by:

*   levels of background noise;
*   reverberation times and sound absorption;
*   information content of the noise;
*   noise transmission between spaces;
*   speech intelligibility;
*   personal control and intermittency of the noise.

Different types of office spaces, such as workstation clusters, social spaces, executive suites, conference rooms and boardrooms, will have specific acoustic requirements, depending on their function.

Impact of acoustic environment on productivity

Of all the aspects of IEQ, noise levels are most often the cause of greatest occupant dissatisfaction in office environments (Jensen and Arens 2005). The major sources of dissatisfaction include:

- speech interruptions, such as people talking over the phone, in adjacent areas and corridors;
- equipment noise;
- excessive background noise from HVAC and lighting systems;
- lack of conversational privacy;
- lack of personal control over noise levels;
- space being acoustically too 'lively' or too 'dead'.

The levels of background noise and speech privacy, and separation between particular types of spaces, have important implications for the work environment and productivity of building occupants. As outlined above, many problems associated with office spaces relate to interruptions by other employees. Distractions due to the sound of speech have been found to be significantly problematic, especially in open-plan offices. This is because speech is more distracting than unintelligible speech or sounds with no information content. Conversely, office spaces with very low background noise can have poor levels of speech privacy, which can also hinder communication. It is a challenge for designers to find the correct compromise between privacy and intelligibility for specific contexts. A detailed review of acoustic satisfaction in open-plan offices is given in Navai and Veitch (2003).

Opportunities for improving acoustic environment

In essence, noise can be controlled by eliminating the source, isolating the source or masking the unwanted sound. Following these basic principles, strategies for creating a high-performance acoustic environment include:

- identifying noise sources and establishing appropriate criteria for background noise, transmission of noise between spaces and speech privacy levels;
- separating noise-sensitive and noise-producing areas, including provision of opportunities for privacy and concentration, when needed, in open-plan offices;
- considering the impacts of building services on ambient conditions. Steps should be taken to minimize background noise from the HVAC system and other equipment, where necessary, by using passive or active methods;
- selecting appropriate surface finishes to control sound reverberation times;

- limiting transmission of unwanted noise from outside the workplace;
- using sound-masking systems to maintain appropriate balance between speech privacy and intelligibility.

Luminous and visual environment

The luminous and visual environment quality refers to the totality of the characteristics of a building that impact on occupants' visual perceptions. Occupant perceptions can be affected by the following:

- luminance levels (ambient and task) for different tasks, and their uniformity;
- glare levels and reflections in computer screens;
- levels of personal control through task lighting, shading or dimmers;
- access to daylight and views;
- lighting characteristics, such as colour temperature and ballast flicker;
- visual appeal and colour scheme of interior design.

Impact of luminous and visual environment on productivity

Luminous and visual environment quality can have a significant impact on occupants' abilities to perform tasks, especially if they are visually intensive. Major sources of dissatisfaction include limited access to daylight, inappropriate light levels, excessive glare and lack of control over the environment.

Various studies have shown a link between lighting quality and productivity. Kats (2003) summarized a series of case studies which indicated that productivity gains ranging from 0.7 to 23 per cent were achieved in buildings with higher-quality lighting fixtures and/or access to daylight. Çakir and Çakir (1998) also identified a health benefit from the use of more extensive task lighting. Their study showed a 19 per cent reduction in headaches for workers with separate task and ambient lighting, as compared to those with ceiling-only combined task and ambient lighting.

Opportunities for improving performance of luminous and visual environment

In any office environment, occupants should be able to see easily, comfortably and accurately. The illumination level required to achieve these results will vary, depending on the activity taking place and the characteristics of the occupant. The illumination level required for most spaces and environments is a function of the type of activity or task being undertaken, the importance and difficulty of its visual aspects, and the age and visual capabilities of the occupant.

Strategies for improving the luminous and visual environment are based on maximizing occupant visual comfort. Practical strategies for creating a high-performance environment include:

- providing appropriate and adjustable task lighting and, where possible, adjustment of ceiling lights;
- integrating natural and electric lighting strategies, including suitable arrangement of fittings with respect to building and workspace layout;
- using day-lighting for ambient lighting wherever feasible;
- using high-performance ballasts, lamps, fixtures and controls;
- reducing direct glare from both natural and man-made sources in the field of view – shading can be combined with light redirection to provide for an effective day-lighting strategy while reducing glare;
- providing light on vertical surfaces/walls and light-shelves to increase the perceived brightness of the space;
- providing internal and external views of nature and visually appealing aesthetics;
- avoiding too much visual uniformity, as well as too much visual chaos;
- cleaning windows and lights regularly to maximize daylight and illumination levels.

Case study: CH2

Building description

Council House 2 (CH2) is a ten-storey office building in Melbourne, Australia, which houses around 500 City of Melbourne staff, together with some ground-floor retail space. Occupied in October 2006, CH2 was conceived, designed and built with a substantial focus on setting a new standard for ecologically sustainable office buildings. It has a raft of sustainable technologies and design philosophies incorporated throughout the entire building, services and fitout. A key element of the business case for CH2 was that provision of high levels of IEQ, along with other design features, would result in significant benefits through improved health, well-being and productivity of staff in the building. Many IEQ-related features of CH2 incorporate principles outlined in this chapter, including:

- 100 per cent fresh air ventilation;
- radiant cooling provided by the thermal mass of concrete ceiling panels, and also through chilled panels;
- lighting provided through a mix of high-efficiency recessed luminaries in the ceiling, suspended strip lighting, daylight penetration and extensive task lighting;
- low-toxicity materials used for all furnishings and finishes;
- extensive use of indoor plants.

The interior design was also intended to achieve productivity benefits through increased communication and collaboration between staff. The fitout of CH2 is based on a modern open-plan philosophy, with no enclosed offices and with low adjustable partitions between workstations. Staircases have been designed to encourage staff to walk between nearby floors. There are relatively unobstructed lines of sight throughout each floor, with the only enclosed spaces being the formal meeting rooms. Informal meeting and social spaces are provided throughout the building. Occupants also have access to external balconies, a winter garden, a summer terrace and a rooftop garden.

Productivity of CH2 occupants

Perceived productivity ratings show that CH2 occupants achieved a significant productivity improvement when compared to the levels recorded in the previous accommodation located next door (in CH1), despite some problems with lighting and increased noise levels due to the open-plan layout. Three-quarters of occupants rated the building as having a positive or neutral effect on productivity, compared with just 39 per cent previously in CH1. CH2 was included in the top 20 per cent of Australian buildings for perceived productivity when compared against a benchmark dataset (Building Use Studies benchmarks). This can be expressed as a 10 per cent perceived productivity enhancement compared to previous accommodation. Contextual indicators showed little change for non-building-related influences on productivity.

Based on the results of occupant questionnaires and a program of physical measurements, it was shown that the significant improvement in perceived productivity could be best correlated to variables relating to the 'building overall', such as the building image, quality of design, perceived healthiness and overall comfort. It was shown that other factors, such as experiences in previous accommodation, may also influence the results. In terms of IEQ impacts on productivity, it was concluded that significantly improved thermal comfort and air quality are likely to have had an enhancing effect, while noise from interruptions and perhaps some aspects of the lighting may have been perceived as a productivity hindrance. Full details of this case study and the productivity analysis are given in Paevere and Brown (2008a, 2008b).

Acknowledgements

Some of the material presented in this chapter was developed for 'Your Building', an initiative of the Cooperative Research Centre for Construction Innovation, as well as the original CRC research project 'Regenerating Construction to Enhance Sustainability: Task 3 – Occupant Health, Wellbeing and Productivity'. The author would like to acknowledge the valuable contributions made to the Indoor Environment Module of Your Building by Stephen Brown, Judith Heerwagen, Adrian Leaman and Mark Luther.

Bibliography

ASHRAE (2004) *ANSI/ASHRAE Standard 55–2004: Thermal Environmental Conditions for Human Occupancy*, Atlanta, GA: American Society of Heating, Refrigerating and Air-Conditioning Engineers.

Browning, W. and Romm, J. (1994) *Greening the Building and the Bottom Line: Increasing Productivity Through Energy-Efficient Design*, Snowmass, CO: Rocky Mountain Institute.

Çakir, A.E. and Çakir, G. (1998) *Light and Health: Influences of Lighting on Health and Wellbeing of Office and Computer Workers*, Berlin: Ergonomic.

Evans, G.W. and Johnson, D. (2000) 'Stress and open-office noise', *Journal of Applied Psychology*, 85 (5): 779–83.

Federspiel, C. (2001) 'Estimating the frequency and cost of responding to building complaints', in J.D. Spengler, J.M. Sammet and J.F. McCarthy (eds) *Indoor Air Quality Handbook*, New York, NY: McGraw-Hill.

Fisk, W.J. (2002) 'How IEQ affects health, productivity', *ASHRAE Journal*, 44 (5): 56–60.

Fried, Y., Slowik, L.H., Ben-David, H.A. and Tiegs, R.B. (2001) 'Exploring the relationship between workspace density and employee attitudinal reaction: an integrative model', *Journal of Occupational and Organizational Psychology*, 74 (3): 359–72.

Heerwagen, J. (2000) 'Green buildings, organizational success and occupant productivity', *Building Research and Information*, 28 (5): 353–67.

Heerwagen, J. and Hase, B. (2001) 'Building biophilia: connecting people to nature in building design', *Environmental Design and Construction*, Apr/May: 30–4.

Heschong, L. (2006) 'Windows and office worker performance: the SMUD call center and desktop studies', in D. Clements-Croome (ed.) *Creating the Productive Workplace*, 2nd edn, Abingdon: Taylor & Francis.

Imanari,T., Omori, T. and Bogaki, K. (1999) 'Thermal comfort and energy consumption of the radiant ceiling panel system: comparison with the conventional all-air system', *Energy and Buildings*, 30: 167–75.

International Organization for Standardization (ISO) (1994) *ISO 7730 Moderate Thermal Environments: Determination of the PMV and PPD Indices and Specification of the Conditions for Thermal Comfort*, Geneva: ISO.

Jensen, K. and Arens, E. (2005) 'Acoustical quality in office workstations, as assessed by occupant surveys', in *Proceedings of Indoor Air 2005*, Beijing.

Kaplan, R. (1992) 'Urban forestry and the workplace', in P.H. Gobster (ed.) *Managing Urban and High-Use Recreation Settings*, USDA Forest Service, General Technical Report NC-163, Chicago, IL: North Central Forest Experiment Station.

Kats, G. (2003) *The Costs and Financial Benefits of Green Buildings: A Report to California's Sustainable Building Task Force*, Sacramento, CA.

Kjellberg, A., Landstrom, U., Tesarz, M., Soderberg, L. and Akerlund, E. (1996) 'The effects of nonphysical noise characteristics, ongoing task and noise sensitivity on annoyance and distraction due to noise at work', *Journal of Environmental Psychology*, 16: 123–36.

Klepeis, N.E., Nelson, W.C., Ott, W.R. *et al.* (2001) 'The National Human Activity Pattern Survey (NHAPS): a resource for assessing exposure to environmental pollutants', *Journal of Exposure Analysis and Environmental Epidemiology*, 11: 231–52.

Lahlou, S. (1999) 'Observing cognitive work in offices', in N.A. Streitz, J. Siegel, V. Hartkopf and S. Konomi (eds) *Cooperative Buildings: Integrating Information, Organizations and Architecture: Proceedings of 2nd International Workshop, CoBuild '99*, Pittsburgh, PA.

Leaman, A. and Bordass, B. (2006) 'Productivity in buildings: The "killer" variables', in D. Clements-Croome (ed.) *Creating the Productive Workplace*, 2nd edn, Abingdon: Taylor & Francis.

Loftness, V., Brahme, R., Mondazzi, M., Vineyard, E. and MacDonald, M. (2002) *Energy Savings Potential of Flexible and Adaptive HVAC Distribution Systems for Office Buildings*, Arlington, VA: Air-Conditioning and Refrigeration Technology Institute.

Navai, M. and Veitch, J.A. (2003) *Acoustic Satisfaction in Open-Plan Offices: Review and Recommendations*, Report IRC-RR-151, Ottawa: Institute for Research in Construction, National Research Council Canada.

Paevere, P. and Brown, S. (2008a) *Project: Regenerating Construction to Enhance Sustainability. Task 3: Occupant Health, Wellbeing and Productivity*, Report 2003–028-B-T3–01, Brisbane: CRC for Construction Innovation.

——— (2008b) *Indoor Environment Quality and Occupant Productivity in the CH2 Building: Post-Occupancy Summary*, Report No. USP2007/23, Melbourne: CSIRO.

Purdey, B. (2005) *Green Buildings and Productivity*, BDP Environment Design Guide: GEN 67, Canberra: Royal Australian Institute of Architects.

Smedje, G. and Norback, D. (2000) 'New ventilation systems at select schools in Sweden: effects on asthma and exposure', *Archives of Environmental Health*, 35 (1): 18–25.

Tennessen, C.M. and Cimprich, B. (1995) 'Views to nature: effects on attention', *Journal of Environmental Psychology*, 15: 77–85.

Wyon, D.P. (2004) 'The effect of indoor air quality on productivity and performance', *Indoor Air*, 14 (7): 92–101.

Part VI

Innovation

Capture and implementation

26 Effectively diffusing innovation through knowledge management

Derek Walker

Technology, Design and Process Innovation in the Built Environment provides numerous examples of innovative developments that have emerged from the CRC for Construction Innovation. These *potentially* can make significant inroads into increasing productivity by those delivering built environment infrastructure and maintaining built assets. This *potential* productivity extends to sustainability so that productivity is realized for long-term benefits rather than merely generating transient short-term advantages. However, the word *potential* is stressed here because no benefit (transitory or long-term) will be delivered unless this potential is realized. Good ideas, like well-crafted strategies, are illusionary unless there is the means to translate thought into effective action.

The challenge that faces all organizations is not only to foster creativity and innovation that generates inventions and innovation, but to also ensure that good ideas and improved processes are effectively adopted and adapted.

Part of this chapter describes findings from a CRC research project that investigated improved ways to transfer knowledge within organizations in the built environment. It was concerned with investigating how knowledge is diffused in such organizations, with a particular focus on information communication technology (ICT) diffusion of a groupware application and also of knowledge management (KM) practices in large, highly computer-literate organizations. A summary of its findings can be found in Walker *et al.* (2004), and other papers and book chapters may be referred to for further details (Peansupap and Walker 2005a; Walker 2005; Walker *et al.* 2005, 2007).

In terms of innovation diffusion, the CRC CI research work did achieve wide distribution of its findings, but the participating organizations did not choose to continue with further stages of the study that would have involved benchmarking KM practices across divisions and companies. Reasons given at the time included anticipated difficulties within the main organizations, with various operating units feeling uneasy about being benchmarked using the capability maturity model (CMM) that is discussed and described later. In this sense, the diffusion was successful only in that it efficiently *disseminated* information and knowledge about the ICT innovations under study. It remains, however, incomplete in that adoption and adaptation of models and processes that were generated from the study have not been applied in the target organizations. This situation forms the core focus of this chapter – how to move beyond disseminating knowledge about innovative ideas to getting them used and applied. The CRC project on which this chapter is based (Walker *et al.* 2004) itself provides a very useful case study that helps us better understand how innovation can be more effectively diffused through clearer understanding of how KM dissemination processes can stimulate appropriate action.

This chapter is structured as follows. The following section explicitly defines terms and core concepts. Next, a description of several models of innovation diffusion that were produced from the CRC CI research project is presented. That provides the idealized model of how we should diffuse innovation. A key driver and barrier to innovation diffusion that critically impacts KM practices is workplace culture and the stickiness of knowledge being transferred, so a section on this naturally follows the idealized diffusion models section. Here the principal barriers to innovation diffusion are discussed, as well as how workplace knowledge transfer practices affect the adoption and adaptation of innovation. This leads to presentation of a simple model indicating how innovation in more general terms may be more effectively diffused for practical use through developing appropriate KM practices to support this aim.

The nature of innovation and creativity

Van de Ven (1986: 590) defines *innovation* as 'the development and implementation of new ideas by people who over time engage in transactions with others within an institutional order'. This definition focuses upon four basic factors (new ideas, people, transactions and institutional context). He stresses that an innovation is simply an idea that is perceived as being new to those involved in discovering it, and that the nature of the idea can be technical (product or service) or administrative (process). A key point about innovation is that it exists elsewhere and is introduced – it is new to the person or organization. The process of recognizing an idea's value and how it can be adopted (or more likely adapted to another context, whether radical or marginal) is crucial to the first steps of its diffusion.

Invention has a different connotation. An invention is something new to the world (as opposed to an individual), and thus it is possible to patent an invention. Being inventive is a highly creative process. Amabile (1998: 78) argues that creativity is the overlap area between having expertise (deep knowledge in a domain), having creative thinking skills (flexibility in ways of perceiving things, curiosity and challenging traditional ways of perceiving things), and being motivated (the desire to do something different and often divergent). Even inventions have to be translated into innovations, in the sense that the invention or discovery becomes morphed into a practical application, often by other people than the originator of the invention.

Empathic design is a hybrid of this. Leonard and Rayport (1997) illustrate how empathic design is developed by using examples where new medical instrumentation tools are invented (or innovations radically morphed into something totally new to the world) by collaborative engagement of people working together to solve common problems.

The one constantly underlying theme to emerge is the way that effective innovation is embraced through effective change. This may materialize from minor step changes through much adaptation of existing ideas (new to people), or from radical innovation that stems from invention of entirely new ways of doing something. We may now start to see innovation diffusion as a social process rather than a technical routine, because it involves supporting people to change from doing something one way to another way.

An organization's 'capacity to change' through innovation is of paramount importance for deployment of innovations. This is stressed in the Van de Ven (1986: 590) definition, as well as by the Leonard and Rayport (1997) examples. The propensity to change can be seen to affect the potential rate of diffusion of innovations.

Rogers (2003) has been one of the most cited authorities on innovation diffusion since publication of the first edition of his book over 25 years ago. His (widely accepted) central idea was that there is a natural distribution of acceptance rates of innovation. His data analysis suggests that there will always be a very small group of leading innovators (of the order of 2.5%, or above 2 standard deviations from the mean). An early adopters' group (about 13.5%) follows. An early majority group follows the early adopters up to 1 standard deviation from the mean (34%), followed by a late majority at 1 standard deviation below the mean (34%). Finally, laggards (the remaining 16%) are very slow to adopt an innovation, at 2 standard deviations and beyond from the mean. Figure 26.1 illustrates an example of this taken from the CRC CI research project that illustrates the adoption profile of an ICT groupware application for six major Australian contractors (see Goldsmith *et al.* 2003).

The distribution illustrated in Figure 26.1 has the six organizations mapped onto the innovation diffusion curve. An interesting part of the exploratory research reported upon here was that it prompted fascinating

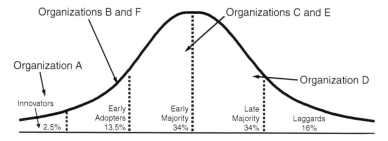

Figure 26.1 Example of an ICT diffusion profile.

questions about how and why these organizations deliberately chose their diffusion strategy, and how that was determined by their organizational culture. Each organization is ICT literate, and each has remained within the top tier of construction contractors in Australia that undertake projects of many hundreds of million and often billions of dollars. Each is a stable commercial entity that makes substantial profits, and so its innovative adoption stance appears to be more of an issue of culture and marketing position than a particular survival response.

Each organization's cultural character is reflected in part by its ICT diffusion profile. Organization A developed its own home-grown intranet for sharing information and knowledge, and can be described as an organization that readily embraces change and is not afraid to be an industry leader piloting new concepts borrowed from other industries. Organizations B and F prefer to adapt technology developed by their competitors and learn from their mistakes (notably A). Organizations C and E were experimenting with the idea of ICT groupware, and were still unsure which direction and application software to adopt. Organization D mainly felt that it should invest in developing its capacity because its main client base demanded that it use this technology to maintain consistency of communication infrastructure.

These organizations exemplify two forms of anxiety about change discussed by Schein (1993) in terms of their motivation to adopt and adapt to change. Schein argues that there are two types of anxiety. Type 1 anxiety occurs when people are reluctant to change or learn something new because the task seems too disruptive or difficult to attempt. The status quo becomes a safe haven for them, and there is little motivation tipping them towards change. Type 2 anxiety, however, occurs when there is fear, shame or guilt about *not* making a change or learning something new. Figure 26.1 suggests that organizations C and E were driven in part by Type 1 anxiety, and that organization D displayed typical Type 2 anxiety behaviour.

The CRC CI research project progressed from the above exploratory analysis and headed off in two complementary directions. One direction led to an in-depth study within one of the above construction organizations to investigate how it managed its KM practices in a business unit that was vital to its continued sustainability and prosperity (Maqsood 2006). The second direction led to development of a model of ICT innovation diffusion based on factor analysis of a sample of 117 skilled and experienced users of a range of ICT applications across three large Australian construction organizations (Peansupap 2004). The two models arising out of those studies are discussed in more detail below.

The first model, illustrated in Figure 26.2, was developed by Maqsood (2006); refer also to Walker and Maqsood (2008: 257–60). It is composed of several simple ideas. The external environment exerts demands for the organization to change and become more innovative through inherent competition presenting opportunities and threats. This exerts a powerful dynamic that drives Type 1 or Type 2 motivational anxiety (Schein 1993).

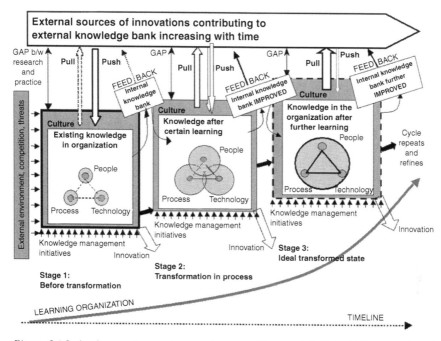

Figure 26.2 An innovation trajectory (source: Maqsood (2006: 134), reproduced with permission).

Opportunities for access to external knowledge repositories exist for firms through their professional bodies, universities, and people sharing and transferring knowledge resources through communities of practice (CoPs). A CoP is a group of people with a shared interest in specific topics who come together formally or informally in real or virtual space to openly discuss and share knowledge about these shared interests (Wenger 1999; Wenger *et al.* 2002). These external innovation sources, such as the CRC CI and its partnered research centres, provide access to research outcomes and other knowledge forms such as theoretical frameworks and case-study data, often with an expectation or hope that the organization will reciprocate by feeding back some of the firm's internal knowledge bank into that source for validation or enhancement through collaborative investigation. Other types of collaborative academic–practitioner groups, such as the rethinking project management group (Winter and Smith 2006), have been described in detail elsewhere (Walker *et al.* 2008a).Thus, we see knowledge push and pull forces being exerted between the external innovation sources and the organization. The organization possesses existing knowledge through its people and processes, and this is enabled by its technology infrastructure. Knowledge and innovation is mainly forced into the organization with much top-level effort as a push force, with very weak pull forces allowing knowledge to be fed back to the outside world for enhancement and validation. This produces a large gap between research and practice so that the organization cannot readily find out about new ideas, and any new ideas generated from the external sources are rarely tested and validated by the organization (or business unit).

The degree to which an organization is oriented towards seeking external knowledge varies according to the type of organization. Figure 26.2 can also be useful in understanding the types of anxiety experienced by organizations relating to their innovation stance. The Miles and Snow (2003) organizational typology of prospector, analyser, defender and reactor can also help explain how and why different organizational types seek knowledge from external sources in different ways. For example, a *prospector* will be seeking knowledge about what external markets fail to currently deliver and what internal innovations it can hone, reshape or further develop to aggressively lead its competitors. An *analyser* would need to carefully seek external knowledge from the market and identify potential for finding a specialized niche market for its products/services, and learning as an early adopter (Rogers 2003) how it can improve upon lead innovators. A *defender* organization would be concentrating to a large extent on knowledge that it can muster to stay in the market and lower costs and raise barriers to entry by others, perhaps through continuous improvement and incremental improvements in its offerings. A *reactor* organization is poor at knowledge absorption, finds itself continually under threat, and perhaps uses external and internal knowledge to mount campaigns to fend off competitors, and spin-doctoring to advertise its offerings in the most favourable light possible and restrict what it may view as dangerous or subversive leakage of proprietary knowledge.

Stage 1, illustrated in Figure 26.2, suggests a low level of organizational learning and limited possibilities for innovation through adoption and adaptation of external ideas. While this may be adequate for the organization's short-term survival, if the organization's internal knowledge capital base is strong, it will inevitably miss signals of external change and turbulence. An organization at this stage is also missing valuable opportunities to innovate through empathic design (Leonard and Rayport 1997; von Hippel *et al.* 1999). People, processes and technology may be poorly aligned so that there are barriers to the generation and flow of knowledge and new ideas. While knowledge may be used, it may be sub-optimally applied – this is indicated by the separation of these three elements and the weak link indicated by dotted lines. Processes may be traditional for the industry with little experimentation and/or they may be somewhat ad hoc. Technology support could include widespread use of groupware and portals centred on data processing rather than any KM activities. In the CRC CI study of three of Australia's leading construction contractors, observed KM initiatives included development of portals for information retrieval, limited within-firm CoPs for ICT use, and enhanced information and knowledge storage through the portal for aspects such as safety reporting, lessons learned and other explicit knowledge repositories (Peansupap 2004; Peansupap and Walker 2005a). People are linked to technology and processes through more formal means in standard ways with little experimentation, and there may be a complacent satisfaction with the way things work. The organizational culture is inward looking and hence the barriers to outside influence are significant, as indicated by the thick boundary lines.

The greater the number and more intense the scope of KM initiatives that are undertaken, the greater is the chance that the organization moves towards or enters Stage 2 development. Experimentation and reflection on experiments undertaken are key elements in an organization increasing its absorptive capacity (to assimilate external knowledge) to incorporate further change and embed learning, thus moving towards being a learning organization (Cohen and Levinthal 1990; Gann 2001; Caloghirou *et al.* 2004). This is because experimentation and proactively seeking external knowledge assists in:

- developing language and meaning (i.e., jargon, 'in-terms' and technical language) that can be shared easily with outside sources;
- being better able to categorize and conceptualize external knowledge and thus in a position to internalize this knowledge and apply it to the workplace;
- being better able to discern patterns and assess the potential value of external knowledge (Cohen and Levinthal 1990).

Stage 2 of the model indicates that the organization has reduced the imperviousness of the boundaries between the organization and the outside world, and has also better integrated its people, processes and technology. There is stronger evidence of knowledge creation, transfer and use for problem-solving, as well as a greater degree of feedback of knowledge to be tested and validated by external sources such as academia, professional associations, clients and competitors (e.g., through undertaking benchmarking exercises). This stage indicates a more or less equal push and pull of knowledge. KM initiatives are more prevalent and intense in scope, and these propel the organization's learning and develop its absorptive capacity. This type of organization is setting out on a path to become a learning organization. Garvin (1993: 80) describes a learning organization (LO) as 'one that is skilled at creating, acquiring and transferring knowledge and at modifying its behaviour to reflect new knowledge and insights'. LOs are active and effective in their KM initiatives that improve their absorptive capacity.

At Stage 3 an idealized condition is attainable: people, processes and technology are completely seamless and embedded, the boundaries are highly permeable, and there is a greater export of knowledge for validation than is being imported. While this appears to many organizations as dangerous and an insecure way of managing critical knowledge assets, it may be that such a state provides greater advantages for rapid innovation and commercialization than its downside of being vulnerable to exploitation from its competitors. As Miles and Snow (1995: 9) indicate, organizations that operate in this way need to have a strong culture, clear governance and cultural protocols and filtering mechanisms to exclude predatory organizations. These organizations are truly LOs, open and flexible to change, and are likely to be able to more rapidly adapt and reinvent themselves to stay well ahead of their competitors even when knowledge about innovative ideas is freely available. This helps them build what Teece *et al.* (1997) refer to as dynamic capabilities that allow them to be highly reflexive and responsive to market changes.

A further model relevant to this chapter developed through the CRC CI research project is illustrated in Figure 26.3, and it indicates how ICT innovation diffusion was taking place within three case-study organizations.

The way that ICT innovation was diffused is influenced by 11 factors. These were derived from factor analysis of the data gathered from the CRC CI study. The factors are illustrated in Figure 26.3, and grouped into four categories: management (F1, F5 and F7), individual (F2, F3, F9 and F10), technology (F4 and F11) and the workplace environment (F6 and F8). Readers interested in more details about the methodology used in this study, and more extensive discussion of the results, should consult the papers published elsewhere (for example, Peansupap 2004; Peansupap and Walker 2005b, 2005c, 2005d, 2006a, 2006b; Walker and Maqsood 2008). The key results of the study that apply more broadly to innovation diffusion through KM initiatives revealed the importance of a work environment that supports the value of being a LO and being involved in general innovation diffusion. Further, the model indicated strong absorptive capacity factors as being important, such as individuals bringing skills with them and having sufficient background knowledge to be able to readily use ICT innovations when provided with opportunities. Important support mechanisms were indicated,

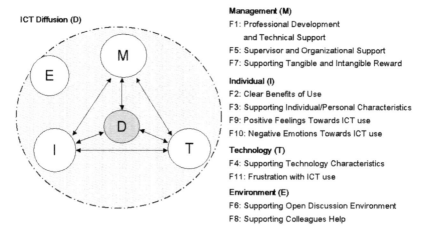

Figure 26.3 Model of ICT innovation diffusion (source: Peansupap (2004: 157), reproduced with permission).

as well as clear management support for the innovations, and providing strategies, resources and implementation plans to facilitate individuals to adopt and adapt innovation. At the individual level (though this was also found to be true at the organizational level), the innovation must be seen as useful and have a clear benefit – adult learners in particular are very focused upon using tools to make their life and job easier (Delahaye 2000). Any introduced innovation must be seen to be user-friendly.

In concluding this part of the discussion, we can comfortably assert that the diffusion of innovation is a socio-technical rather than a purely technical process. We have seen in Figure 26.1 examples of ostensibly similar construction organizations with similar opportunities to adopt and adapt innovative ICT tools available to them all – yet some chose to bravely lead in taking up those opportunities while others tended to be reticent in doing so. Figure 26.2 also helps explain how innovation may be diffused more easily in one organization (or business unit) than another in terms of how well people, processes and technology are integrated and how open the organization is to external knowledge sources. Figure 26.3 helps to better explain how ICT innovation diffusion can be enhanced and improved. The study provided a descriptive discourse about how leading Australian construction organizations adopt and adapt ICT innovation, and also went some way towards providing insights into the forces at work in supporting or inhibiting ICT innovation diffusion. This suggests that other innovative ideas may follow a similar path in these organizations.

Knowledge management and being a learning organization

We can envisage two distinct phases of the diffusion of innovation. The first is that any innovation is, by definition, an idea introduced to an organization (through individuals) that is new to that person/organization. This means that there needs to be a process whereby not only information about the idea but also *knowledge* about how the idea can be usefully and effectively deployed is transferred. Thus, there is a need for KM schemes to be put in place to foster these processes. The second phase relates to the diffusion of innovation throughout the organization. This also requires effective KM processes because in adapting any new innovative idea, the idea undergoes adaptation to suit the local circumstances. Organizations that can effectively manage their knowledge assets can develop characteristics of being a LO. These knowledge assets include highly descriptive knowledge (information about things), procedural knowledge (how to deal with using that information within the organizational context) and cultural knowledge (how to influence people and organizations to effectively create, share, transfer and use knowledge that is necessary to successfully use an innovation).

Nonaka (1991) introduced a model of knowledge transfer and exchange (SECI) whereby individuals in small groups share tacit knowledge through *socialization* and as this tacit knowledge is explained it becomes *externalized* into explicit knowledge that through being *combined* with existing explicit knowledge becomes *internalized* by the individual and reframed again as personal tacit knowledge. This way of looking at knowledge generation and use sees both data and knowledge as being both inert and being actively refined. Others have advanced this view of the *knowledge generation and use* process from both the group and individual perspective. Crossan *et al.* (1999), for example, offer a model described as the '4 Is' – intuiting, interpreting, integrating and institutionalizing. Intuition is tacit knowledge, this is made explicit through interpreting it relative to its context, the knowledge becomes combined and integrated with the pool of knowledge and this becomes organizationally internalized as a whole. Thus, KM is about the creation, sharing, reframing and use of knowledge. Lawrence *et al.* (2005) added to this model by considering the role of power in the process. This better explains how the dynamics of the process operates. They argue that individuals influence groups, and groups force the organization to internalize knowledge into routines and processes, and once that happens this knowledge becomes institutionalized through culture and governance to discipline groups and individuals.

Knowledge, information and data about best practices for diffusing an innovation, for example (what they are, how they may be applied, and how this resource should be stored, accessed and used), are very important issues to be addressed if an organization wishes to remain competitive and become a LO. Creating an environment where knowledge assets are valued and taking the care to properly value and manage organizational routines such as best practices, innovation and creativity can be argued as vital in today's knowledge-based economy. LOs require groups of people to interact intellectually, and one useful way of facilitating this is to encourage people with a shared passion about how to improve various aspects of a business to form CoPs (Wenger 1999; Wenger *et al.* 2002). Discussing how best to shape and encourage a CoP to develop is beyond the scope of this chapter, but readers could refer to Wenger *et al.* (2002) to learn more about this, and practical examples of CoPs in a project management environment can be found in Jewell and Walker (2005) and Peansupap and Walker (2005a).

One interesting feature of knowledge is that it is notoriously difficult to transfer – it is what some researchers have described as being 'sticky' (von Hippel 1990). Szulanski (1995, 1996, 2003) identified seven reasons for knowledge stickiness:

1   The source lacks motivation (unwillingness to share knowledge)
2   The source lacks credibility (lacks authority, expertise or is perceived as unreliable or untrustworthy)
3   The recipient lacks motivation (doesn't care)
4   The recipient lacks absorptive capacity (has not the background to perceive cause and effect links, lacks underpinning knowledge or experience in experimentation to know how to use the knowledge)

5 The recipient lacks retentive capacity (forgets vital details)
6 There is a barren organizational context (the culture or governance structure inhibits knowledge sharing)
7 There is an arduous relationship between source and recipient (lack of empathy, trust or commitment to collaborate in the task of sharing knowledge).

Szulanski (1996) concluded from testing his model that, contrary to conventional wisdom that blames primarily motivational factors, the major barriers to internal knowledge transfer are:

• knowledge-related factors, such as the recipient's lack of absorptive capacity (reason 4);
• causal ambiguity (reason 4);
• an arduous relationship between source and recipient (reason 7).

These factors relate to the ability of individuals and organizations to build and sustain innovation capacity. They also help to explain potential barriers to organizations moving from stages 1 through 3 of the innovation trajectory (Figure 26.2), and the factors affecting ICT innovation diffusion (Figure 26.3). These factors also help to explain how an organization's culture, as well as national cultural traits, can influence the workplace environment; moreover, how able that workplace is to encourage individuals to share knowledge or how a leadership team may influence that workplace culture. It is highly important to gain an understanding of the impact of culture (national and organizational), as culture largely shapes perceived 'strategies and tactics' – that is, politics associated with decisions. Lawrence *et al.* (2005) have discussed the impact of power and influence and politics in shaping the way that knowledge transfer takes place, and Rowlinson *et al.* (2008) also provides a useful overview.

One of the main objectives of the CRC CI research project was to gain a better understanding of the forces at work that influence the success of KM initiatives. This ended up as focusing on what is referred to as the 'people infrastructure'. This infrastructure is supported by a leadership strategy that recognizes how technology is merely a support mechanism, and that a nurturing workplace culture is the key to developing an LO. The next section briefly outlines the model that was developed for improving KM in organizations that aim to better diffuse a wide variety of innovation – from technical and administrative to market development.

A model KM initiative that fosters innovation diffusion

The key goal of being a LO or an innovative organization is that competitive value is delivered through having a knowledge advantage. This advantage sets the framework for creating, sharing and transferring knowledge needed not only to learn about new innovations but also to test and trial ideas and thereby generate new knowledge about the innovation and how it can be effectively used within the organization's operating business context.

The key to this model's usefulness is not so much that it promotes sound KM initiatives and fosters a LO mentality; rather, its purpose is to facilitate action. Knowledge without action provides a dormant or latent competitive competence.

Figure 26.4 illustrates the CRC CI model described by Walker (2005: 16). Before proceeding further with this section, it would be wise to define what is meant by a knowledge advantage (K-Adv). The term K-Adv was used in the research as an abbreviation, but was never registered as a brand name. Walker (2005: 16) argues that 'an organization's K-Adv is its capacity to liberate latent creativity and innovation potential through effective management of knowledge both from within its organizational boundaries and its external environment'.

The K-Adv requires a coordinated approach in addressing leadership actions to:

- *establish and deploy a vision* of what knowledge as a strategic asset (K-Adv) means to the organization;
- support the *people infrastructure* necessary to effectively use knowledge in their business activities;

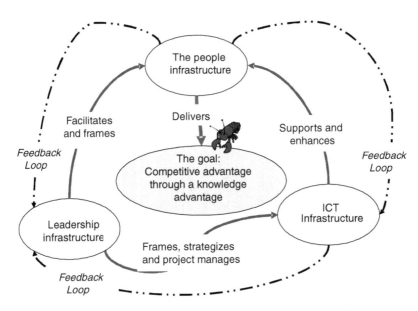

Figure 26.4 A conceptual model for knowledge advantage delivery (source: © Walker (2005: 16)).

- provide the necessary *enabling information and communication technologies (ICT) infrastructure* to do so.

This model features support and feedback that allows this coordinated approach. Figure 26.5 illustrates a more detailed view of how this model can be practically applied.

The starting point for an organization gaining a K-Adv is through its strategic leadership. There has to be a high-level leadership team that must first appreciate the value of KM and organizational learning as a critical strategic asset. Once this is in place, the leadership group then needs to find a way to engage with all organizational stakeholders who hold knowledge that can be harvested, shared, traded or collaboratively developed and refined. The K-Adv leadership infrastructure framework comprises two processes: envisioning, and vision realization. Vision has been shown to be vital in motivating people to perceive a preferred future (Clegg *et al.* 2006; Christenson 2007). This is an active process. First, it involves seeking out stakeholder knowledge about innovation and for these people to help shape and develop the content of the preferred vision of the innovation's application. This requires prioritization, and is where leaders and stakeholders can jointly decide, or help to steer, a strategic direction for the innovation's use. That vision should be clearly articulated (Christenson and Walker 2004). Leaders also need to develop a vision realization plan and ensure that the vision is achieved. This involves the use of standard project management processes such as planning, resourcing and coordinating, and also for maintaining focus on the vision.

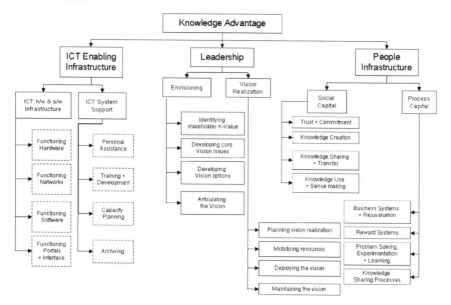

Figure 26.5 A detailed K-Adv model (source: © Walker (2005: 21)).

The important point to remember about leadership infrastructure activities is that the leadership team must champion the processes described above. Without this strategic leadership intent, the necessary infrastructure (human and ICT support) will not be made available and any bottom-up approach to drive the model will be stymied because essential resources will be viewed as a luxury.

People are anchored as the core enabler of the K-Adv model for the simple reason that it is people who make things happen. *Thought leaders* define an organization in terms of social structures and social capital and the way that these interact, in contrasting ways to the command and control power model of leaders influencing followers (Nahapiet and Ghoshal 1998; Cohen and Prusak 2001; Inkpen 2005). Organizations must deploy the people infrastructure's social capital, and support this through the organization's process capital to enable people to deliver a K-Adv. People generate knowledge about innovations that are then shared, transferred, tested, reframed and reabsorbed into the organization. This requires developing a supportive workplace environment for knowledge sharing, as noted by the models developed by Peansupap (2004) and Maqsood (2006). It also requires social capital formation through CoPs, and that needs a plan and process for developing and maintaining trust, commitment and KM processes. The process capital that is highlighted in Figure 26.5 as supporting the people infrastructure relates to business processes that may need to be re-engineered and fine-tuned to help people communicate, gain access to needed resources and be administratively supported rather than being dragged down by bureaucracy – this poses particular challenges to organizations. The culture of the organization also interacts with this process capital element through the organization's reward systems, processes on how people actually share knowledge and engage in collaborative problem-solving. All require processes and routines that are supported by a helpful workplace culture.

The third infrastructure illustrated in Figure 26.5 comprises the ICT support systems in place. Innovation diffusion involves not just technical communication support, such as manuals and online help systems, but also groupware and other collaborative tools to enable people to easily communicate with each other. It will include, too, advanced IT tools such as Building Information Models and innovative visualization and simulation tools (Duyshart 1997; Walker *et al.* 2008b). There is a strong need for a robust groupware ICT platform to connect people in order for this to be helpful to those experimenting with innovations and trying to communicate various issues, problems and discoveries. CoP software provides just one of many collaborative tools that may be useful (for examples of these, see Duyshart *et al.* 2003; Jewell and Walker 2005). A powerful ICT platform is of little use unless it is well supported, and so the second key sub-element of the ICT infrastructure is the ICT support processes and assets that are put in place; as indicated in Figure 26.5, these comprise people support as well as technology support.

The K-Adv tool was envisaged to also serve as a capability maturity modelling (CMM) tool (Walker and Nogeste 2008). CMM tools have been in existence for several decades. Paulk *et al.* (1993) are the authors most frequently cited as having developed the idea. However, it has been adopted by others in project management (Project Management Institute 2003) and, 15 years ago, in the construction industry (Construction Industry Development Agency 1994), though it was never sustainably developed in that industry. Possibly the same reasons apply as was the case with (unsuccessfully) attempting to extend the CRC CI project to test and fine-tune the CMM aspects of the K-Adv concept that were developed. Challenging organizations to be open and to abandon secrecy and the 'management prerogative' is difficult. Walker and Nogeste (2008) explain in more depth how the CMM concept can be used. Others have tested the approach – for example, Manu and Walker (2006) applied this in assessing social capital and knowledge transfer for an aid project in Vanuatu. Figure 26.6 provides a hypothetical result in which an audit of an organization, using the CMM models developed in the CRC CI research project reported upon in this chapter, is presented showing the 'as is' audited position and the desired state at some future time 'X'.

Figure 26.6 shows existing maturity levels of developing the leadership infrastructure for envisioning at level 2, and vision realization also at level 2. Level 2 represents a level of preactive initiation where plans are being developed but are not yet finalized or being put in place. The social capital element of the people infrastructure is indicated as audited at a level 3, active adoption, where planned activities to build social capital are in place. The process capital element is indicated as being currently audited at level 4, indicating proactive adoption. This may imply that planned actions are being deployed and fine-tuned to incrementally improve this aspect. The physical ICT infrastructure of hardware, software and portals, for example, has been audited as being proactively deployed and improved

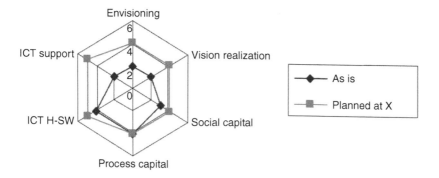

Figure 26.6 CMM hypothetical example.

at level 4 maturity. However, ICT support appears to be deployed at a preactive level 2. The planned 'to be' situation is for full level 4 for all elements except the ICT infrastructure, which is planned to be at level 5 – that is, totally embedded as a regularly reviewed and improved routine adopted into the organizational culture.

The most important value that a tool such as the K-Adv CMM offers in improving innovation diffusion is that it rigorously integrates the three main infrastructures in a planned way to deploy and diffuse innovation. Moreover, it entails providing KM initiatives and applications that improve the people infrastructure, as well as demanding ICT support to help this be ubiquitous and fully supported. It also stresses the initial leadership needed to establish a K-Adv plan and to ensure that plans are realized. The gap analysis is important, because it triggers action to be taken to progress the organization from its 'as is' situation to its 'preferred future'.

Conclusions

The purpose of this chapter has been to indicate how innovation can be effectively diffused by organizations through appropriate KM processes that facilitate firms and business units becoming learning organizations. Innovation was defined and explained, and KM principles were introduced and their relevance explained. A key thread in this chapter has been that innovation diffusion is an active process – indeed, it could be described as needing a proactive process and certainly not a reactive one. The models presented in Figures 26.2 and 26.3 help provide a context for the way that knowledge about innovations can be used and adapted to suit the local situation. This led to a discussion on 'sticky knowledge' that reinforced the relevance and validity of the conceptual model for delivering a knowledge advantage, presented in Figure 26.4, and that helped to explain the detailed K-Adv model illustrated in Figure 26.5. Finally, a hypothetical example was used to illustrate how the K-Adv tool can be used to audit how innovation diffusion can be planned for and deployed, together with a visualization tool for benchmarking results (Figure 26.6).

The most important message running through this chapter is that while the first step in diffusing innovation is dissemination of information and knowledge about the advantages that the innovation can bestow, the real work lies in ensuring that identified innovations are used, and that knowledge about their use and how to best use that innovation is effectively managed. KM in this sense is merely another tool in triggering action and in ensuring that adequate feedback knowledge and information is circulated within an organization to those who need it. KM is a tool in helping organizations to be more innovative and become learning organizations.

Finally, it must be stressed that the value of innovation deployment and diffusion is that it helps organizations to become more reflexive, flexible and ultimately more sustainable in developing and maintaining their dynamic capabilities and competencies so that they can better serve their stakeholders. In this light, innovation diffusion is seen as a sustainable value generator.

Bibliography

Amabile, T.M. (1998) 'How to kill creativity', *Harvard Business Review*, 76 (5): 76–87.

Caloghirou, Y., Kastelli, I. and Tsakanikas, A. (2004) 'Internal capabilities and external knowledge sources: complements or substitutes for innovative performance?', *Technovation*, 24 (1): 29–39.

Christenson, D. (2007) 'Using vision as a critical success element in project management', Doctor of Project Management thesis, School of Property, Construction and Project Management, RMIT University, Melbourne.

Christenson, D. and Walker, D.H.T. (2004) 'Understanding the role of "vision" in project success', *Project Management Journal*, 35 (3): 39–52.

Clegg, S.R., Pitsis, T.S., Marosszeky, M. and Rura-Polley, T. (2006) 'Making the future perfect: constructing the Olympic dream', in D. Hodgson and S. Cicmil (eds) *Making Projects Critical*, Basingstoke: Palgrave Macmillan.

Cohen, D. and Prusak, L. (2001) *In Good Company: How Social Capital Makes Organizations Work*, Boston, MA: Harvard Business School Press.

Cohen, W.M. and Levinthal, D. (1990) 'Absorptive capacity: a new perspective on learning and innovation', *Administrative Science Quarterly*, 35 (1): 128–52.

Construction Industry Development Agency (1994) *Two Steps Forward, One Step Back: Management Practices in the Australian Construction Industry*, Sydney: Construction Industry Development Agency.

Crossan, M.M., Lane, H.W. and White, R.E. (1999) 'An organizational learning framework: from intuition to institution', *Academy of Management Review*, 24 (3): 522–37.

Delahaye, B.L. (2000) *Human Resource Development: Principles and Practice*, Brisbane: John Wiley.

Duyshart, B. (1997) *The Digital Document: A Reference for Architects, Engineers and Design Professionals*, Oxford: Architecture Press.

Duyshart, B., Mohamed, S., Hampson, K.D. and Walker, D.H.T. (2003) 'Enabling improved business relationships: how information technology makes a difference', in D.H.T. Walker and K.D. Hampson (eds) *Procurement Strategies: A Relationship-Based Approach*, Oxford: Blackwell.

Gann, D. (2001) 'Putting academic ideas into practice: technological progress and the absorptive capacity of construction organizations', *Construction Management and Economics*, 19 (3): 321–30.

Garvin, D.A. (1993) 'Building a learning organization', *Harvard Business Review*, 71 (4): 78–90.

Goldsmith, P.W., Walker, D.H.T., Wilson, A. and Peansupap, V. (2003) 'A strategic approach to information communication technology diffusion: an Australian study', paper presented at the 2003 Construction Research Congress, American Society of Civil Engineers, Honolulu, 19–21 March.

Inkpen, A.C. (2005) 'Social capital, networks, and knowledge transfer', *Academy of Management Review*, 30 (1): 146–66.

Jewell, M. and Walker, D.H.T. (2005) 'Community of practice perspective software management tools: a UK construction company case study', in A.S. Kazi (ed.) *Knowledge Management in the Construction Industry: A Socio-Technical Perspective*, Hershey, PA: Idea Group Publishing.

Lawrence, T.B., Mauws, M.K., Dyck, B. and Kleysen, R.F. (2005) 'The politics of organizational learning: integrating power into the 4I framework', *Academy of Management Review*, 30 (1): 180–91.

Leonard, D. and Rayport, J.F. (1997) 'Spark innovation through empathic design', *Harvard Business Review*, 75 (6): 102–13.

Manu, C. and Walker, D.H.T. (2006) 'Making sense of knowledge transfer and social capital generation for a Pacific island aid infrastructure project', *The Learning Organization*, 13 (5): 475–94.

Maqsood, T. (2006) 'The role of knowledge management in supporting innovation and learning in construction', PhD thesis, School of Business Information Technology, RMIT University, Melbourne.

Miles, R.E. and Snow, C.C. (1995) 'The new network firm: a spherical structure built on a human investment philosophy', *Organizational Dynamics*, 23 (4): 5–18.

—— (2003) *Organizational Strategy, Structure and Process*, Stanford, CA: Stanford University Press.

Nahapiet, J. and Ghoshal, S. (1998) 'Social capital, intellectual capital, and the organizational advantage', *Academy of Management Review*, 23 (2): 242–66.

Nonaka, I. (1991) 'The knowledge creating company', *Harvard Business Review*, 69 (6): 96–104.

Paulk, M.C., Curtis, B., Chrisses, M.B. and Weber, C.V. (1993) 'Capability maturity model, version 1.1', *IEEE Software*, 10 (4): 18–27.

Peansupap, V. (2004) 'An exploratory approach to the diffusion of ICT innovation in a project environment', PhD thesis, School of Property, Construction and Project Management, RMIT University, Melbourne.

Peansupap, V. and Walker, D.H.T. (2005a) 'Diffusion of information and communication technology: a community of practice perspective', in A.S. Kazi (ed.) *Knowledge Management in the Construction Industry: A Socio-Technical Perspective*, Hershey, PA: Idea Group Publishing.

—— (2005b) 'Exploratory factors influencing ICT diffusion and adoption within Australian construction organizations: a micro analysis', *Journal of Construction Innovation*, 5 (3): 135–57.

—— (2005c) 'Factors affecting ICT diffusion: a case study of three large Australian construction contractors', *Engineering, Construction and Architectural Management*, 12 (1): 21–37.

—— (2005d) 'Factors enabling information and communication technology diffusion and actual implementation in construction organizations', *Electronic Journal of Information Technology in Construction*, 10: 193–218.

—— (2006a) 'Information communication technology (ICT) implementation constraints: a construction industry perspective', *Engineering, Construction and Architectural Management*, 13 (4): 364–79.

—— (2006b) 'Innovation diffusion at the implementation stage of a construction project: a case study of information communication technology', *Construction Management and Economics*, 24 (3): 321–32.

Project Management Institute (2003) *Organizational Project Management Maturity Model (OPM3) Knowledge Foundation*, Newtown Square, PA: Project Management Institute.

Rogers, E.M. (2003) *Diffusion of Innovation*, 5th edn, New York, NY: Free Press.

Rowlinson, S., Walker, D.H.T. and Cheung, F.Y.K. (2008) 'Culture and its impact upon project procurement', in D.H.T. Walker and S. Rowlinson (eds) *Procurement Systems: A Cross-Industry Project Management Perspective*, Abingdon: Taylor & Francis.

Schein, E.H. (1993) 'How can organizations learn faster? Lessons from the green room', *MIT Sloan Management Review*, 34 (2): 85–92.

Szulanski, G. (1995) *Appropriating Rents from Existing Knowledge: Intra-Firm Transfer of Best Practice*, Fontainebleau, France: Institut Européen d'Administration des Affaires.

—— (1996) 'Exploring internal stickiness: impediments to the transfer of best practice within the firm', *Strategic Management Journal*, 17: 27–43.

—— (2003) *Sticky Knowledge: Barriers to Knowing in the Firm*, Thousand Oaks, CA: Sage.

Teece, D., Pisano, G. and Shuen, A. (1997) 'Dynamic capabilities and strategic management', *Strategic Management Journal*, 18 (7): 509–33.

Van de Ven, A.H. (1986) 'Central problems in the management of innovation', *Management Science*, 32 (5): 590–607.

von Hippel, E. (1990) ' "Sticky information" and the locus of problem solving: implications for innovation', *Management Science*, 40 (4): 429–39.

von Hippel, E., Thomke, S. and Sonnack, M. (1999) 'Creating breakthrough at 3M', *Harvard Business Review*, 77 (5): 47–57.

Walker, D.H.T. (2005) 'Having a knowledge competitive advantage (K-Adv): a social capital perspective', keynote address, International Conference on Information and Knowledge Management in a Global Economy, Lisbon, 19–20 May.

Walker, D.H.T. and Maqsood, T. (2008) 'Procurement innovation and organizational learning', in D.H.T. Walker and S. Rowlinson (eds) *Procurement Systems: A Cross-Industry Project Management Perspective*, Abingdon: Taylor & Francis.

Walker, D.H.T. and Nogeste, K. (2008) 'Performance measures and project procurement', in D.H.T. Walker and S. Rowlinson (eds) *Procurement Systems: A Cross-Industry Project Management Perspective*, Abingdon: Taylor & Francis.

Walker, D.H.T., Anbari, F.T., Bredillet, C., Söderlund, J., Cicmil, S. and Thomas, J. (2008a) 'Collaborative academic/practitioner research in project management: examples and applications', *International Journal of Managing Projects in Business*, 1 (2): 168–92.

Walker, D.H.T., Aranda-Mena, G., Arlt, M. and Stark, J. (2008b) 'E-business and project procurement', in D.H.T. Walker and S. Rowlinson (eds) *Procurement Systems: A Cross-Industry Project Management Perspective*, Abingdon: Taylor & Francis.

Walker, D.H.T., Maqsood, T. and Finegan, A. (2005) 'The culture of the knowledge advantage (K-Adv): an holistic strategic approach to the management of knowledge', in A.S. Kazi (ed.) *Knowledge Management in the Construction Industry: A Socio-Technical Perspective*, Hershey, PA: Idea Group Publishing.

—— (2007) 'A prototype portal for use as a knowledge management tool to identify knowledge assets in an organization', in A. Tatnall (ed.) *Encyclopedia of Portal Technology and Applications*, Hershey, PA: Idea Group Publishing.

Walker, D.H.T., Wilson, A. J. and Srikanthan, G. (2004) *The Knowledge Advantage (K-Adv) for Unleashing Creativity & Innovation in Construction Industry*, Brisbane: CRC for Construction Innovation. Online. Available at HTTP: <www.construction-innovation.info/images/pdfs/2001–004-A_Industry_Booklet.pdf>.

Wenger, E.C. (1999) 'Communities of practice: the key to knowledge strategy', *Journal of the Institute for Knowledge Management*, 1 (1): 48–63.

Wenger, E.C., McDermott, R. and Snyder, W.M. (2002) *Cultivating Communities of Practice*, Boston, MA: Harvard Business School Press.

Winter, M. and Smith, C. (2006) *Rethinking Project Management: Final Report*, Manchester: Engineering and Physical Sciences Research Council.

27 The business case for sustainable commercial buildings

Kendra Wasiluk and Ralph Horne

It is now widely accepted by most climate scientists and governments that human-induced climate change is real and occurring, with recent extreme weather events, sea-level rise and climate trends causing severe damage to human life and supporting ecosystems (IPCC 2001; G8 Presidency 2005; Lowe 2005; Newton 2008; Flannery 2005). Greenhouse gas emissions from the use of energy in buildings and transport are a key contributor to climate change (Fien 2004; Hamnett and Freestone 2000; Gleeson *et al.* 2000). In Australia, 23 per cent of such emissions are a result of energy demand in the building sector (CIE 2007). However, the environmental impact of buildings extends well beyond their operational energy use and associated greenhouse gas emissions.

In OECD countries, buildings consume 30 to 50 per cent of available raw materials and generate about 40 per cent of waste to landfill (OECD 2003). The built environment can also have an impact on occupant health, as people spend on average approximately 90 per cent of their time indoors (Paevere and Brown 2007). From an economic and income-generating perspective, the building and construction sector in many countries constitutes more than half of the country's real capital (Huang *et al.* 2005). In Australia, the construction of buildings alone accounts for almost 10 per cent of GDP and has been growing at a rate of 3 per cent per annum (Department of Environment and Heritage 2006: 54). If all of the residential and commercial activity within the built environment is taken into account, the building sector houses the majority of Australia's economy – nearly 60 per cent of GDP (CIE 2007: 8). As a result, the sector has significant environmental, economic and social impacts both locally and globally. In order to make strides towards a sustainable future, the building sector is a key part of the challenge.

The total commercial building stock in Australia is approximately 20.5 million m^2, with approximately 14.5 million m^2 located in central business districts (PCA 2007). Most performs poorly against sustainability benchmarks, and there is huge scope for improving the sustainability of the existing commercial building stock. Entrepreneurs in the sector are already responding to environmental pressures, primarily on a voluntary basis. Some government agencies have used their office accommodation guidelines as an incentive to push compliance with voluntary building rating tools, particularly for energy efficiency (AGO 2007; Victorian Government Property Group 2005). However, this is generally limited to new buildings and larger occupancies. The Australian Building Codes Board has also recently introduced national regulations with regard to energy efficiency, through Section J in the Building Code of Australia.

Supporters of sustainability in commercial buildings are often challenged to justify their motivations to their board or executive management, clients, customers and agents, frequently against widespread perceptions that sustainable buildings cost more, hold inherent risks and do not add to portfolio value. Due to the infancy of many such buildings, some of the arguments used to support the business case are often weakened by a lack of documented evidence related to actual outcomes. Another key issue is that many of the benefits of investment in sustainability often accrue to an organization's intangible assets. However, intangibles cannot be dismissed, as that is where most of the market value of companies actually sits – an exceptionally high 77 per cent of stock exchange (ASX 200) value in Australia as of 31 December 2006 (AMP Capital 2007). These intangible factors, such as employee and customer relationships, supply chain and intellectual property, can be extensively linked to a company's performance on environmental and social issues.

This chapter presents the business cases for sustainability – the rationale for why individuals and companies should make decisions in favour of sustainability when owning, developing, designing, building, occupying or managing commercial buildings. It addresses both profit performance and value, and the broader additional indirect and/or intangible benefits that accrue through factoring sustainability into decision-making.

This chapter will enable individuals, whatever role they play across the industry, to:

- understand and value the benefits of sustainable practices, products and processes related to commercial buildings;
- gain a greater understanding of the motivations, benefits and value of sustainability to others in the industry;
- communicate sustainability to their clients, customers and work colleagues in relevant terms and at critical stages of their decision-making throughout the complete cycle of conceiving, financing, planning, developing, designing, building and occupying commercial buildings.

Definition of a sustainable commercial building

A sustainable commercial building can be defined as one with planning, design, construction, operation and management practices that minimize the impact of development on the environment. It is also economically viable, and enhances social amenity for its occupants and the community

(McCartney 2007). The Green Building Council of Australia (GBCA 2006: 16) defines a green building as 'one that incorporates design, construction and operational practices that significantly reduces or eliminates the negative impact of development on the environment and occupants'.

Definition of a business case

A business case is a key form of advice used by executive decision-makers. It is a key element in the decision-making and budgeting process, providing a commercial assessment for a project, policy or program proposal requiring a commitment of financial or human resources. A business case should present a detailed summary of the business benefits, market impact, financial benefits and potential risks of undertaking a new venture. Within the strategic framework of the company's mission, vision and values, it sets out:

- the problem or situation addressed by the proposal;
- the features and scope of the proposed initiative;
- options considered and the rationale for choosing the solution proposed;
- the proposal's conformity with existing policies, regulations and processes;
- an implementation plan;
- expected costs of the proposal;
- anticipated outcomes and benefits of the proposal;
- expected risks associated with the proposal's implementation.

A business case should always:

- contrast the cost, benefits and risks associated with the proposed solution against the option of doing nothing (the base case);
- be justified in terms of the organization's strategic objectives and priorities;
- demonstrate how the outputs from the project would generate the outcomes required by the organization.

The business case for sustainable commericial buildings

In Australia the business case for sustainable commercial buildings is in a rapid state of transition, as sustainability objectives are becoming increasingly relevant for all groups in the commercial building industry. Davis Langdon (2006a, 2006b) reported a 10 per cent increase in the number of respondents indicating that green objectives are 'already an issue' in their business over a six-month period in 2006 (January to July). Respondents believed that green objectives are being driven by informed clients, and it is this client demand that has increased the overall demand for green buildings. Clients are aware that meeting corporate sustainability objectives, for themselves or their tenants, is a long-term 'value add' for their asset. A number of other factors are driving the increased uptake of sustainability in the commercial building sector, including government regulation, stakeholder demands, environmental imperatives, cost efficiencies and brand value.

Consideration of profit and value

As all projects require a budget, choices need to be made to distinguish between competing claims for the allocation of resources. To fully evaluate those claims and the benefits that can accrue from investments in sustainable commercial buildings, it is critical to create a business case that looks beyond the traditional project-cost model towards one that recognizes the value that such buildings can add to an organization. The link between the costs and the added value (tangible or intangible) that may result from the investment is of particular importance, recognizing that 'cost savings' do not mean the same thing as 'value', and that cost savings do not necessarily add value to a building.

The argument has become not about profit or value, but about profit *and* value. Therefore, the business case needs to provide decision-makers with a sound understanding of all the ways in which a sustainable approach to commercial buildings will help the organization achieve business success.

A range of literature exists on the business case for sustainability in general and for other industry sectors (Salzmann *et al.* 2005; Schaltegger and Wagner 2006; Steger 2004; Steger *et al.* 2007; SustainAbility 2002; SustainAbility and UNEP 2005; Willard 2002), and two recent publications specifically address the business case for sustainable commercial buildings (Kats 2003; GBCA 2006). Other studies have outlined the 'value' derived from investment in sustainable commercial buildings (Reed and Wilkinson 2006; RICS 2005; Saxon 2005). From this literature comes a host of 'value drivers' which motivate organizations to incorporate sustainability into their decision-making. These value drivers include the following:

- boost employee productivity;
- enhance health and well-being of occupants;
- reduce liability;
- benefit your community;
- be quicker to secure tenants;
- command international recognition;
- attract and retain staff;
- achieve lower tenant turnover;
- command higher prices/rents;
- provide export opportunities;
- produce infrastructure cost benefits;
- cost less to operate and maintain;
- reduce insurance rates;

- attract grants, subsidies and other inducements;
- create jobs;
- reduce risk;
- attract government tenants;
- reduce capital costs;
- provide higher returns on asset and increase property values.

This list demonstrates that there are numerous business case arguments for the value of sustainable commercial buildings. Not all factors are relevant for every business case, and generally one or two key factors may be particularly relevant in a given business case. It is important to understand that each business case is unique, due to:

- different motivations and drivers of value across industry groups, often related to the group's role in the building life cycle;
- the strategic objectives of each organization (i.e., the business case can vary between organizations within the same industry group);
- the values and beliefs of the decision-makers at all levels.

Thus, there is not just *one* business case for sustainable commercial buildings – the business cases can vary considerably across the sector.

Your Building business case module – four fundamental questions

The *Your Building* web portal, <www.yourbuilding.org>, is an online Australian resource regarding sustainable commercial buildings. It provides information for all those involved across the building life cycle – from investors, owners and occupiers to developers, builders, designers and facility managers. One of its major features is the significant work on the business cases for sustainability designed to explain why individuals and companies should make decisions in favour of sustainability when owning, developing, designing, building, occupying or managing commercial buildings.

The business case module poses four key questions underlying business cases for sustainable commercial buildings, intended to capture a range of value factors:

1   How can a sustainable commercial building contribute to my company's profit?
2   How can a sustainable commercial building affect my company's risk?
3   How can a sustainable commercial building contribute to my company's business continuity?
4   How can a sustainable commercial building help me realize my personal values and beliefs?

Within these four fundamental questions, the benefits sought by, and accruing to, each industry stakeholder group (and, in fact, each company) are different, and so each group will have its own business case. In this chapter, each question is outlined along with a summary table showing how sustainable commercial buildings may provide benefits in each instance (the value drivers) and the industry group for which they may be relevant. For each question, one value factor has been selected and explored in further detail.

The *Your Building* web portal presents additional information for each of the four questions, including:

- relevant value drivers (listed and explained);
- typical areas of cost saving, value add and risk management;
- Australian and international case study examples;
- whose business case benefits?

Sustainable commercial buildings and profit

The commercial property and construction industry has begun to warm to the sustainable building agenda, 'not necessarily because its members have miraculously developed an insatiable urge to save the planet, but because they have begun to see a viable new investment opportunity' (Building Design+Construction 2006: 5).

A mindset of cutting upfront costs has long been established as standard practice when developing, designing, constructing, managing and occupying commercial buildings. However, sustainable commercial buildings require a different kind of budget approach – one that takes into account life cycle costs – as the majority of potential savings will likely result after construction. Less then 10 per cent of the total life cycle cost of a building lies in construction, while 60 to 85 per cent lies in the ongoing expenses – including continuous maintenance, and energy and repair costs. However, while the majority of cost savings accumulate over the life of the building, shorter-term financial gains are also achievable by the developer or builder by using a 'least first cost' approach.

Allocation of cost saving benefits is an important business case consideration: one of the barriers to sustainability uptake continues to be the disassociation of benefits and costs. The developer and owner pay the capital costs of sustainability features (passive or active) – costs which may not flow through to increased rental levels – while many of the operational savings accrue to tenants by way of reduced occupancy costs.

When preparing a business case for sustainable commercial buildings, it is also important to identify and communicate where there are opportunities for increasing bottom line profit, and to identify areas where profit may be reduced.

Table 27.1 summarizes the profit motivated business case factors, and the industry groups that are impacted by each of them.

Table 27.1 Sustainable commercial buildings and profit value factors

Sustainability and profit	Owners	Developers	Designers	Builders	Managers	Occupiers
Reduced commissioning, operating and maintenance	✓					
Energy savings	✓✓	✓✓			✓	✓✓
Reduced capital costs of mechanical system, as control systems reduce need for over-sizing	✓✓	✓✓				✓✓
Water savings	✓✓	✓✓			✓✓	✓✓
Waste reduction and disposal savings	✓✓	✓✓			✓✓	
Reduced development costs	✓✓	✓✓			✓	
Accelerated planning approval process	✓✓		✓			
Lower carrying costs	✓✓	✓✓				
Compressed schedule	✓✓	✓✓				
Green premium (higher return on asset)	✓✓	✓✓				
Tenancy benefits	✓✓	✓✓				
Reduced vacancy rates	✓✓				✓✓	
Higher NOI for gross leases	✓✓				✓✓	
Financial incentives	✓✓	✓✓				✓
Avoided energy infrastructure investment	✓✓	✓✓				
Improved indoor environment quality	✓✓				✓	✓✓
Lower churn, turnover, tenant inducements	✓✓				✓✓	✓✓
Ability to attract and retain employees						✓

Source: Wasiluk (2007), reproduced with permission.

Example 1: value driver – improved indoor environment quality

Indoor environment quality is an area of great interest to researchers and the industry more generally. While empirical evidence is difficult to locate, there is an expectation that improved indoor environment quality contributes to improved productivity and reduced employee and tenant turnover, due to healthier and more attractive working conditions. This may result in lower costs for owners and occupiers. Typical areas of cost savings include:

- lower churn;
- increased ability to attract and retain employees;
- improved occupant productivity;
- reduced absenteeism;
- reduced turnover and tenant inducements.

Lower churn

The costs of churn are often neglected when the life cycle costs of a building are estimated. A survey by the IFMA (1997) indicated that, on average, 44 per cent of occupants move within a given year, with churn rates increasing over time. Data collected on churn costs in an actual government office building with 1,500 workstations (the Rachel Carson State Office Building in Harrisburg, Pennsylvania) indicated that a high-performance sustainable building could save over $800,000 annually, based on a large building with a 25 per cent annual churn rate. This equates to a saving of $2,250 per person moved (Loftness *et al.* 2002).

To reduce churn costs, many buildings now include features such as raised floor systems and movable partitions, in place of suspended ceilings and permanent walls. Underfloor HVAC systems, reduced ductwork, modular power cabling and telecommunications/data systems, and movable partitions reduce the need for complex construction during alterations, which reduces churn costs and allows spaces to be modified quickly. Some of these design features may also have flow-on sustainability and cost benefits by reducing the waste, the time and the disruption to business that can be associated with less flexible and adaptable systems.

Increased ability to attract and retain employees

The costs associated with staff turnover are estimated to be 50 per cent of salary (for mid-level managers), and therefore there is a bottom line benefit for businesses to invest in and retain their staff (CABE and BCO 2005: 11). When employees leave, they take with them firm-specific tacit knowledge that is not easily valued or quantified. However, as with a number of other savings to an organization's operating costs, it may not be possible to identify a cause and effect relationship between reduced staff turnover and occupancy of a sustainable commercial building.

Improved occupant productivity

Employee salaries and benefits represent the largest proportion of business costs (see Figure 27.1), and improvements in employee productivity are claimed to produce cost savings for occupants that will surpass savings in their operating costs (CABE and BCO 2005). For example, it is claimed that an increase of just 1 per cent in employee productivity can nearly offset a company's entire annual energy bill. Accordingly, a large proportion of the research related to the profit-driven business case centres on the paybacks of improved employee productivity, and reduced illness and absenteeism.

Sustainable commercial buildings and business continuity

To survive, any business must continue to perform and be profitable in a competitive and changing environment. While the major cause of business failure is financial, competitive position may be eroded – even to the extent of business failure – by other factors, such as poor reputation, lack of brand recognition, inability to attract and retain quality employees, and inability to attract and retain new customers and new business. Business continuity value factors examine the contribution that sustainable commercial buildings can make to these non-financial elements of the competitive advantage that is necessary for business continuity.

Traditionally, decisions in building projects have been cost- and time-based, with a view to achieving short-term benefits and paybacks, while tending to overlook those non-financial factors that are critical to longer-term business survival. Yet integrating the management of these non-financial issues – often classed as intangible assets – with the economic

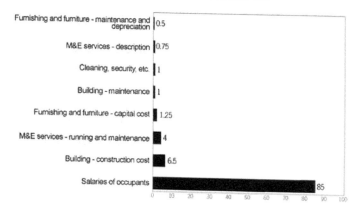

Figure 27.1 Breakdown of occupiers' business costs across a 25-year lease (source: adapted from CABE and BCO (2005)).

502   K. Wasiluk and R. Horne

purpose of the company is important, and can be considered a 'constitutional requirement of the management of any profit-orientated company' (Schaltegger and Wagner 2006).

There is growing awareness of the contribution that sustainable buildings can make to creating and maintaining value in intangible assets which can, in turn, enhance a company's competitive position. Some of that value derives directly from the physical characteristics of a sustainable commercial building, while some may result from associated changes to organizational culture, behaviours, policies, structures or processes.

Table 27.2 summarizes business-continuity motivated business case factors, and the industry groups that are impacted by each of them.

Example 2: value driver – improved corporate profile and community relations

Sustainable commercial buildings can provide branding, marketing and publicity benefits, leading to competitive advantage for companies, particularly those that take an early leadership position. Value is typically created by:

- living corporate values through the building asset;
- enhanced marketability;
- enhanced publicity.

Living corporate values through the building asset

Public organizations have always been accountable to the public for their licence to operate. However, the public, investors and clients are increasingly interested in the social and environmental performance of both private and public companies, and are increasingly demanding that their operations be more accountable and transparent. Buildings are the largest visual symbols of an organization's corporate presence and culture, and can link into a company's sustainability profile (CABE and BCO 2005), therefore contributing to brand recognition and the value of reputational assets.

Enhanced marketability and publicity

Many case-study examples exist where stakeholders have derived marketing benefits from free and unsolicited coverage in major publications because of their association with sustainability (Building Design+Construction 2006: 15). In April 2007, Melbourne City Council's CH2 was featured on the ABC TV show *Catalyst* in a segment entitled 'Council House Two: the eco-office block of the future', while companies such as Colliers International, Bovis Lend Lease, Kador Group, Szencorp and NAB have all been publicized for their involvement in, or ownership of, sustainable commercial buildings (Southgate 2007). Companies may also receive public recognition when their

Table 27.2 Sustainable commercial buildings and business continuity value factors

Sustainability and business continuity	Owners	Developers	Designers	Builders	Managers	Occupiers
Improved occupant productivity	✓	✓			✓	✓
Reduced occupant complaints	✓	✓		✓	✓	✓
Improved corporate profile and community relations	✓	✓	✓	✓	✓	✓
Living corporate values through building asset	✓	✓		✓	✓	
Enhanced marketability	✓	✓	✓	✓	✓	✓
Enhanced publicity	✓	✓	✓	✓	✓	✓
Ability to attract and retain employees	✓				✓	
Tenancy benefits	✓				✓	
Ability to attract and retain tenants	✓				✓	
Reduced vacancy rates	✓				✓	
Responding to client demands	✓	✓	✓	✓	✓	
Ability to attract new business		✓	✓	✓		✓

Source: Wasiluk (2007), reproduced with permission.

buildings achieve sustainability ratings under designated building ratings schemes, while others have received publicity from winning green awards.

Sustainable commercial buildings and risk

In the commercial building industry there are traditionally a number of types of risks facing industry groups, including:

- economic risk of overall investment quality;
- financial risks, such as access to capital, unforeseen events or budget over-runs and delays;
- market risks, such as damage to reputation, ability to attract and retain tenants and employees, and economic cycles;
- social risks, such as occupant health and safety;
- legal risks, including contractual disputes and legislative changes.

Faced with this large scope of risk, the building and construction market is generally said to be 'risk averse' because it seeks to reduce, minimize and carefully manage risk. An aversion to increasing risk is claimed to be slowing the acceptance and uptake of sustainability-related innovation, 'because of an understandable reluctance to accept new methods without proof that they work' (RICS 2005: 13).

However, value in each of the above-mentioned areas can be derived from reducing risks through the adoption of sustainability initiatives, extending to both compliance and beyond compliance activities. Potential also exists for systematic early identification and attention to addressing long-term issues such as resource depletion, fluctuations in energy costs, product liabilities and pollution and waste management.

Understanding the language of risk in the business case for sustainable commercial buildings means appraising not only what the risks of incorporating sustainability are, but also what the risks are if sustainability is *not* incorporated. This adds an extra layer of scrutiny to every decision made about a building.

Table 27.3 summarizes risk-motivated business case factors, and the industry groups that are impacted by each of them.

Example 3: value driver – future-proofing

Sustainability implies an investment beyond that required by compliance. However, there are few (if any) companies that would take the risk of developing, owning or renting a building that merely complied with legislated requirements. The debates around climate change and cost internalization of energy, water and emissions mean that legislation is changing rapidly. Furthermore, environmental benchmarking of the built environment is on the rise in Australia and, with the new Section J BCA (Building Code of

Table 27.3 Sustainable commercial buildings and risk value factors

Sustainability and risk	Owners	Developers	Designers	Builders	Managers	Occupiers
Improved environment quality	✓					✓
Future liability					✓	✓
Sick building syndrome	✓				✓	✓
Employee expectations						✓
Reputational capital	✓	✓	✓	✓	✓	✓
Insurance	✓				✓	
Risk of new materials and methods	✓	✓	✓	✓	✓	
Future-proof buildings	✓	✓			✓	
Regulatory requirement	✓	✓			✓	✓
Performance disclosure and building ratings	✓	✓	✓		✓	

Source: Wasiluk (2007), reproduced with permission.

Australia) energy-efficiency guidelines coming into effect, extends beyond voluntary tools such as Green Star, NABERS (National Australian Built Environment Rating Scheme) and ABGR (Australian Building Greenhouse Rating). These guidelines will affect all asset classes, including older commercial assets. New sustainability inclusions in the Property Council of Australia quality matrix will also impact not only on new but also on existing buildings. The introduction into the market of new buildings with higher sustainability performance standards threatens the future value of older stock, which will need to be upgraded to satisfy fast-evolving market expectations. In Australia, all state governments and the Commonwealth government have minimum ABGR requirements for leased or owned office accommodation, and most also have Green Star requirements.

As a consequence, to begin a building project – either a new building or a refurbishment – or to rent a building that does not meet or exceed today's sustainability benchmarks risks being functionally obsolete from the day it opens, and economically disadvantaged for its entire lifespan. Similarly, developing and owning buildings that will not be able to cope with the growing demand for sustainable commercial assets opens owners to greater risk of losses on their investment.

In contrast, a sustainable commercial building that exceeds minimum legislated requirements may be future proofed against risk through being safeguarded against future energy and water price increases and downturns in the real estate market, the impact of which, according to Davis Langdon Australia (2007), 'should not be underestimated'. Future-proofing of assets and investments so that they anticipate market trends and minimize the need for future large-scale refurbishments and upgrades is a risk-reduction strategy that may help to stabilize asset value.

Sustainable commercial buildings and personal values and beliefs

A diverse set of personal values and beliefs underlies each individual's commitment to sustainability, and will undoubtedly have an impact on how they factor sustainability into their business planning and decision-making. Personal motivations may influence – or bias – decision-making towards sustainability, even where organizational policies and processes prescribe a particular approach. For example, a developer with a personal commitment to sustainability is more likely to factor sustainability considerations into their decision-making. Thus, factors that lie behind decision-making about sustainable commercial buildings will be a complex amalgamation that may not be well documented or transparent. Furthermore, any organization that wishes to embed sustainability in its strategic and organizational decision-making will benefit from making sustainability part of its cultural values.

Table 27.4 summarizes personal value and belief motivated business case factors, and the industry groups that are impacted by each of them.

Example 4: value driver: ability to attract and retain employees

As the baby-boom generation reach retirement age, the smaller workforce of generation X (people born from the 1960s through to the late 1970s) and generation Y (people born from the early 1980s through to the late 1990s) will replace them. By 2011, the Australian workforce will have been transformed to include a significant increase in the apparently more environmentally conscious generation X and Y workers, with generation Y numbers expected to double from 20 to 40 per cent of the workforce (Colliers International 2006). Increasing industry group participation in sustainable commercial buildings will allow the next generation of staff to live their personal values in their work environment. In seeking to attract and retain these employees, it is important to understand their personal values and beliefs, and how these can be realized through involvement in sustainable commercial buildings.

Younger generations are more likely to report greater environmental concerns, with Australians aged 18 to 24 twice as likely to rate the

Table 27.4 Sustainable commercial buildings and personal values and beliefs value factors

Sustainability and personal values and beliefs	Owners	Developers	Designers	Builders	Managers	Occupiers
Improved corporate profile and community relations	✓	✓	✓	✓	✓	✓
Living corporate values through building asset	✓	✓	✓	✓	✓	✓
Ability to attract and retain employees	✓	✓	✓	✓	✓	✓

Source: Wasiluk (2007), reproduced with permission.

environment as the most important social issue than those aged 55 and over (ABS 1998). Generation Y are said to have less loyalty to companies, and are always seeking promotions, better opportunities and more varied job roles. Many feel personally responsible for trying to make a difference in the world, and they often consider environmental and social issues when looking for work (Jayson 2006). Generation Y has also been called 'generation give', for their personal commitments to social and environmental issues (Sheedy 2007).

Sustainable business strategies can also align an organization with the intrinsic values and beliefs of its current and prospective employees. Survey results from the 2006 BCI Australia report found that 'being part of an industry that values the environment' is the top personal driver for all stakeholder groups. Over 70 per cent of all stakeholder groups surveyed responded that this was a key reason why they are involved in sustainable building, making this altruistic motive 'the most significant driver of green building involvement in the AEC (architecture, engineering and contracting) and building owner community in Australia today' (BCI and GBCA 2006: 23). Satisfaction from 'doing the right thing', providing a 'better place to live', proving that development can be done in a different way, educating people about environmental issues and helping to make the world a better place for their grandchildren are some of the other personal values and motivations of individuals in the commercial building industry (Wilson *et al.* 1998).

Barriers to a sound business case

Potential barriers to creating and implementing a sound business case for sustainable commercial buildings may include:

- lack of understanding and knowledge across industry groups;
- perceptions of additional costs;
- deficient valuation techniques;
- lack of hard data.

Each of these can have an impact on how the business case is formulated, presented and received.

Lack of understanding and knowledge across industry groups

Given the recent nature of much of the terminology and interest regarding the benefits and costs of sustainable buildings, there is an understandable knowledge gap when it comes to understanding how to own, develop, design, construct, manage and occupy sustainable commercial buildings. This may make it difficult for one industry group to convince another of the business case, and for the widespread uptake within in each group.

Common knowledge gaps across all industry groups which can have an impact on understanding the business case for sustainable commercial buildings (RICS 2005: 7) include:

- a general lack of awareness in the market;
- a shortage of environmentally sustainable development knowledge, research and resources;
- sustainability strategies not being widely understood;
- the steep learning curve required for developers, owners and consultants;
- a lack of construction companies with experience in sustainable construction;
- a shortage of engineers with experience in designing and operating green building systems.

Perceptions of additional costs

There is a perception that sustainable commercial buildings cost significantly more than traditional buildings. Although a shift is occurring, the discourse on the costs and values of sustainable design is dominated by these perceived additional costs, in part because methods for calculating costs are more highly developed and more readily accepted than methods for assessing less tangible benefits and values, externalities and future costs/benefits.

A number of key Australian and international publications have sought to dispel myths about the costs of sustainable commercial buildings (GBCA 2006; RICS 2005; Kats 2003), with a recent study by Davis Langdon

Australia (2007) finding that the initial impact of sustainability measures on construction costs is likely to be in the order of 3 to 10 per cent. Nonetheless, the debate about the cost of sustainability, and its effect on profit, remains a key barrier in the commercial building industry. Davis Langdon Australia (2006a: 7) concluded that this 'perception of extra costs will diminish as the design strategies become the norm'. This means that the cost of 'business as usual' will gradually be revised to align with sustainable outcomes. In the meantime, additional value propositions are needed to justify any incremental increases in construction costs for sustainable buildings.

Deficient valuation techniques

Current valuation techniques lack the ability to reflect the additional capital and operating costs associated with sustainability that may be involved for owners and developers. These costs may not flow through into increased rents and building valuations, and so may reflect negatively on predicted profitability, even though operating cost reductions are of benefit to the occupiers. As a consequence, sustainability continues to be overlooked as a significant factor in valuation models used by most owners, investors and valuers. This situation may change in the near future, but at this point in time, valuation techniques continue to be a barrier to the development of a successful business case for sustainable commercial buildings.

Lack of hard data

A lack of quantitative data, particularly around intangibles (such as the contribution of a sustainable commercial building to productivity, employee attraction and retention, and organizational reputation), is another barrier to assembling the business case. For example, because measuring the 'soft' benefits of sustainable buildings may necessarily be qualitative in nature, and because it is often difficult to establish cause and effect of various strategies, convincing people that sustainable design can improve productivity, reduce operating costs and future-proof an investment has proven to be difficult. Yet those assembling businesses' cases are increasingly being required to provide less anecdotal evidence and to substantiate their claims. In addition, only a handful of sustainable commercial building projects exist in Australia at present, and many of these have only been recently completed, which means that long-term data on costs and performance are not yet available.

Future directions for the sustainable building business case

While a range of literature exists on the business case for sustainability, and specifically the business case for sustainable commercial buildings in Australia (GBCA 2006; Wasiluk 2007), there is no agreed methodology to link, capture and measure the value created as a result of adopting sustainability principles in the commercial building sector. Studies have outlined the 'value' derived from sustainable commercial buildings (Reed and Wilkinson 2006; RICS 2005; Saxon 2005), but much of the hypothetical value identified is claimed by many to be 'intangible' or difficult to quantify. Intangibles, though not always recognized, have always been a driver of corporate performance (Cohen and Low 2001). As stated earlier, they cannot be dismissed, as that is where most of the market value of companies may actually reside.

There is a pressing need for a set of widely accepted metrics by which corporate leaders and the investment community can account for the non-financial factors that profoundly affect value creation in the modern enterprise.

Conclusion

The business case for sustainable commercial buildings indicates why and how individuals and companies may make decisions in favour of sustainability when owning, developing, designing, building, occupying or managing commercial buildings. In Australia this business case is in a rapid state of transition, as sustainability objectives are becoming increasingly relevant for all industry groups in the sector. A number of other factors are driving the increased uptake of sustainability, including government regulation, stakeholder demands, environmental imperatives, cost efficiencies and brand value.

The *Your Building* web portal is the key online Australian resource for sustainable commercial buildings. It provides information for all those involved across the building life cycle – from investors, owners and occupiers to developers, builders, designers and facility managers. One of its major features is the significant work on the business cases for sustainability. This explores the process involved in making decisions in favour of sustainability when owning, developing, designing, building, occupying or managing commercial buildings. It addresses both profit performance and value, and the broader additional indirect and/or intangible benefits that accrue through factoring sustainability into decision-making.

Bibliography

ABS (1998) *Australian Social Trends 1998: People and the Environment: Attitudes and Actions and People's Concerns About Environmental Problems*, Canberra: Australian Bureau of Statistics. Online. Available at HTTP: <www.abs.gov.au/Ausstats/abs> (accessed 28 May 2004).

AGO (2007) *Energy Efficiency in Government Operations (EEGO): Green Lease Schedule*, Canberra: Australian Greenhouse Office, Department of the Environment and Water Resources. Online. Available at HTTP: <www.environment.gov.au/

settlements/government/eego/publications/pubs/eego-fs2.pdf> (accessed 31 October 2007).

AMP Capital (2007) *SRI Funds Deliver Once Again*. Online. Available at HTTP: <www.ampcapital.com.au/K2DOCS/47B68ADC-9252–43DC-855E-3AEB3 AFEE931/20070925SRImedianstudy.pdf?DIRECT> (accessed 1 April 2008).

BCI and GBCA (2006) *Green Building Market Report*, Sydney: Building and Construction Interchange Australia and Green Building Council of Australia.

Building Design+Construction (2006) *Green Buildings and the Bottom Line*, Oak Brook, Ill. Online. Available at HTTP: <www.bdcnetwork.com/contents/pdfs/whitepaper06.pdf> (accessed 5 March 2007).

CABE and BCO (2005) *The Impact of Office Design on Business Performance*, London: Commission for Architecture and the Built Environment and British Council for Offices. Online. Available at HTTP: <www.cabe.org.uk/AssetLibrary/2191.pdf> (accessed 1 March 2007).

CIE (2007) *Capitalising on the Building Sector's Potential to Lessen the Costs of a Broad Based GHG Emission Cut*, Canberra: Centre for International Economics. Online. Available at HTTP: <www.asbec.asn.au/files/Building-sector-potential_Sept13.pdf> (accessed 5 January 2008).

Cohen, P. and Low, J. (2001) 'The value creation index: quantifying intangible value', *Strategy and Leadership*, 29 (5): 9–15.

Colliers International (2006) *Lifeblood: Sustaining the Value of Australian Business*, Sydney: Colliers International.

Davis Langdon Australia (2006a) *The Davis Langdon Sentiment Monitor: Construction Sentiment (May 2006)*, Sydney. Online. Available at HTTP: <www.davislangdon.com/upload/StaticFiles/AUSNZ%20Publications/Sentiment%20Reports/Construction%20Industry%20Sentiment%20Report.pdf> (accessed 5 March 2007).

—— (2006b) *The Davis Langdon Sentiment Monitor: Construction Sentiment (November 2006)*, Sydney. Online. Available at HTTP: <www.davislangdon.com/upload/StaticFiles/AUSNZ%20Publications/Sentiment%20Reports/Sentiment%20Report%20Two.pdf> (accessed 5 March 2007).

—— (2007) *The Cost and Benefit of Achieving Green Buildings*, Sydney. Online. Available at HTTP: <www.davislangdon.com/upload/StaticFiles/AUSNZ%20Publications/Info%20Data/InfoData_Green_Buildings.pdf> (accessed 1 May 2007).

Department of Environment and Heritage (2006) *Scoping Study to Investigate Measures for Improving the Environmental Sustainability of Building Materials*, Canberra: Australian Greenhouse Office. Online. Available at HTTP: <www.greenhouse.gov.au/buildings/publications/pubs/building-materials.pdf> (accessed 10 September 2007).

Fien, J. (2004) 'Beyond the city edge', in E. Charlesworth (ed.) *City Edge: Case Studies in Contemporary Urbanism*, Oxford: Architectural Press.

Flannery, T. (2005) *The Weather Makers: The History and Future Impact of Climate Change*, Melbourne: Text.

G8 Presidency (2005) *Gleneagles Plan of Action: Climate Change, Clean Energy and Sustainable Development*. Online. Available at HTTP: <www.britishembassy.gov.uk/Files/kfile/PostG8_Gleneagles_CCChangePlanofAction.pdf> (accessed 19 October 2005).

GBCA (2006) *The Dollars and Sense of Green Buildings 2006: Building the Business*

Case for Green Commercial Buildings in Australia, Green Building Council of Australia. Online. Available at HTTP: <www.gbcaus.org/> (accessed 2 February 2007).

Gleeson, B., Low, N., Elander, I. and Lidskog, R. (2000) *Consuming Cities: The Urban Environment in the Global Economy After the Rio Declaration*, New York, NY: Routledge.

Hamnett, C. and Freestone, R. (2000) *The Australian Metropolis: A Planning History*, Sydney: Allen and Unwin.

Huang, T.-L., Noguchi, T. and Kanematsu, M. (2005) *An Assessment System for Eco-building Materials in Japan*, Tokyo: Building Material Engineering Lab, University of Tokyo. Online. Available at HTTP: <http://bme.t.utokyo.ac.jp/bmd/papers/bmd_material_environment/EASEC9_Paper.pdf>.

IFMA (1997) *Benchmarks III*, Houston, TX: International Facility Management Association.

IPCC (2007) *Climate Change 2007*, Geneva: Intergovernmental Panel on Climate Change.

Jayson, S. (2006) 'Generation Y gets involved', *USA Today*, Oct. Online. Available at HTTP: <www.usatoday.com/news/nation/2006–10–23-gen-next-cover_x.htm> (accessed 21 May 2007).

Kats, G. (2003) *Green Building Costs and Financial Benefits*, Westborough, MA: Massachusetts Technology Collaborative. Online. Available HTTP: <www.cape.com/ewebeditpro/items/O59F3481.pdf> (accessed 30 September 2006).

Loftness, V., Brahme, R., Mondazzi, M., Vineyard, E. and MacDonald, M. (2002) *Energy Savings Potential of Flexible and Adaptive HVAC Distribution Systems for Office Buildings*, Arlington, Va.: Air-Conditioning and Refrigeration Technology Institute. Online. Available at HTTP: <www.arti-21cr.org/research/completed/finalreports/30030-final.pdf> (accessed 1 March 2007).

Lowe, I. (2005) *A Big Fix: Radical Solutions for Australia's Environmental Crisis*, Melbourne: Black Inc.

McCartney, D. (2007) *Definition of Sustainable Commercial Buildings*, Brisbane: CRC for Construction Innovation. Online. Available at HTTP: <www.yourbuilding. org/display/yb/Definition+of+sustainable+commercial+buildings> (accessed 31 March 2008).

Newton P.W. (ed.) (2008) *Transitions: Pathways Towards Sustainable Urban Development in Australia*, Melbourne: CSIRO and Dordrecht: Springer.

OECD (2003) *Environmentally Sustainable Buildings: Challenges and Policies*, Paris: OECD.

Paevere, P. and Brown, S. (2007) *Indoor Environment, Productivity and Sustainable Commercial Buildings*, Brisbane: CRC for Construction Innovation. Online. Available at HTTP: <www.yourbuilding.org/display/yb/Indoor+environment,+productivity+and+sustainable+commercial+buildings> (accessed 31 October 2007).

PCA (2007) *Australian Office Market Report July 2007*, Sydney: Property Council of Australia. Online. Available at HTTP: <www.propertyoz.com.au/ppt/Media Presentation077.ppt#330,2,Office> (accessed 6 October 2007).

Reed, R. and Wilkinson, S. (2006) 'Green buildings: issues for the valuation process', paper presented at Queensland Property Conference, Australia Property Institute, Gold Coast, 27–28 October.

RICS (2005) *Green Value: Green Buildings, Growing Assets*, Victoria, BC: Royal Institute of Chartered Surveyors. Online. Available at HTTP: <www.rics.org/greenvalue> (accessed 5 March 2007).

Salzmann, O., Ionescu-Somers, A. and Steger, U. (2005) 'The business case for corporate sustainability: literature review and research options', *European Management Journal*, 23 (1): 27–36.

Saxon, R. (2005) 'Be valuable: a guide to creating value in the built environment', Constructing Excellence. Online. Available at HTTP: <www.constructingexcellence.org.uk/> (accessed 28 February 2007).

Schaltegger, S. and Wagner, M. (2006) 'Capturing the relationship between sustainability performance, business competitiveness and economic performance', in S. Schaltegger and M. Wagner (eds) *Managing and Measuring the Business Case for Sustainability*, Sheffield: Greenleaf.

Sheedy, C. (2007) 'Generation give', *Voyeur*, May. Online. Available at HTTP: <www.virginblue.com.au/products/voyeur/may07/index.php?section=Generation%20give> (accessed 20 May 2007).

Southgate, L. (2007) 'Better building blocks', *The Australian Way: Qantas In-Flight Magazine*, Feb.: 58–61.

Steger, U. (ed.) (2004) *The Business of Sustainability: Building Industry Cases for Corporate Sustainability*, Basingstoke: Palgrave Macmillan.

Steger, U., Ionescu-Somers, A. and Salzmann, O. (2007) 'The economic foundations of corporate sustainability', *Corporate Governance: The International Journal of Business in Society*, 7 (2): 162–77.

SustainAbility (2002) *Developing Value*, London: SustainAbility.

SustainAbility and UNEP (2005) *Buried Treasure: Uncovering the Business Case for Corporate Sustainability*, 2nd edn, London: SustainAbility and United Nations Environment Programme. Online. Available at HTTP: <www.sustainability.com/downloads_public/insight_reports/buried_treasure.pdf> (accessed 1 April 2007).

Victorian Government Property Group (2005) *Victorian Government: Office Accommodation Guidelines 2005*, Melbourne: Department of Treasury and Finance. Online. Available at HTTP: <www.dtf.vic.gov.au/ CA25713E0002EF43/WebObj/VGPGOfficeAccommodationGuidelines2005/$File/VGPG%20Office%20Accommodation%20Guidelines%202005.pdf>.

Wasiluk, K.L. *et al.* (2007) *Introduction to Business Cases for Sustainable Commercial Buildings*. Online. Available at HTTP: <www.yourbuilding.org/display/yb/Introduction+to+business+cases+for+sustainable+commercial+buildings>.

Willard, B. (2002) *The Sustainability Advantage: Seven Business Case Benefits of a Triple Bottom Line*, Gabriola Island, BC: New Society. Online. Available at HTTP: <www.sustainabilityadvantage.com/> (accessed 29 February 2007).

Wilson, A., Seal, J.L., McManigal, L.A., Lovins, L.H., Cureton, M. and Browning, W.D. (1998) *Green Development: Integrating Ecology and Real Estate*, Brisbane: John Wiley.

28 Innovation drivers for the built environment

Karen Manley, Mary Hardie and Stephen Kajewski

This chapter has been written for construction industry professionals, and provides information to assist businesses in improving their innovation performance. It focuses on the factors that promote innovation in the built environment at firm, project and industry levels, and is based on data gathered from a large-scale survey of nearly 400 firms in the Australian construction industry conducted in 2004, and 12 in-depth case studies of innovation on Australian construction projects between 2003 and 2005. It was found that firm and project level innovation is driven primarily by the actions of clients and the challenges thrown up by emergent project crises. In the longer term, industry-level innovation is driven by businesses responding to emerging social and economic trends.

Three priorities for business planning are suggested by this research. First, relationships are required with leading-edge clients in order for a business to be exposed to the pressure to perform beyond business-as-usual. Second, relationships are required with firms capable of successfully engineering joint approaches to innovation as a means of accommodating the risks involved in a challenging project. Finally, relationships with employees are needed to maximize the impact of training programs aimed at enhancing managerial and social skills to address non-technical obstacles to innovation.

This chapter provides a more fine-grained analysis of construction innovation drivers than is typically available. An industry-focused perspective is adopted to provide value to construction businesses alienated by more abstract and conceptual treatments of the topic. The chapter focuses on the drivers of innovation processes, and suggests that businesses can improve the success of their innovation programs by understanding these forces and incorporating insights into their strategic plans. The innovation benefits documented indicate that businesses can ensure their reputations and ongoing competitive advantage through more concerted efforts to maximize their innovation potential.

Background

An extensive body of literature indicates that innovation has frequently been found to have strong links with economic performance and growth. According to Gann (2003: 553):

> A sophisticated research, development and innovation system has emerged in OECD countries over the past 100 years. This has created many of the most significant advances in engineering, manufacturing, healthcare, leisure, defence, education and the service sector. Innovation is widely recognised as a driving force of economic growth, providing the means by which firms compete, and explaining in part why industries thrive or decline.... This is particularly the case in the production and use of the built environment, which provides much of the fixed capital infrastructure required by modern society.

Gann describes a 'process of intensification' that is occurring as construction firms strive to maintain profitability while addressing newly important priorities such as the environmental and social consequences of built environment processes. This is evidenced by the increasing interest in the theory and practice of innovation among governments, industry bodies, researchers and academics. Creating an atmosphere conducive to innovation is seen as important both for individual organizations and for the industry as a whole. Flynn *et al.* (2003: 417) point out that 'The ability of an organisation to grow is dependent upon its ability to generate new ideas and to exploit them effectively for the long-term benefit of the organisation.'

The connection between innovation and profitability is a consistent theme in the literature. Van der Panne *et al.* (2003: 310) note that 'It is widely recognized that innovation is key to the economic performance of firms. Innovative firms grow more quickly and make higher profits.' Innovation is important to the growth of firms, and in aggregate impacts the growth of nations. Steele and Murray (2004) report that corporate expenditure on innovation and R&D can be said to largely sustain global economic growth. In addition they note that: 'The benefits of investing in innovation are now being realized by many of the forward thinking (innovator/early adopter) organizations' (Steele and Murray 2004: 322).

The literature suggests that businesses can gain a competitive edge by encouraging creativity, identifying talent and having formal structures for adopting and diffusing innovations. More specific advice for construction businesses is provided in this chapter, based on a research study undertaken by the authors and their associates for the CRC for Construction Innovation (Manley 2005). The next section outlines the nature of this study and its methods.

The research study

The study comprised two main components: an innovation survey of construction businesses in Australia, and 12 case studies of project-based innovation in Australia.

Survey

The purpose of the survey conducted in 2004 was to examine innovation levels, types, strategies, drivers, obstacles and impacts. It covered the non-residential building and civil sectors in the states of New South Wales, Queensland and Victoria. The industry was defined broadly to include five groups: main contractors, trade contractors, consultants, suppliers and clients from the public sector who regularly commission work.

Overall, 1,317 questionnaires were distributed and 383 useable responses were returned, equating to a response rate of 29 per cent. The survey covered 'key' organizations, which were defined as government clients, members of eight selected industry associations, and consultants and contractors appearing on the pre-qualification lists of the clients. The industry associations chosen were identified through an industry workshop in Brisbane in 2004 as those that made the most significant contribution to construction projects. Table 28.1 summarizes key survey data.

Case studies

The 12 case studies conducted between 2003 and 2005 were based on the flagship Egan demonstration projects undertaken in the UK (Constructing Excellence 2006). Each involved between three and six personal interviews, of approximately one hour's duration, across senior representatives of the main project stakeholders. The purpose was to elicit the benefits of innovation and to highlight the nature of successful implementation strategies. The cases were nominated to the program through referrals from clients and through a public call for nominations. Approximately 100 cases were considered, all involving innovation on a specific construction project. The selection of the final 12 was based on:

- the existence of significant measured benefits, or the clear potential to assess such benefits;
- the likely usefulness of the study in highlighting innovation challenges;
- the likely level of cooperation from project stakeholders in completing the study;
- the likely industry interest in the type of innovation nominated;
- the desirability of including case studies from different Australian states;
- the desirability of a balance between civil and building studies.

Table 28.1 Key survey data

Industry sector	Firms sent survey forms	Completed survey forms returned	Response rate	Population size, by number of firms	Population definition	Percent population sampled	Sampling method
All sectors	1,317	383	29%	3,476	–	38	–
Main contractors: non-residential building and civil	300	93	31%	1,122	Pre-qualified firms	32	Random
Consultants: non-residential building and civil	409	130	32%	1,549	Pre-qualified firms/assoc. members	26	Random
Trade contractors: electrical, communication, air conditioning, mechanical	236	74	31%	346	Major assoc. members	68	Census
Suppliers: glass, plaster, asphalt, steel	328	63	19%	415	Assoc. members/ Yellow Pages	79	Various
Public sector clients: non-residential building and civil	44	23	52%	44	Agency managers	100	Census

The case studies are listed below, and will be referred to by number through-out this chapter:

1   Energy Cost Savings in 5-Star Office Building;
2   Clever Planks at Sports Stadium;
3   Port of Brisbane Motorway Alliance;
4   Fire Engineering at National Gallery of Victoria (NGV);
5   Fibre-Reinforced Polymer Bridge Deck;
6   Ground Penetrating Radar (GPR) and Defective Bridge Beams;
7   Managing Stormwater with Storage Gutters and Infiltration;
8   Saving On-Site Remediation Costs;
9   Post-tensioned Steel Trusses for Long Span Roofs;
10 Twin-Coil Air Conditioning at the Art Gallery of South Australia;
11 Better Project Outcomes with Relationship Management and 3D CAD;
12 Using Recycled Tyres to Construct an Access Road over Saturated Terrain.

Tables 28.2 and 28.3 show the key features of the case studies.

The 12 case studies comprised three sporting stadiums, two bridges, two art galleries, two commercial buildings, two very different roads (a motorway and an access track) and one case of contaminated land. The projects ranged in value from $13,000 to $112 million, and in all cases the benefits achieved were significant.

Drivers of construction innovation in Australia

This section reviews descriptive data provided by the research study to consider agents of innovation, the role of clients, crises-driven innovation, the business environment and sources of innovative ideas.

Agents of innovation

The study indicated the importance of businesses maintaining active per-sonal contacts with related businesses, industry associations and research centres for driving innovation. Value-adding relationships were found in four domains:

- on individual construction *projects*, particularly with the client;
- along the *supply chain*, especially with manufacturers and consultants;
- with *technical support providers*, especially universities and industry associations;
- at *different organizational levels*, including within a particular organi-zation or industry, or across related industries, and interstate or overseas.

The BRITE (Building, Research, Innovation, Technology and Environment) case studies and survey highlighted the importance of these relationships.

Table 28.2 Key data, case studies 1–6

	Study 1	Study 2	Study 3	Study 4	Study 5	Study 6
Project name	William McCormack Place	Lang Park Sports Stadium	Port of Brisbane Motorway	National Gallery of Victoria, Australian Art Building	Coutts Crossing Bridge	Cattle Creek Bridge
Location	Cairns, Qld	Brisbane, Qld	Brisbane, Qld	Melbourne, Vic.	Coutts Crossing, NSW	Near Mackay, North Qld
Project description	4,568-m³ public building	52,500-seat world-class stadium	5-km, 4-lane motorway, with 12 major new bridges	Iconic public building, 11,000 m³	Repair of 12-metre length of 90-metre long timber bridge deck	Identification and repair of faults in 200 new concrete bridge beams
Budget	$17.5 million	$280 million	$112 million	$65 million	$1 million	$1 million
Innovation summary	Chilled water thermal storage tank and moisture absorbing thermal wheel	Precast prestressed polystyrene voided concrete planks	Project delivered under an alliance contract	Fire engineering enabled use of unprotected steel	Fibre-reinforced polymer (FRP) bridge deck	Ground penetrating radar to find defects in bridge beams
Main benefits achieved	37% saving in energy costs	8% saving in cost of grandstand steelwork	10% of project cost saved, 30% of time saved	5% of project cost saved	75% saved in transport costs	50% of project cost saved

Table 28.3 Key data, case studies 7–12

	Study 7	Study 8	Study 9	Study 10	Study 11	Study 12
Project name	Gladesville Road Centre	Imago Site	Stadium Australia	Art Gallery of South Australia	Adelaide Oval	Tomago All-Weather Access Road
Location	Hunters Hill, NSW	East Perth, WA	Sydney, NSW	Adelaide, SA	Adelaide, SA	Tomaree, NSW
Project description	Stormwater management at a small community building	Remediating 5,800 m³ contaminated land	Two 3,500 m³ roofs over sports stadium ends	Upgrading air-conditioning system at an art gallery	Redeveloping the eastern grounds of a sports stadium	16-km road through saturated ground
Budget	$13,000	$1.8 million	$10 million	$100,000	$22 million	$4 million
Innovation summary	Managing stormwater with storage gutters and infiltration	Saving site-remediation costs through a new waste disposal method, sprinkler and wheel washer	Post-tensioned steel trusses to create long span roofs	Twin-coil air-conditioning to improve energy efficiency	Relationship based contract and 3D CAD to efficiently deliver complex project	A permeable road pavement meeting strict environment and community requirements
Main benefits achieved	26% saved in mains water demand	13% of project cost saved	50% less steel weight; 25% less roof erection time	30% saved in energy consumption	50% saved in prefabrication time	15% of project cost saved

For example, two case studies (3, 11) involved innovative forms of contract driven by clients, the intention of which was to provide incentives for cooperative behaviour between team members on the *projects*. The impressive outcomes on those two projects (see Tables 28.2 and 28.3) are based on innovation driven by the high-quality and durable relationships that ensued. These projects employed, in one case, an Alliance Contract, and in the other, a modified 'C21' contract combined with relationship workshops.

Within the *supply chain*, the case studies suggest that manufacturers (2, 5, 10, 12) and consultants (1, 2, 4, 6, 8, 9, 11) were the most innovative groups. Manufacturers have a more consistent stream of work compared to construction businesses, which typically face discontinuities arising from project-based production. Manufacturers are therefore better able to maintain ongoing R&D programs, and R&D is a key input to effective innovation. On the other hand, consultants are the practical problem-solvers in the industry, using their creativity to develop design innovations.

The survey findings highlight the same industry groups as 'encouragers' of innovation, as shown in Table 28.4.

Clients play the lead role in encouraging innovation, having been nominated by nearly 60 per cent of the industry, while over 50 per cent of industry participants nominated consultants, particularly architects and engineers. Manufacturers also emerge as playing a critical role, with nearly 50 per cent of the industry nominating them.

Table 28.4 Industry group by per cent of survey respondents perceiving them to encourage innovation, Australian construction industry, 2004

Encouragers of innovation	%
Large/repeat clients	59
Architects	55
Engineers	51
Manufacturers	46
Building designers	44
Main contractors	43
Developers	38
Project managers	38
One-off clients	27
Trade contractors	27
Other suppliers	26
Organizations that set industry standards	26
Quantity surveyors	19
Funders	15
Government regulators	11
Letting agents	7
Insurers	5

The key role of clients, consultants and manufacturers in encouraging innovation highlights the significance of establishing strategic relationships with each of them, particularly for access to their innovation assets, which include 'demanding requirements', 'creativity' and 'R&D' respectively. Such assets will assist businesses seeking to build a successful innovation program.

The role of *technical support providers* in driving innovation, particularly through testing and validation activities, was also reflected in the case studies (1, 2, 5, 7, 9, 10, 12), suggesting the value of relationships with universities and other research centres.

Another way to think about relationships is to look at their operation at *different organizational levels*. Industry-level relationships have been covered above, while the importance of relationships *within an organization*, and particularly the treatment of employees, was highlighted in all the case studies. For example, case study 8 showed that employees are a key source of innovation ideas, particularly when:

- they have a direct and secure employment relationship with the organization;
- the organizational climate is one in which failure is tolerated as an acceptable means of advancing creativity and innovation;
- there are organizational incentive programs to encourage employees to share ideas.

Businesses might also focus on relationships beyond their own industry, state and country if they want to maximize innovation opportunities, as was demonstrated in most of the case studies (1, 2, 3, 4, 5, 6, 7, 9, 12).

In summary, clients, manufacturers, consultants, technical support providers and employees are key agents of innovation. Hence, businesses wishing to improve their innovation performance are advised to focus on building and maintaining relationships with these groups, particularly clients.

Role of clients

The survey results show that Australian repeat public sector clients can play a significant role in promoting firm-level innovation, as 60 per cent of respondents nominated them as 'encouragers' of innovation. Compared to other industry groups, such clients had the highest rate of investment in R&D, the highest rate of adoption of advanced practices and technologies, and the best return on innovation, and played a key role in providing ideas for innovation. Hence, it may be that an effective innovation program for construction firms should focus, where possible, on cultivating deeper and broader relationships with repeat public sector clients.

The clients covered by the survey are 'leading-edge' clients with high levels of technical competence, with challenging needs and with extensive experience (Morrison *et al.* 2004). They are also more likely to use value-based tender selection than other clients. The innovation opportunities of a business are influenced by the level of sophistication of the clients for which it works – the more demanding, technically competent and experienced the client, the more likely it is to stimulate innovation in projects, by demanding outcomes that exceed business-as-usual practice.

The BRITE case studies identified specific roles played by leading-edge public and private sector clients in driving project-based innovation, as shown in Table 28.5.

The clients in these case studies promoted innovation by setting challenging energy targets, designing new forms of contract, undertaking R&D, networking with specialist experts and organizing demonstrator projects.

Crises-driven innovation

In the three cases in Table 28.5 where client behaviour was not driving innovation (4, 6, 8), crises during projects were responsible. The innovation was possible because a cooperative team approach was adopted, leading to a clear focus on finding solutions. The blame-shifting that is often characteristic of the traditional adversarial approach was avoided, and more creative 'best for project' responses were generated. For businesses wishing to improve their innovation performance, this suggests the value of maintaining robust industry relationships and the flexibility to respond to changing project circumstances.

The business environment

Looking beyond the project level, the case-study data can also be examined to reveal broad-based innovation drivers. Such analysis revealed the following influences on construction innovation:

- user needs such as the increasing requirement from building tenants for reduced operating costs, which is reflected in changing demands from building owners and design innovation (1, 7, 10);
- regulatory regimes where performance-based building codes offer greater opportunity for innovation than prescriptive approaches (Gann *et al.* 1998) (4);
- trade conditions reflecting growing internationalization which offers opportunities for Australian construction businesses to expand their markets globally (6, 10);
- social values where increasing concern about the environment is reflected in government energy targets (1, 7, 10).

Surveillance of these meta-trends is an important task for construction businesses wishing to maximize their innovation outcomes.

Table 28.5 Drivers of innovation on case-study projects

Study 1	Study 2	Study 3	Study 4	Study 5	Study 6
Client wanted to minimize whole-of-life costs	Client's contract provided innovation incentives	Client wanted improved time/cost/quality	During the project, time and cost started to blow out and needed containment	Client attracted by the weight and corrosion benefits of a new material	During bridge repair, faults were found in new concrete beams

Study 7	Study 8	Study 9	Study 10	Study 11	Study 12
Client (local council) wanted to educate community re water-saving technologies	During site remediation highly toxic materials were unexpectedly found	Client had tight time-line; needed to keep stadium operating during construction	Client needed to gain better control of fluctuating humidity and temperature levels	Client had tight time-line; wanted better than usual time/cost/quality outcomes	Client constrained by restrictive community and environmental requirements and needed new ideas

Sources of innovation

The research study also collected data on the industry groups that provide new ideas for innovation. These data reinforce the importance of relationships with the types of industry participants emphasized earlier. The case studies show that consultants, clients and manufacturers are prominent sources of innovation ideas for the industry, as shown in Table 28.6.

The role of consultants in providing ideas is highlighted in seven of the studies (1, 2, 4, 6, 8, 9, 11), manufacturers were a key source in five (2, 5, 7, 10, 12) and clients in four (3, 5, 6, 11). These case-study findings identify the same three groups as the survey findings regarding innovation 'encouragers'.

In addition to the central role played by particular industry groups in providing ideas for innovation, businesses should bear in mind the importance of idea generation through in-house staff, industry associations, conferences and previous projects. These sources were highlighted in the survey, with nearly 70 per cent of industry participants nominating in-house staff as a prominent source.

In summary, the evidence presented in this section shows that innovation in the construction industry is typically either sponsored by client needs, in which case businesses can proactively suggest innovations; or driven by crises during projects, in which case, creativity in the context of harmonious project relations is required for effective reactive innovation. In the longer term, innovation in the industry is driven by businesses that ensure their competitive advantage by responding to changing social and economic trends. Overall, the research study has emphasized the importance of construction businesses responding to these trends by managing relationships with consultants, clients, manufacturers and their own employees.

Conclusions

This chapter has quantified the benefits that can flow from successful innovation, and shown how it is driven at firm, project and industry levels. For businesses wishing to improve their innovation performance in response to these drivers, the single most important strategy is to focus on relationship-building. The learning required for effective innovation is maximized via enhanced relationships and shared understandings, which give rise to a knowledge-multiplier effect and the development of collaborative advantage. Enhancing relationships involves thinking about the firms, industry associations, government departments, research centres and industries that are likely to have knowledge that would complement the in-house knowledge base. Then, decisions need to be made about the most appropriate level, mode and frequency of contact. Resulting relationships may be informal, or they may evolve into specialized arrangements involving, for example, facilitated workshops, working groups, reference groups,

Table 28.6 Industry group providing ideas for innovation on case-study projects

Study 1	Study 2	Study 3	Study 4	Study 5	Study 6
Mechanical/electrical consultant	Structural engineering consultant and plank manufacturer	Client	Fire engineering consultant and architect	Pre-existing research group comprising clients and manufacturers	GPR consultant who had pre-existing relationship with client
Study 7	Study 8	Study 9	Study 10	Study 11	Study 12
Manufacturer	Environment consultant and main contractor	Steel design consultant/specialist sub-contractor	Manufacturer	Client, architectural and engineering consultants	Manufacturer

memorandums of understanding, partnerships, alliances, joint ventures, joint R&D, joint publications, joint presentations or joint patenting.

Three priorities for business planning are suggested by this study. First, relationships are required with leading-edge clients in order for a business to be exposed to the pressure to perform beyond business-as-usual. Second, relationships are required with related firms to successfully engineer joint approaches to innovation as a means of accommodating the risks involved. Finally, relationships with employees are needed to maximize the impact of training programs aimed at enhancing managerial and social skills to address non-technical obstacles to innovation.

Bibliography

Constructing Excellence (2006) Website. Online. Available at HTTP: <www.constructingexcellence.org.uk/resources/demonstrationprojects/default.jsp> (accessed 20 June 2007).

Flynn, M., Dooley, L., O'Sullivan, D. and Cormican, K. (2003) 'Idea management for organisational innovation', *International Journal of Innovation Management*, 7 (4), 417–42.

Gann, D.M. (2003) 'Guest editorial: innovation in the built environment', *Construction Management and Economics*, 21 (6), 553–5.

Gann, D.M., Wang, Y. and Hawkins, R. (1998) 'Do regulations encourage innovation? The case of energy efficiency in housing', *Building Research & Information*, 26 (5), 280–96.

Manley, K. (2005) *BRITE Innovation Survey*, Brisbane: CRC for Construction Innovation.

Morrison, P.D., Roberts, J.H. and Midgley, D.F. (2004) 'The nature of lead users and measurement of leading edge status', *Research Policy*, 33 (2), 351–62.

Steele, J. and Murray, M. (2004) 'Creating, supporting and sustaining a culture of innovation', *Engineering, Construction and Architectural Management*, 11 (5), 316–22.

van der Panne, G., van Beers, C. and Kleinknecht, A. (2003) 'Success and failure of innovation: a literature review', *International Journal of Innovation Management*, 7 (3), 309–38.

29 Seeking innovation

The construction enlightenment?

Peter Brandon

The past 40 years (roughly one working lifetime) have been a period of rapid technological progress across a wide range of human activity. It has included the development and miniaturization of the computer, the development of the Internet, massive improvement in communication, significant understanding in medicine, vast increases in our knowledge in many areas, and enormous improvements in infrastructure such as transport. This has been coupled with major reductions in cost, placing them within range of many more people than was ever envisaged. One newspaper was quoted as saying that there is more knowledge in a single edition of the New York *Sunday Times* than was acquired by the average individual living 400 years ago. It is difficult for anyone living in the developed world today to understand what it must have been like to live at that time. Our dependence on technology, enhanced by education, has transformed our lives.

But what about construction? There has been incremental innovation over the centuries, but the industry has tended to stay within the craft paradigm where it is confident it can deliver (Sebestyen 1998). The citizens of the past would have little difficulty in understanding and participating in many of the processes the industry adopts today, at least in some of the major sectors such as housing. Perhaps the service engineering aspects would cause problems, but the rest would be instantly recognizable. The industry has refined its methods, of course, and it is almost as efficient as it can be whilst it uses existing approaches. The innovations of recent years have seen a greater specialization, with new professions springing up to support the standards and give assurance to the public. Knowledge has become identified with these bodies, and they in turn have created systems to maintain standards, usually by examination, which has had the effect of making the knowledge and practice inaccessible to those who do not pursue the professional pathway. By default, the boundaries between professions and trades have become less porous, and a series of silos of knowledge have grown up, often protected by law. The innovation which encouraged efficiency by specialization has also led to fragmentation of the industry, which in many cases can lead to an adversarial approach in which the best interests of the client can be lost.

For many years, academics and leading industrialists have been advocating change for the benefit of all the industry. This has largely fallen on deaf ears except in some specialist sectors, for example, where the design is driving a technological revolution. However, is this about to change? Are we at a point where the pressures from clients, the competitiveness of firms and the transformation of other parts of society is such that the industry must change and adopt some of the practices from other manufacturing industries? There are dangers in suggesting such a move, as it could be that another false dawn is about to be launched! Is there a parallel which could be seen to be the model for innovation that might give us a clue that something dramatic might happen?

One such model might be the revolution that occurred between 200 and 500 years ago, which led to a change in the way humans thought about themselves and their environment and to the transformation of the developed world. This prolonged period (by today's standards) has been labelled the Age of Reason by historians, and was a period rich with new knowledge. It provided the foundation for our own technological advance in all areas of society. It included such well-known names (and discoveries) as William Gilbert (Earth magnetism), Thomas Harriot (algebra), William Oughtred (the slide rule), John Napier (logarithms), William Harvey (blood circulation), William Gascoigne (the micrometer), Isaac Newton (calculus) and many others. What these visionaries did was to provide the basis for the creation of knowledge based on observation and enquiry. In order to do their work, they needed to measure and develop tools for examining and testing ideas. For many centuries, much of the perceived knowledge had been handed down by the state or the church without challenge. Now was the opportunity to challenge and examine whether it was true. The outcome of these advances was a series of tools for scientists to use in their enquiry. These included:

- measurement methods and tools;
- scientific methods based on observation and hypothesis testing;
- reductionism, whereby the problem could be reduced to a sufficiently small size for examination.

These, in turn, allowed an increase in precision and the development of rational thought where reason held sway over superstition and revelation. By challenging the views of the establishment it also led to the querying of the social status quo and encouraged revolutionary thinking, where the individual had a voice and the opportunity to express his or her opinion. This change is reflected in the change in thinking advanced by the philosophers in a new age, the Age of Enlightenment.

Philosophy can be defined as 'The academic discipline concerned with making explicit the nature and significance of ordinary and scientific beliefs and investigating the intelligibility of concepts by rational argument concerning their presuppositions, implications and interrelationships' (*Collins English Dictionary*, 2000). It seeks to answer the reason why we believe certain things. Once humankind found itself able to establish its knowledge without deference to established powers, then our belief systems changed. Enlightenment reflected that change in view, although many would argue that it has claimed too much. Can the rational thought of mankind explain the meaning of life? Nevertheless, it has dominated the thinking of the Western world ever since. In 1784, in *What Is Enlightenment?*, the philosopher Immanuel Kant suggested the following:

> Enlightenment is man's release from his self-incurred tutelage. Tutelage is the incapacity to use one's own understanding without the guidance of another. Such tutelage is self-imposed if its cause is not lack of intelligence, but rather a lack of determination and courage to use one's intelligence without being guided by another.

The proponents of enlightenment included such luminaries as Voltaire, Gibbon, Paine, Hume, Rousseau and Berkeley, who together started a series of changes in thought which have stayed with us until today, for example:

- a move from faith based on edicts of the state and church to reason;
- a challenge to established tradition;
- a change in values which gave more credence to rigour, rationality and geometric order.

It could be argued that these changes provided a liberal stance against superstition and intolerance which forms the basis of our modern understanding of freedom. In so doing, it provided the basis for democracy, religious tolerance, and acceptance of scientific method as a way of establishing at least some aspects of 'truth'.

The key point is that the technology of the day transformed the thinking of vast numbers of people, and this in turn created a society in which change could occur through examination, challenge and rational ideas. New thinking was established which was built on rational thought. Today we face a similar revolution in our thinking processes which is challenging us in all aspects of our lives, including the way we think about ourselves and each other. The rise of the machine is making us consider what we are and why we exist in a way that has not been undertaken at this scale ever before.

Construction is no exception to these changes. As a key component of our quality of life and the way we behave, it must be impacted upon by the change in thinking of the society which it serves. It must reflect the new demands of that society, and one of those aspects must be the tools which society expects to be used to achieve its requirements.

However, an examination of the industry would suggest that not much has changed over several centuries, and some would claim that it has acted more like the establishment of old and resisted change to protect its own established traditions. Authors who have pointed to this problem include Woudhuysen and Abley in *Why Is Construction So Backward?* (2004) and, with a more positive spin, Cole, Lorch and others in *Buildings, Culture & Environment: Informing Local and Global Practices* (2003). There is a strong argument which suggests that tradition has value especially when applied across a major industry such as construction. The education of its participants, the legal and financial frameworks and the processes involved are embedded in its culture. It would be costly to change, and we tamper with this at our peril.

Despite the reluctance to move from traditional practice, many authors have been forecasting and expecting change to happen for the last 40 years. If the question is asked, 'If we knew what we know today, would we design an industry that acts in the way it does today?', the answer would almost certainly be no. Whether we really have a better alternative that we could lay on the drawing board is less clear. However, the recent increase in foresight and agenda setting for research in construction (Flanagan and Jewel 2003) suggests that there is a feeling that something must be done. The Australian Cooperative Research Centre for Construction Innovation undertook a major exercise in visioning the future which has received widespread interest (Hampson and Brandon 2004). Through its members and engaging with around a thousand professionals from all major disciplines and site staff, it identified certain aspects of technology as a key enabler in providing a better and more effective industry.

There is, therefore, a potential parallel with the Age of Enlightenment:

- technology has created a new understanding of how things can be done in the industry, harnessing advances found in other fields of endeavour;
- it has opened new opportunities for advancement which were not there before the technology was developed;
- already, it is beginning to change relationships between people and the power structures of the industry;
- it has begun to change the way we think about ourselves and our industry.

All these things were forecast as the computer revolution began to develop. Christopher Evans, in *The Mighty Micro: The Impact of the Computer Revolution* (1979), devoted a whole chapter to the decline of the professions. He argued that professions gained their status because they were exclusive repositories of information whose members had made it difficult for others to access their knowledge. Consequently, when the computer enabled knowledge to be generally available (an early expectation of the Internet), then this would undermine their ability to protect their status. It can be debated as to how far this is true, but nevertheless, the ability of the general public to engage with any form of knowledge is undeniable. It took several hundred years for the Age of Reason to impact fully, and we do not know where this current revolution will lead.

Questions can be asked as to whether we are approaching an enlightenment for construction – for example:

- Are we at a tipping point or watershed for construction innovation in a similar way to that experienced following the development of science?
- Is there a focus for construction research which could transform the industry?
- Where might the innovation emerge from that might generate a major transformation?

If we are to answer the above, we might need to answer more specific enquiries related to the way we do our construction business. For example, we might compare the recent past with the enlightened future:

- Instead of an industry based on a craft tradition – can we base it on rational thought where the primary question is 'Should we be doing it this way?'
- Instead of tradition being valued more than reason, based on current knowledge – new improved approaches take precedence over past methods.
- Instead of the industry taking its cue from its own past – it embraces the advances made in other industries and is more outward looking.
- Instead of change being derided – change outside of the current paradigm is encouraged.

The new technologies

The range of technologies is so great that it is not possible to deal with more than just a few in this chapter. They range from processes to materials, from communications to free-form design, and from new financial models to enhanced visualization of both product and manufacture. However, in most of the developments it is information technology that has enabled the potential for change to take place. A quick look at the advanced design and construction schools of leading universities would reveal the use of building information models, virtual reality, augmented reality, web-based communication and a whole series of other devices, such as Second Life and Facebook, which suggest that even our perception of the real world is changing. Going further afield, the use of computers to jack into the brain to aid those with hearing or eye defects seems to be a forerunner of what might become more widespread, leading possibly to new lifestyle changes which we can design for ourselves.

It is probably true to say that no one knows with complete confidence where we are heading with the developments in information technology. If the introduction of the Internet, which was not foreseen even 40 years ago (but which has transformed many aspects of human behaviour), was such an innovation, then who knows what the future might bring? Nevertheless, it is incumbent on those who lead the thinking of the industry to anticipate what might happen and how this might influence what we do now. We have an obligation to ask, 'Where should we look for an improvement in construction innovation?' The rest of this chapter will attempt to explore this in strategic terms through just three examples, namely:

- management, which has the ability to intervene and make change happen;
- information technology, which can provide both a vision and support for change;
- engagement with the people and stakeholders who are the ultimate beneficiaries of the innovation.

However, the theme through all these is the need to reduce the interfaces between people, and between people and machine.

Innovation in management: dissolving the interfaces

The last 200 years at least have seen a change in the management structure of the industry. It has gone from the design and control of the construction process under the auspices of the designer/engineer to a series of specializations which have been created to deal with the complexity of projects and, to some extent, their scale. These changes were the result of innovative thinking at the time, and yielded useful results. As time progressed, however, they became fossilized around the roles people were expected to play, often reinforced by the professions that were created as they became established, and a new problem emerged. The impact was to create an adversarial position in which each role defended its boundaries, giving rise to potential conflict and to a blame culture when things started to go wrong. The role of management is now a major one within the industry, but it exists as a separate role because the interfaces between the participants have increased and someone has to manage them. It is paradoxical that the more we manage, the more complex the process appears to become. Ferry and Holes (1967) found that a single item is measured up to 16 times during the course of a contract, usually by different people, but not always. This is just a simple example of the interface problem, but it illustrates that trust begins to diminish (otherwise we would all accept one person's view) and we need to assure ourselves that our understanding is right.

If an interface is created, then the following can happen (see Figure 29.1):

- the two parties need to collaborate;
- they therefore need agreed methods of communication between them, often in the form of documentation;
- they may well need to negotiate who does what and when and for what price;
- in order to understand what is in the mind of these parties as an agreement, a contract is drawn up to define it; this might involve another profession;
- the engagement between the two parties may have public interest, and hence the regulatory authorities then provide a new regulation to cope with it;
- the process becomes so complex that one side decides that it is sensible to employ a manager to deal with it;
- the other side then realizes that it might lose out and does the same;
- in order to facilitate the two managers, a project manager is introduced to keep them in order!

Of course, this is a caricature of what happens, but it seems to be recognizable by many people in the construction industry.

In actual fact, the situation is slightly worse than that above. Throughout the process of construction there are a large number of interfaces where knowledge is passed from one participant to another. To do this, models

Figure 29.1 The problem with interfaces...

are created to convey information such as drawings, bills of quantities and specifications. These, by the very definition of model, do not convey the whole of the information which the participant who creates the model actually knows. However, the next person takes the simplified information, enlarges it from his or her own understanding and then creates another model which is passed on to the next consultant, and the process begins again. Through this process, knowledge is created, simplified and passed on in a different form. The result is a knowledge entropy, where knowledge of one type is being lost and knowledge of another type is being created. Any one participant only has partial knowledge of the project, and this has the potential for misunderstanding, leading in some cases to litigation.

The innovation required is to dissolve the interfaces wherever possible and to provide a better shared common understanding of the meaning and processes which are required to design and build the building without error.

Innovation: the technological support

It is not possible to do without management of some sort, but where it resides is critical. If it is within one organization or entity which can act without formal boundaries, then it may well be possible to move away from adversarial positions, hence the increase in 'design and build' and multidisciplinary consultancy practices.

However, it is one thing to say we will aim to get rid of the interfaces, but another to find a way of achieving it. The past 40 years have seen giant steps taken towards connectivity between people and organizations through technology which facilitates the breaking down of the interface barriers. Emerging over this period has been the concept of the building information model (BIM), whereby there is a common model available to all approved participants which can be updated and with internal integrity as decisions are made by designers and other consultants. The process needs managing, but can be done in real time providing certain protocols are agreed. Work in this area includes that by Martin Fischer at CIFE, Stanford University (Fischer 2008), Ghassan Aouad and others at Salford University (Aouad *et al.* 2005) and Martin Riese (Riese 2008). A major European project, *Divercity*, engaging several organizations across the EU, also experimented with partners cooperating together across geographical boundaries. One of the most impressive uses of such a model occurred in the Stata building at MIT, engaging Frank Gehry Associates and Gehry Technologies together with Skanska and much of the major supply chain (Joyce 2004). The model became the reference for all activity and design evaluation. It also took a big step forward towards paperless design. In addition, it engaged the steel fabricators in manufacturing the steel members direct from the engineers' drawings, and the site erection team used laser technology for setting out in order to ensure compliance with the model. These sophisticated models are not yet in general use, but the trend is towards adoption. Interestingly, the regulatory authorities have in some instances not kept pace with the technology (e.g., requiring 2D drawings for projects where this is inappropriate), and it is easier to adopt these new practices where the construction industry is not so heavily regularized.

These changes are not only assisting the process as it stands, but also challenging the way we think about the project, its design and the engagement of people within it. Do we need the design team which we have traditionally required? Who is in charge of coordination, and is it the person driving the model? How should the regulatory authorities act if they are not to be the people who stop innovation? What is the limit to design, now that the algorithms have been developed for complex shapes, and new materials have extraordinary performance characteristics which allow them to be built?

Although this is happening through those architects who have to have these tools in order that their designs can be built, they are paving the way for the rest of the industry. Often, it is the clients who are demanding these new methods to avoid the massive extra costs and overruns which seem to be the province of large projects. One Island East, a multi-storey building in Hong Kong for Swires Properties has a very sophisticated BIM model by Gehry Technologies using CATIA software, and this was driven by a client who recognized that it would avoid a significant proportion of cost and time, as well as stress. The savings are self-evident. The model found 2,000 clashes before tender, and over a two-year period, as the design developed, it identified and dealt with a further 150 to 200 clashes a week. It also coped with major changes in design which would have been impossible or very expensive if undertaken manually. It is now thought that savings of up to 25 per cent might be available in post-occupation costs as well.

However, to get this approach into the mainstream requires a radical change of thinking. The technology is beginning to create that change which goes far beyond the concept of just adopting a technology to support design and construction activity. It goes to the very root of why we do things in the way we do, and why we behave in the way that we have done for centuries. Like the Age of Enlightenment, these are issues which will transform our industry and our understanding of the role that each of us will play. In time the change will be even more significant, as we hand more and more of the design function to machines with intelligence and more of the assembly process to robots. These moves are already being suggested by many of the software gurus, including Grady Booch, chief scientist of IBM (Booch 2007), who believes that within 30 years we will see the rise of the machine, where we will move from dependence upon machines to being driven by them. This, of course, raises all sorts of ethical and moral questions related to our adoption of the technology. These, in turn, will make us question why we do things in this way, and what our behaviour should be. The Age of Enlightenment based on technology is going to be tackled by the wider public in the first instance, but it will be a much shorter period than the four or more centuries seen previously. The speed of change may be so fast that in itself it will create problems which are not yet identified, and may even exaggerate the gap in the two-speed world between the developed and the underdeveloped peoples which has been forecast in both science and fiction, resulting in severely destructive social stresses.

In the meantime, we are developing new technologies which allow people to meet in a more natural way across traditional geographical boundaries. Tele-collaboration is a research topic which could be of great interest to construction. The industry exists by creating virtual teams which are increasingly global. Although building modelling is a big step forward, the human/computer interface still requires significant research to make it natural. In real life we pick up body language, and especially eye gaze, which allows us to interpret the words and actions of others. The teams at Salford University are investigating ambient technologies and methods of collaboration which will make them more accessible. These include hidden computers with more natural methods of communication, and augmented reality where virtual spaces are combined or overlapped with real space to examine or instruct behaviour – for example, the use of RFID tags, together with personal data appliances, to examine whether the service engineering has been built according to planned design (see CoSpaces website <www.cospaces.org/>) in the model.

Space can be shared in a number of different ways (see Figure 29.2):

- by observation, which involves the use of a flat screen and is the normal way we video-conference at the present time;
- by tele-immersion, where virtual objects can be shared across a virtual table or a workbench;
- by collaborative virtual environments, where avatars can meet and act, controlled by external persons, as in the virtual environment of Second Life;
- by immersive collaborative environments, where a person can engage with others, or with other spaces, immersed in a virtual world using a cave.

At Salford University a reconfigurable cave is being created which has the flexibility to undertake all these approaches. In time, it will allow participants in the design and construction process to have an appropriate method of communicating with each other which is as near real life as we can make it. The barriers of space and time may well begin to disappear, and in a global industry this must be significant.

Innovation: the shift to public participation and democratic responsibility

So how will the technology which is beginning to remove some of the geographical interface problems be used? This is a much more difficult issue to

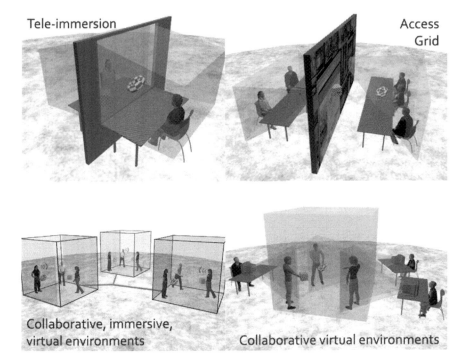

Figure 29.2 Different ways of sharing space in tele-collaboration (source: Professor David Roberts and Dr Robin Wolff, Centre for Virtual Environments and THINKLab Octave, University of Salford).

anticipate. Predicting the impact of technology on human behaviour is a matter of much speculation and, as the introduction of the Internet has shown, humans adapt their actions to suit the technology in unforeseen ways.

One further considerable interface is that between the public and the business community in making decisions. Understandably, the people wanting to take business decisions and investing their own money in the impact of that decision also want to retain control. This has been easy to maintain until now, as it was not a practical proposition to engage the stakeholders in a detailed assessment of the proposed development. Often this was left to representatives of people such as planning authorities, who exercised some control over events within their jurisdiction. In large infrastructure projects within democratic regimes, enquiries have been set up which have allowed the public to make representations for or against a proposal. These often delayed planning consent, but at the same time provided endorsement of the project by stakeholders once the project had been authorized. However, the nature of these enquiries – often extremely formal, with experts and consultants providing submissions – created barriers which the public found intimidating. More recently, government agencies have provided websites on key issues (sometimes identified by the public themselves in the form of a motion) where people can express their opinion or even provide a petition online. One such example was a UK government website which asked for opinion on a road-charging scheme to replace road tax, and several hundred thousand people quickly responded with a resounding 'No' to the perceived proposal. It was quietly dropped, at least for the time being. Increasingly, technology is being used by planning authorities to demonstrate to residents the impact of a new scheme, and some of the Virtual Reality work at Salford has been used to explain more clearly to planning authorities the impact of a proposed development (e.g., a new prison which could be unsightly) or to engage residents in planning a regeneration scheme encouraging them to explain how crime, for example, can be designed out of the scheme. They know the area best, and can make informed suggestions.

Another use of technology was used by media researchers at Salford to develop a game, *Plasticity*, which allowed schoolchildren to have a say in the development of Bradford city centre. This was based on a game engine which allowed the children to remove buildings and replace them from a fixed menu of buildings which could then be rendered in various ways. The purpose was to engage all the people affected – in this case, the children – in the redevelopment, so that they felt some ownership of what was eventually proposed.

Public participation is a key feature of Principle 10 of Agenda 21 (United Nations Report: Conference on Environment and Development), the agreed blueprint for local authorities to provide sustainable development. The principle states that:

- individuals should have an opportunity to participate in decisions which affect them;
- democratic decision-making allows a more sensitive choice to be made related to the needs of the community;
- by engaging the general public, values can be shared and an agreed solution can be achieved.

The new technologies allow a better understanding of what is proposed in a way which most people can understand. They break down the barriers between the lay person and the expert in certain fields, and allow the values of a society to be brought to bear within the process. There have been many examples in the past where authorities have ridden roughshod over the people, and the people have risen up to stop the proposal. One example is the introduction of the high-speed train through Turin, which has been delayed for 18 years because those affected were not engaged from the start and a 'not in my back yard' syndrome was allowed to develop. Large-scale protests then followed. Engagement through better communication and participation may have avoided such a delay. In long-term issues such as sustainable development, it is absolutely critical that the public are involved so that the proposals can take into account their values and engage in dialogue which might influence those values for many years to come.

This may seem a long way from the concept of enlightenment, but, if increased democratization takes place, it changes the power structure within society and may well affect everyone's thinking and behaviour. It will start with the models we create in our brain, and will continue to what we propose and what behaviours we adopt. Many will be afraid of the impact, many will welcome it, and there is a large debate to be had just on the benefits or otherwise of such a shift. The point is that technology which enables this change has provided a platform for new thinking in a not dissimilar way to the Age of Reason followed by the Age of Enlightenment of a few centuries ago.

There is, of course, a very strong potential downside to adoption of this technology. To gain knowledge, the technology may be intrusive. To provide information, a machine may need to interpret raw data, and it may not be evident who provided this or what that interpretation was. Sometimes, extensive use of routines are hidden in software which have value judgements within them (Brandon 2005). These may not be known to the user and are, therefore, unchallengeable in the way that normal human debate allows adaptation and evolution of thought. These are serious issues with which society has to grapple. At the moment we appear to be drifting towards a society where machines are beginning to rise (Booch 2007) at the expense of human engagement and modification. This is the opposite of a technology which opens up opportunities for humans to participate in the issues which concern them. In moving in this direction, we must be aware of the negative potential of the technology we are adopting.

Summary

This chapter has attempted to explore the concept of a new Age of Construction based on technology, and to draw a parallel with the so-called Age of Reason and Age of Enlightenment. It has recognized that, in order to gain a change in thinking, there needs to be a platform of new knowledge and tools which provide the fertile ground for ideas to develop. Much of the current literature on innovation (Dodgson *et al.* 2005) seems to take this as a given before innovation can take place. These authors are right, but it is worth exploring in some depth to discover whether we can identify those technologies which may have the major impact on our world, the way we behave and the development of processes and products which might be adopted in the construction industry. If we can focus on these, we might speed up the innovation process for the benefit of all concerned or we might be able to avoid the pitfalls that such technologies might bring as a side-effect to their major intent.

Earlier in this chapter a number of questions were asked.

Are we at a watershed or tipping point which is similar to the Age of Enlightenment?

The answer is almost certainly yes, but it is important to remember that the Age of Enlightenment took place over a period of 200 years or more. Although we are likely to be faster at adopting ideas, a period of 50 years, say, would be a great improvement but is still a long time. It is the cultural aspects which need to adapt before the technologies can be fully harnessed, and these are much more difficult to change.

Is there a focus for research which might transform our industry?

Here we are into the area of speculation, but it is possible that a focus on removing many of the interfaces in the construction and manufacturing processes may yield major benefits. This applies to man/machine as well as people-to-people interfaces, and may also be an issue with machine-to-machine understanding in the future. Construction has spent 200 years creating silos of knowledge as part of an innovation process to build more efficiently; it now needs to unlearn these processes and think in a different way, and the integrating technologies will aid this transformation. It is now possible to do what was impossible when the industry had to work within the inadequacies of human thought and limbs.

Is there innovative thinking that we can identify which might provide the transformation?

Again, the answer is almost certainly yes. Removing the interfaces is the start, but it is not just in new forms of procurement, such as alliancing or partnering, that this will occur. These do remove the formality of the interfaces, but it is technologies such as CAD/CAM where major advances may occur, including the greater development of off-site manufacture, yielding efficiencies and better quality control. In this chapter, use of communication technologies which begin to dissolve the geographical barriers has been suggested as ways to improve the interface between people working in the current technologies. Construction will not be immune from the transformations occurring in other aspects of society and the cultural changes that this might bring. It is already clear that new approaches to democratization are on the horizon, and no one quite knows where this will lead, even in the short run. Participation will be demanded, but at what level and with what authority? It is being encouraged in areas such as sustainable development, but will this extend to routine business decision-making?

It is not possible to consider any new technology without addressing its negative impact. As we develop more and more powerful machines and begin to provide intelligence in these machines which can rival or exceed our own, then our perception of ourselves and our ability to solve world problems will change. These are uncharted waters, but no citizen can absolve himself or herself from the impact this will make. Harnessing technology for good must be the mantra of us all, and perhaps the technology can help us to identify what is good and what is bad. For construction, there is a long way to go before this impact is felt, but it is vital that we do not lose sight of our benign objectives for humankind.

Bibliography

Aouad, G., Lee, A. and Wu, S. (eds) (2005) *Constructing the Future: nD Modelling*, Abingdon: Taylor & Francis.

Booch, G. (2007) *The Promise, the Limits, and the Beauty of Software: The BCS/IET Manchester 2007 Annual Turing Lecture*, 22 January. Online. Available at HTTP: <http://intranet.cs.man.ac.uk/Events_subweb/special/turing07/> (accessed May 2008).

Brandon, P. (2005) ' "If … then": the Achilles Heel for knowledge management', in F.L. Ribeiro, P.D.E. Love, C.H. Davidson, C.O. Egbu and B. Dimitrijevic (eds) *Information and Knowledge Management in a Global Economy*, vol. 1, Lisbon: IST.

Cole, R.J. and Lorch, R. (eds) (2003) *Buildings, Culture & Environment: Informing Local and Global Practices*, Oxford: Blackwell.

Divercity Project: A Virtual Toolkit for Construction Briefing, Design and Management (2003), Salford: University of Salford.

Dodgson, M., Gann, D. and Salter, A. (2005) *Think, Play, Do: Technology, Innovation and Organization*, Oxford: Oxford University Press.

Evans, C. (1979) *The Mighty Micro: The Impact of the Computer Revolution*, London: Gollancz.

Ferry, D.J. and Holes, L.G. (1967) *Rationalization of Measurement*, London: Royal Institution of Chartered Surveyors.

Fischer, M. (2008) 'Reshaping the life cycle process with virtual design and construction methods', in P. Brandon and T. Kocatürk (eds) *Virtual Futures for Design, Construction & Procurement*, Oxford: Blackwell.

Flanagan, R. and Jewel, C. (2003) *A Review of Recent Work on Construction Futures*, London: CRISP Commission 02/06, Construction Research and Strategy Panel.

Hampson, K. and Brandon, P. (2004) *Construction 2020: A Vision for Australia's Property and Construction Industry*, Brisbane: CRC for Construction Innovation.

Joyce, N.E. (2004) *Building Stata: The Design and Construction of Frank O. Gehry's Stata Center at MIT*, Cambridge, MA: MIT Press.

Riese, M. (2008) 'One Island East, Hong Kong: A case study in construction virtual prototyping', in P. Brandon and T. Kocatürk (eds) *Virtual Futures for Design, Construction & Procurement*, Oxford: Blackwell.

Sebestyen, G. (1998) *Construction: Craft to Industry*, London: E. & F.N. Spon.

United Nations Report: Conference on Environment and Development (1992) *Annex 1: Rio Declaration on Environment and Development, Principle 10*, Rio de Janeiro. Online. Available at HTTP: <www.un.org/documents/ga/conf151/aconf15126–1annex1.htm> (accessed May 2008).

Woudhuysen, J. and Abley, I. (2004) *Why Is Construction So Backward?*, Chichester: John Wiley.

Index

Printed and bound by CPI Group (UK) Ltd, Croydon, CR0 4YY

22/10/2024

01777621-0018